Nanothermites

Series Editor
Bernard Dubuisson

Nanothermites

Eric Lafontaine
Marc Comet

ISTE

WILEY

First published 2016 in Great Britain and the United States by ISTE Ltd and John Wiley & Sons, Inc.

ISTE Ltd
27-37 St George's Road
London SW19 4EU
UK

www.iste.co.uk

John Wiley & Sons, Inc.
111 River Street
Hoboken, NJ 07030
USA

www.wiley.com

Library of Congress Control Number: 2016941697

British Library Cataloguing-in-Publication Data
A CIP record for this book is available from the British Library
ISBN 978-1-84821-837-6

Contents

Introduction

Thermites are combustible substances, usually little known to the general public, prepared by physically mixing powders of metal oxide and metal. The particular chemical characteristic of thermites lies in the nature of their constituents, which are considered by common sense to be non-combustible.

The displacement of the oxygen contained by metal oxides by aluminum was discovered by the Russian chemist Nikolay Beketov in 1865, but it was only between the late 19th and early 20th Century that the German chemist Johannes Wilhelm Goldschmidt patented the formulation of aluminothermic compounds, which were then intended for welding metal parts [GOL 07]. The mixtures prepared by Goldschmidt consisted of metal oxides or sulfides that were reduced by metals with a marked electropositive character, such as aluminum, calcium or magnesium. It is worth noting that the first thermites were manufactured by the same industrial processes that made use of molten salts electrolysis to obtain the metals used as fuels in these compounds came to maturity. The Hall–Héroult process for producing aluminum by electrochemical reduction of a molten cryolite bath dates back to 1886. Several years later, in 1897, Herbert Henry Dow founded the famous "Dow Chemical Company", which manufactured magnesium by the electrolysis of molten magnesium chloride. The considerable amount of electrical power required for melting and breaking down the salts employed as reducing metal precursors required a source of abundant and inexpensive electricity. The invention of the dynamo in 1868 by the Belgian physicist Zénobe Théophile Gramme, and then the use by Aristide Bergès, in 1882, of "white coal" to activate it, marked the dawn of the age of industrial production of electricity.

The analysis of the historical context provides an explanation as to why thermites, despite their seeming chemical simplicity and the unsophisticated process used to prepare them by powder mixing, emerged quite late in the history of pyrotechnics.

The term "thermite" was coined by Goldschmidt to denote the reactive compositions he had developed. This term is justified by the very significant amount of heat released during these combustions. The Larousse dictionary defines thermite as "a mixture of metal oxides and fine-particle aluminum powder, whose highly exothermic combustion is used in aluminothermic welding". This highly restrictive definition should be broadened to allow for taking into account the wide variety of compositions whose reaction modes are similar to aluminothermic reactions. In light of recent scientific advances in this field, thermites may be defined as "energetic compositions formed of reactive constituents that have a high proportion of metal elements, whose self-propagating reaction is accompanied by significant heat release".

The classical definition of thermites reflects the fact that aluminothermic mixtures were for a long time the main representatives of this particular family of energetic materials. The mixtures of micron-sized powders of aluminum and metal oxides are insensitive to various forms of stress: flame, impact, friction and electrostatic discharge. It is very difficult to ignite micron-sized aluminothermic mixtures by means of a simple flame, and the only way to reliably and rapidly activate the reaction is to use a more sensitive pyrotechnic ignition composition [COM 06a]. The reaction is accompanied by a shower of sparks, but most of the combustion products remain in condensed form, either solid or liquid. Due to the difference in density, the melting metal separates from slags, which consist essentially of alumina. By cooling, the drop of metal forms a nugget that remains encased in its ceramic layer. Rail welding is done by means of a device that uses the effect of gravity to enable flowing of the molten metal resulting from the reaction.

Due to the transfer of the significant amount of heat stored in the liquid metal to the matter it is in contact with, micron-sized thermites can be used as incendiary substances. While flowing, the incandescent metal drops become subdivided into droplets whose oxidation in contact with air provides additional energy. The strong exothermicity of aluminothermic reactions is also taken advantage of in the field of demolition to perform thermal shearing of massive metal structures used as reinforcement. As these examples show, the uses of micron-sized thermites are quite limited and they mainly consist of using the significant amount of heat generated by the aluminothermic reaction in order to melt objects or set them on fire.

Aluminothermic reactions are highly exothermic and propagate slowly and without oxygen inflow. The oxidation–reduction reaction is characterized by the transfer of oxygen contained in the metal oxide toward the aluminum, a highly oxophilic metal. This reaction is difficult to activate, and the ignition of micron-sized aluminothermic compositions takes place at a temperature nearing alumina's melting point (~2,053°C).

Nanothermites are manufactured starting from the same chemical compounds as their ancestors, the thermites. The only difference is the smaller size of the particles that compose them, which is only 5–1,000 nm. The mixtures containing at least one nanostructured reactive species are sometimes called “nanothermites”, but it seems more accurate to assign this designation to mixtures whose constituents are all sub-micron sized (<1,000 nm). The term “superthermite”, which is sometimes employed in literature, refers rather to reactivity than to structure [PIE 10]. In the English language scientific literature, nanothermites are also frequently referred to as “metastable interstitial (or intermolecular) composites”.

The study of nanothermites began some 20 years ago in the large national laboratories in the United States, and most likely at about the same time in Russia. Approximately one century would therefore elapse between the invention of thermites and their nanosized formulation. In reality, the fabrication of nanothermites was possible only starting with the moment when aluminum was produced in the form of stable nanoparticles and in sufficient quantity. Once again, it is worth noting that the history of thermites is closely linked to the history of the metal fuels that they contain.

Nanothermites ignite at lower temperature than thermites, thus they are more sensitive to ignition than the latter. On the other hand, nanothermites react so rapidly that the behavior of some of them makes them more similar to primary explosives than to combustible substances. The release of a similar amount of heat in a much shorter time confers higher reactive power to nanothermites compared to thermites.

Although research on nanothermites is relatively recent, it has already shown that these new materials have exceptional pyrotechnic properties. What the authors of this work aim at is not only drawing up a state of the art of this fascinating field of pyrotechnics, but also suggesting paths worth exploring by future research.

Chapter 1 refers to methods for the synthesis of metals and metal oxides that are in a highly divided state. The “fundamental bricks” that serve to formulate nanothermites must have the smallest possible size in order to facilitate interfacial contacts in the energetic composition. This chapter will be of particular use to scientists willing to themselves manufacture the nanoparticles they employ for nanothermite formulation.

Chapter 2 describes the main methods for the preparation of nanothermites, which often consist of mixing, in a more or less ordered manner, the nanoparticles of metal fuel and oxidizer.

Chapter 3 details the experimental study of nanothermites and their constituents. The properties of the fuels and oxidants most frequently employed are described here, as well as the methods used to characterize the reactivity and morphology of nanothermites.

Chapter 4 approaches nanothermites from the original perspective of security: pyrotechnic security, neutralization and toxicological risk.

1

Elaboration of Nanoparticles

Many methods have been developed for the preparation of metal powders depending on their desired basic characteristics, such as size, size distribution, morphology and specific surface area, to which other properties can be added, such as chemical reactivity and electrostatic stability. These methods for producing powders can be grouped in three large families. The first group is based on solid-phase methods, which consist essentially of mechanical milling and mechanosynthesis or reactive milling.

The second one is based on methods by which powders are elaborated in the liquid phase:

– sonochemical synthesis;

– microemulsion synthesis;

– solvothermal synthesis;

– sol-gel synthesis.

The third group includes methods for elaborating powders in the gas phase:

– inert gas condensation;

– explosion of metal wires;

– thermal plasmas;

– laser ablation;

– pyrotechnic synthesis.

This chapter does not claim to provide an exhaustive approach to all of the nanoparticle synthesis methods, as it more specifically focuses on various implementations for obtaining nanosized metal or metal oxide particles that may present an interest for the mixture-based formulation of nanothermites.

1.1. Solid-phase elaboration

1.1.1. *Mechanical milling*

For thousands of years, man has resorted to various milling methods in order to reduce the particle size of materials. However, in the 1960s, Benjamin developed a new method, high energy milling, which allowed for obtaining materials with nanoscale organization [BEN 70, BEN 74].

Starting from the 1980s, this technique was rapidly developed, as it allowed for obtaining structural states that are difficult to obtain by other synthesis methods, and even impossible to obtain by classical methods, such as melting solidification.

As an example, the preparation of the amorphous alloy type of compounds [WEE 88] can be mentioned. The works of Gaffet *et al.* showed that injected mechanical power was the parameter controlling the crystalline to amorphous transition in nickel alloys. This is obtained in a planetary mill by independently varying the rotation speeds of the disc and satellites [GAF 91]. A further example is to obtain supersaturated solid solutions from immiscible elements at thermodynamic equilibrium [YAV 92] or metastable crystalline phases [KOC 96, KOC 93].

Two terms are commonly employed in the literature to refer to high energy milling. The term mechanosynthesis is used when powders of different nature are milled together to finally obtain materials presenting new alloy compositions and/or new structures, or when chemical reactions are activated by the mechanical energy transferred during the phenomena triggered by milling [SUR 01, GLU 08].

The term mechanical milling or mechanomilling is employed when high energy milling is used not only to reduce the size of powder grains but also to modify the structure and/or microstructure of powders.

1.1.1.1. *Principle*

High energy milling is a method that permits us to obtain in solid phase ultrafine and homogeneous powders by exerting mechanical stress on a material. This consists of introducing one or several materials in a sealed chamber that contains one or several impactors, which are generally spherical, and shake everything in a more or less forceful manner. Under the effect of the mechanical energy transferred during collisions, the powder grains are subjected to very strong plastic deformations and go through a sequence of fractures and cold welding.

The plastic deformation rate increases enormously under milling [DEL 97], which leads to a significant increase in the hardness of the material with milling duration [KIM 95].

Nevertheless, a material's toughness cannot increase indefinitely with decreasing grain size because the reinforcement mechanism relies on the pile-up of dislocations at the level of obstacles, such as grain boundaries. Therefore, Hall–Petch law is valid as long as the grain size of the nanocrystalline material can sustain the dislocation pile-up.

$$\sigma_c = \sigma_0 + \frac{k}{\sqrt{d}} \quad \text{(Hall–Petch law)} \qquad [1.1]$$

where σ_c is the yield strength, d is the size of crystal grains and σ_0 and k are material-dependent constants.

Nieh and Wadsworth [NIE 91] propose a mechanism that describes the dislocation pile-up at grain boundaries and carry forward a relation that allows us to estimate the critical distance between two dislocations.

$$Lc = \frac{3Gb}{\pi(1-v)H} \qquad [1.2]$$

where G is the shear modulus, v is Poisson's ratio, b is the Burgers vector, h is the material's hardness and Lc is the critical distance between two dislocations.

Nevertheless, the decrease in grain size is also limited by the rate of dislocation recombination during milling. Fetch *et al.* [FEC 90] have shown that dislocations induced by mechanical milling combine and annihilate starting from a certain level of constraint, which results in a decrease in dislocation density. This effect is all the more important for the materials with low melting points, such as aluminum. Thus, for these materials, dislocation density is instead controlled by the recombination rate rather than by the deformation energy as a result of milling. The opposite is observed for materials with high melting point. Grain size is instead controlled by plastic deformation. When equilibrium is reached, the new deformations that can occur are grain boundary slidings that do not influence nanostructure [KIM 95, ZHA 01].

Eckert *et al.* have shown that the minimal grain size induced is inversely proportional with the melting temperature for the group of metals with face-centered cubic crystal structure. This seems quite clear for the four metals with the lowest melting points (Al, Ag, Cu and Ni) [ECK 92]. The other metals having a face-centered cubic crystal structure and higher melting points, which were studied (Pd, Rh, Ir) along with the metals with centered cubic and hexagonal close packed crystal structure [FEC 90], seem to have a grain size that is constant with the melting temperature.

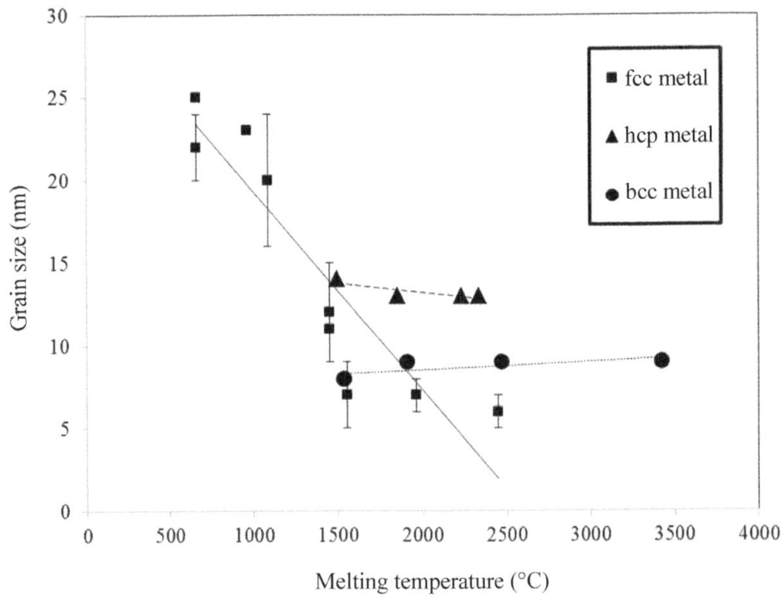

Figure 1.1. *Grain size as a function of melting temperature according to data from [ECK, 92, FEC 90, OLE 96]*

Ball milling and associated methods provide effective means for producing particles from bulk starting materials, and in the case of brittle materials it is possible to obtain a size reduction in a range below 100 nm.

Works conducted by Svrcek *et al.* aimed at the fragmentation of crystalline silicon by means of a planetary mill have allowed us to obtain particles of sizes between 2 and 6 nm, as well as silicon clusters of around 16 nm. The latter are disassembled by adding several drops of 30% ammonia, the size of particles then being brought to around 4 nm [SVR 05]. Russo *et al.* have obtained nanoparticles with a mean diameter of 55 nm, but also in a small quantity [RUS 11].

In the case of surfactant-assisted direct milling of metal particles, Chakka *et al.* have milled iron and cobalt for 50 h in the presence of around 10% by weight mixture of oleic acid and heptane, thus obtaining more or less spherical particles with sizes between 3 and 9 nm [CHA 06]. In slightly different milling conditions (milling duration of 20 h, same surfactant, but with a concentration of 50% by weight), Poudyal *et al.* have obtained mainly nanoplatelets with a diameter ranging between 5 and 30 µm and a thickness ranging from 20 to 200 nm, but also a small fraction of nanoparticles of the order of 6 nm [POU 11].

This approach may prove ineffective for ductile and malleable materials because the particles are not easily fractured and are cold welded. However, by using a surfactant (oleic acid) diluted at 3 or 5% in a polar solvent (acetonitrile), during milling under argon atmosphere in a planetary ball mill (Retsch PM 400) for 3 h, McMahon *et al.* obtained nanoparticles of aluminum, iron and copper with sizes indicated in Table 1.1 [MCM 14].

Metal	Type of balls	Size of particles
Al	Aluminum balls 8 mm diameter	bimodal 5–10 nm (24% by weight) 20–50 nm
Fe	Mild steel balls 3 mm diameter	10–20 nm
Cu	Copper cylinder 6.35 mm diameter, 6 mm length	Bimodal 250 nm (7% by weight) 500–900 nm

Table 1.1. *Size of nanoparticles obtained by mechanical milling of ductile metals*

1.1.1.2. *The main types of mills*

In order to decrease grain size by mechanical milling or mechanosynthesis, various types of mills can be employed, such as:

– vibratory mill;

– attritor;

– ring mill;

– planetary mill.

1.1.1.2.1. Vibratory mill

Vibratory mills consist of a vial that is set in vibration motion at high frequency. Among them, we can distinguish a type of mill with only one vibration axis and only one milling ball and another type of mill with high-frequency vibration along the three axes. There are many models, which are marketed by several companies such as Fritsch (Pulverisette 0 or 23, for example) or Spex, with the 8000 M shaker mill model.

In this type of mill, mechanical effects are essentially obtained by the balls' collisions with powder. It operates at oscillation frequencies ranging from several dozen to several thousand hertz, and it can mill powder quantities ranging from 10 to 20 g, and within quite short periods of time (of the order of 24 h) [SUR 01]. This type of mill is highly energetic in comparison with the attritor.

1.1.1.2.2. Attritor

The attritor was the first mill used by Benjamin [BEN 70] to obtain metal alloys by milling. It can be vertical [GIL 83] or horizontal [ZOZ 97]. It is composed of a cylinder body that contains balls and a central shaft equipped with paddles perpendicular on the shaft axis. The rotation of this shaft sets balls into motion. A characteristic of this device is the large number of balls used (around 1,000 balls with diameters ranging from 0.2 to 1 cm). The rotation speeds are relatively low and can reach 250 rpm. In this type of mill, powder milling is essentially carried out by frictions among balls or with the body wall [SUR 01].

1.1.1.2.3. Ring mill

The ring mill is a variant of the attritor. It was developed by the team of Senna [HAM 96] and marketed by the Japanese company Nara Machinery Co under the brand name Micros. Its operation relies on a pile-up of zirconium oxide rings on six vertical axes fixed on a rotating disc. The diameter of the central hole of the rings is larger than that of the axis, thus during the rotation of the assembly the off-centered rings rub the wall of the recipient due to the centrifugal force developed. Similarly to the (vertical or horizontal) attritor with balls, the milling of powders is essentially obtained here by friction. Depending on its size, the capacity of this type of device ranges from 0.4 to 33 L and rotation speed varies between 250 and 3,000 revolutions per minute [AVV 01].

1.1.1.2.4. Planetary mill

This mill consists of a rotating platform featuring one to four spinning vials. This device owes its name to the planet-like movement. Due to the simultaneous rotation of platform and vials, centrifugal forces are induced inside the vials, and they act alternatively in the same and opposite directions. This results in collisions and frictions between balls and the powder.

This type of mill is used for the laboratory synthesis of materials, as it allows for controlling various milling parameters. The capacity of this type of mill ranges from several grams to a hundred grams. The rotation speed of the main disc can vary between 100 and 1,100 revolutions per minute. The main planetary mills are marketed by Retsch company, with the PM range of products or by the Fritsch company with the "Pulverisette" range, some of which were developed starting from the G5 and G7 mills conceived by Gaffet *et al.* [GAF 95].

1.1.1.3. *Milling parameters*

Each milling depends on many parameters that directly influence the morphological and microstructural characteristics of the final product [GAF 04, SUR 01]. The main parameters that can be subjected to variation are the following:

– energy transferred during milling;

– milling duration;

– nature of the milling media;

– size of balls;

– balls-to-powder mass ratio;

– filling ratio;

– milling atmosphere;

– control agents;

– temperature.

1.1.1.3.1. Transferred energy

A mill's energy varies from one type of mill to another. In principle, high energy leads to more rapidly obtaining the final product.

Several teams have taken an interest in the study of modeling and simulation of phenomena that occur during milling. It is in particular worth citing the works of McCormick *et al.* [HUA 97a, DAL 96, HUA 97b], Gaffet *et al.* [ABD 95, CHO 97, ABD 96], Courtney *et al.* [MAU 94, MAU 95a, MAU 95b, MAU 96a, COO 95, MAU 96b, MAU 92, COU 96, RYD 93] or those of Hashimoto and Watanabe [WAT 95, HAS 90].

Modeling the phenomena that are happening in a mill, such as kinematics, mechanisms through which mechanical energy is transferred to the material to be milled, as well as its reaction to this transfer, is a particularly complex task. It is however possible to distinguish phenomena occurring at the local and global level. The first level refers to interactions between balls and powder, while the second relates to the mill's dynamics and kinematics. Gaffet *et al.* have added a third level, which takes into account the structural evolutions of materials subjected to mechanical stress and their evolutions as an energy of energy transfer [CHO 97].

In the vertical vibratory mills with one ball, impacts are frontal, at normal incidence; as for the vibratory mills with several balls, their kinematics are more complex, being characterized by quasi-normal impacts to the walls.

Studying a Ni_xZr_y alloy, Chen has defined milling intensity as the momentum transferred ($M_b \times V_{max}$) to the unit mass of powder (M_p) per unit time (f):

$$I = \frac{M_b V_{max}}{M_p} f \quad [1.3]$$

with

$$V_{max} = 2\pi A f_{bol} \quad [1.4]$$

where M_b is the mass of the ball, M_p is the mass of powder, V_{max} is the ball velocity relative to the vial at the moment of impact, f is the frequency of the collisions, A is the amplitude of vibration and f_{bol} is the frequency of vial vibration [CHE 92].

The kinetics of the planetary or horizontal mills depends on a significant number of parameters such as the speed of rotation of vials and platform, load ratio and filling ratio.

Mio *et al.* have studied the influence of various parameters, and in particular the effect of the speed ratio and rotational direction on the balls' impact energy [MIO 02]. The latter increases with platform speed and reaches an optimal value beyond which impact energy decreases.

This so-called critical speed V_c for which impact energy is optimal can be written as:

$$V_c = \sqrt{\frac{R_p}{R_b - r_b} - 1} \quad [1.5]$$

where R_p is the platform radius, R_b is the vial radius and r_b is the ball radius.

This evolution of impact energy as a function of speed ratios can be explained by balls moving in various modes. Four modes can be distinguished as the speed ratio increases. The "balancing" mode means that balls and powder are moving as a whole. The "cascade" mode means a rolling of balls that leads to a relative movement of balls arriving in the "high" section of the pot, which cascade back to the "low" section of the pot by successive small scale collisions. In addition to this impact mechanism, balls are also sliding on the pot walls. This sliding phenomenon has a direct effect on the balls' starting frequency [DAL 96].

The "cataract" mode describes the situation where balls fill the whole space of the pot and collide with one another. This results in a high number of collisions and the sliding phenomenon is almost absent.

The "rolling" mode occurs at high speeds. Due to the centrifugal effect, balls remain glued to the inner wall of the pot, no ball gets away from the wall to strike the opposite side of the pot and friction effects are weak.

Another important point highlighted by the works of Mio *et al.* is that impact energy is much higher when the pots and platform rotate in opposite directions [MIO 02].

1.1.1.3.2. Milling duration

This parameter defines the minimum time needed for the system to reach a state of equilibrium. It depends on the type of mill used, the way balls act upon powder (collision or friction) and for metals that are sensitive to hydrogen-induced brittle fracture, on the atmosphere [ECK 92]. In effect, Eckert *et al.* have reported that face-centered cubic metals have a tendency to stick to the milling tools during milling under an argon atmosphere, and thus milling effectiveness decreases [FEC 90]. Finally, milling duration depends on the temperature at which it takes place [BOR 97]. At the beginning of the process, cold welding will predominate and can lead to an increase in the particle size [SUR 01]. Then, the fracturation phenomenon will become dominant, and as a result the size of particles will decrease. This will rapidly take place, then will slow down and will finally stop, thus reaching a state of equilibrium [SUR 01, LEE 98]. It generally takes several hours, but it may also take several dozen hours of milling to reach equilibrium [KHA 06, KHA 10, REV 05]. It is also worth noting that the longer the duration, the higher the contamination of powders by the milling media. On the other hand, it is possible to obtain undesirable phases if milling continues beyond what is researched, as shown, for example, by Suryanarayana [SUR 95] on a titanium and aluminum alloy.

During the milling phase, the balls' collisions or frictions with powder generate local heating of the material. In order to minimize heating, a cycle-based approach is often adopted by alternating milling and pause phases [KHA 08, KLE 05].

1.1.1.3.3. Nature of media

The choice of hardness, and therefore of the nature of the material that balls and vials are made of, needs to take into account the powder to be milled. The use of low-hardness material will not allow for the powder milling, while using material with too high hardness will lead to premature wear of the balls and vials, and consequently to a contamination of the milled powder. Materials currently employed are steel nuances [ELE 15, CHE 99] or chromed steel [CUK 01], yttrium oxide stabilized zirconium for the jars [YEN 96] and yttrium oxide or magnesia partially stabilized zirconium [YEN 96] for the balls [BEG 97], agate [TON 91, STU 01, LI 03], corundum [BIL 07], tungsten carbide [PAB 99], WC-Co tungsten carbides [STU 01] and sapphire [ELE 05]. Uses of other materials such as titanium, copper or copper-berylliums, ceramics or silicon nitride are also reported by Suryanarayana [SUR 01].

When a ductile material such as aluminum is milled, balls and vials are generally made of steel [ECK 92, KHA 08, REV 05].

1.1.1.3.4. Size of balls

In most cases, the balls used have the same diameter and are of the same nature [PAB 99]. Shin *et al.* have studied the effect of the size of balls with a constant load ratio on the size of powder particles obtained after milling. They have shown that the optimum diameter of the ball decreases as the rotation speed increases. At a given rotation speed, this is interpreted as an incidence of two opposite effects: the first effect, unfavorable to milling efficiency, is that kinetic energy is lower for smaller size balls. The second effect is that an increase in the number of balls, and thus of collisions, favorably influences milling efficiency [SHI 13].

Vaezi *et al.* have shown that using a mix of balls with various diameters (here, 5 and 10 mm) is more efficient than using balls of the same diameter (5 or 10 mm) at constant load ratio [VAE 12]. The explanation they propose for this phenomenon is that in a mix, the smaller balls can fill the space between the larger ones. Due to this rearrangement of balls, the number of collisions increases. Moreover, as the load ratio is constant, the weight of each large ball equals that of eight smaller balls. This means a higher number of collisions take place, while the balls' kinetic energy is not significantly reduced. This leads to an optimized number of collisions to collision energy ratio for the considered range (rotation speed, load ratio, filling ratio, etc.).

Another hypothesis can be proposed based on the works of Takacs and Pardavi-Horvath [TAK 94], who have shown that during the mechanosynthesis of iron particles produced by the reaction of iron oxide with zinc, the latter, due to its ductility, is deposited on the surface of milling media. Such behavior facilitates cold welding phenomena. To solve this problem, they have used a mix of balls of various diameters. Even though this may effectively enable us to avoid zinc deposits on walls and balls, no explanation is proposed. In light of these results, Suryanarayana suggests that various sizes of balls generate shear forces. These forces would facilitate powder removal from the surface of vials and balls and would therefore favor fragmentation to the detriment of cold welding, thus improving milling yield [SUR 01].

1.1.1.3.5. Balls-to-powder mass ratio

Balls-to-powder mass ratio, sometimes called load ratio, is an important variable in the milling process. It is defined as the ratio between the balls' mass and powder mass. In the literature, we can equally find low ratios of 1:1, and very high values such as 220:1 [SUR 01]. Generally speaking, a ratio between 10 and 20 to 1 is more commonly used for powder milling in a low capacity mill, such as a SPEX mill or a planetary mill of the Fritsch Pulverisette type [PAB 99, AVE 03].

Load ratio influences the time needed to reach a particular phase state and/or size of the milled powder. The higher the load ratio, the shorter the time needed. An increase in the load ratio is equivalent to an increase in the number of balls, if their mass is on average identical, which involves an increase in the number of collisions per time unit. A higher quantity of energy is transferred to the powder, which more rapidly reaches its final state. We should, however, be aware of the fact that a high quantity of energy transferred within a short period of time can lead to an increase in temperature, which can influence the material's final state.

1.1.1.3.6. Filling ratio

The filling ratio of the pot with balls is another important factor in mechanical milling. A low level of filling will result in low milling yield, while on the other hand, if the filling ratio is too high, there will not be enough space for the balls to move freely. This will reduce the impact energy and will thus decrease process efficiency. An optimal filling ratio is obtained when 50–60% of the pot volume is filled [SUR 01, GAF 04].

1.1.1.3.7. Milling atmosphere

Milling can be performed under various atmospheres, either of neutral gas (argon, helium) or reactive gas [SUR 01, GAF 04], in order to obtain hydrides in a hydrogen atmosphere or to embrittle some metals [ECK 92], oxides under atmospheres of air or oxygen diluted in a neutral gas, or even nitrides when milling takes place under ammonia or nitrogen atmosphere [MIK 92].

The type of atmosphere can also influence the material's final structure. To take an example, Ogino *et al.* have shown that in the case of Cr-Cu alloys milled under nitrogen or argon atmosphere that contains air, nitrogen atoms were absorbed by the alloy, thus increasing the thermodynamic stability of grain boundaries and amorphous phases compared with the centered cubic phase. Absorption thus facilitates grain refining and causes amorphization. No alloy amorphization is observed in the case of pure argon atmosphere [OGI 90].

Similar works conducted by Li *et al.* highlight the fact that oxygen accelerates the titanium amorphization kinetics [LI 07].

1.1.1.3.8. Control agent

As previously noted, powder particles are alternately cold welded and then fragmented during milling. A control agent can be added during milling in order to modify the welding/fragmentation ratio under given experimental conditions [SUR 01]. The agent acts to lower the surface energy of the powder particles and thus to reduce the cold welding phenomenon, to the benefit of fracturing, the intention being to reduce the size of the grains resulted after complete milling [GAF 04, ECK 92].

Control agents can be solid, liquid or gas [SUR 01, ECK 92]. When liquid agents are added, the particles obtained will have a more homogeneous size distribution compared to adding a solid agent [SUR 01]. The main agents used are stearic acid [LIU 15, WOL 96, RAM 12], alcohols [HUA 97c, LEE 96, RAM 12] or even saturated hydrocarbons (paraffin, hexane, etc.) [AME 95, SUZ 95] or unsaturated hydrocarbons (benzene [IMA 96, LEE 96] and toluene [PAB 99, PAB 96]).

Other more exotic agents are employed as control agents, such as mineral or organic salts of sodium [SHI 99], classical solvents such as tetrahydrofuran [IMA 96] or solid substances such as graphite [MOR 90]. As far as graphite is concerned, it may be interesting not to eliminate it upon milling completion, and to incorporate everything in the formulation of a nanothermite, for the purpose of reducing its sensitivity to electrostatic discharge, for example.

The agent to powder mass ratios is normally within the range from 0.5 to 3% [KHA 06, KHA 08, KLE 05] but in some cases they may reach several hundreds of percent [REV 05] in relation to the load to be milled.

1.1.1.3.9. Milling temperature

Another important parameter in determining the material's final state is the milling temperature. When temperature is high, the size of crystallites increases, but their constraints and solubility in solid state are reduced [HON 94]. The works of Zhou *et al.* referring to aluminum milled at 90 K temperature have shown that grain size reaches an average value of 26 nm after a milling duration of only 8 h, while it took nearly 100 h of milling at 298 K for the size of the grains to become stabilized at around 25 nm. This shows that, for a similar level of milling energy, cryomilling is more efficient than milling at ambient temperature [ZHO 04].

1.1.1.4. *Mechanosynthesis*

Chemical reactions may take place during mechanical milling. In this case, the process is called reactive milling, mechanochemistry or mechanosynthesis.

Chemical reactions take place at the interface of particles that are subject to continuous renewal during milling. This technique presents the major advantage that the reaction takes place at a much lower temperature (it may often reach ambient temperature) than the thermodynamic equilibrium temperature. Reactions can occur either in stable state or during self-propagated combustions. When the latter is the case, due to the strong increase in local temperature, the resulting particles' diameters are often in the micrometer range, while the particles resulted from self-combustion have diameters in the nanometer range. In order to prevent autoignition of the mixture during milling, experimental milling parameters can often be adjusted so that the impact energy decreases and/or an inert diluent can be added to slow down reaction kinetics.

As previously described in section 1.1.1.3.7, during mechanosynthesis reactions solid matter can react with the milling atmosphere when gases are reactive. To take an example, Liu *et al.* have used mechanosynthesis to obtain tantalum and niobium nitride by milling either pure tantalum or pure niobium grains of around 100 μm in size. They have used for this purpose a planetary mill with a pot and balls (d = 17 mm) made of stainless steel, a balls-to-powder load ratio of 20:1, under nitrogen atmosphere at 2.5 atmospheres. After a milling duration of 80 h, the result was the formation of an NbN phase whose grain size ranged from 10 to 30 nm. By comparison, Ta_2N forms more rapidly than niobium nitride, as the reaction is nearly completed after a milling duration of 16 h. Even though an amorphous phase emerges when milling duration exceeds 30 h, Ta_2N is the only crystalline phase formed throughout milling. The authors advance the hypothesis that in the Ta-N system the amorphous phase emerges due to iron contamination originated in the milling medium [LIU 99].

Solid–liquid reactions may occur when a solid is milled in a reactive liquid, for example, in the case of the synthesis of nanosized crystallized silicon particles by reaction between liquid silicon tetrabromide and magnesium pellets in a Spex 8000 mill, under argon atmosphere [SAN 07] or by the reaction of silicon tetrachloride with lithium pellets in the presence of lithium chloride used as control agent to avoid a combustion reaction [CHA 14b]. The silicon obtained in this latter example is amorphous. It has been recently shown that mechanochemically induced solid–solid reactions could also be used to produce nanosized metal and metal oxide particles.

When adequate conditions in terms of chemical reaction, the stoichiometry of initial reagents and milling parameters (rotation speed or collision frequency, size and number of balls, load ratio, nature and quantity of control agent, etc.) are chosen, mechanochemical treatment can be used for the synthesis of nanosized particles dispersed in a soluble salt matrix. It is possible to obtain weakly agglomerated nanoparticles in the desired phase by selectively eliminating the matrix phase by washing the obtained powder with appropriate solvents.

Reports of the syntheses of nanosized metal and oxide particles obtained by mechanosynthesis are presented in Table 1.2.

Metal	*d* (nm)	Conditions	Reference
Al	25–40	$AlCl_3 + 3\ Li \rightarrow Al + 3\ LiCl$ Control agent: LiCl load/agent mass: 0.25 Glen Mills Turbula T2C, B/P load ratio: 35/1 Atmosphere: argon, 3 h, steel balls, Washing nitromethane + $AlCl_3$	[PAS 09]
	40–50	$AlCl_3 + 3\ Na \rightarrow Al + 3\ NaCl$ Control agent: NaCl load/agent mass: 0.3 Glen Mills Turbula T2C, B/P load ratio: 55/1 Atmosphere: argon, 4 h 40, steel balls	
Cu	30	$CuCl_2 + 2\ Na \rightarrow Cu + 2\ NaCl$ Control agent: NaCl load/agent mass: 1 SPEX 8000, B/P load ratio: 3/1 Atmosphere: argon, 16 h, steel balls ($d = 4.8$ mm) Washing with deionized water, then methanol under ultrasounds	[DIN 96a]
Co	10–50	$CoCl_2 + 2\ Na \rightarrow Co + 2\ NaCl$ Control agent: NaCl load/agent mass: 0.5 SPEX 8000, B/P load ratio: 3/1 Atmosphere: argon, 10 h, steel balls ($d = 4.8$ mm) Washing with deionized water, then methanol under ultrasounds	[DIN 96b]
Fe	15–20	$FeCl_3 + 3\ Na \rightarrow Fe + 3\ NaCl$ SPEX 8000, B/P load ratio: 3/1 Atmosphere: argon, 16 h, steel balls ($d = 4.8$ mm) or 24 h ($d = 3.2$ mm) or 3.3 h ($d = 6.4$ mm) Washing with deionized water, then methanol under ultrasounds	[DIN 96c]
	30–35	$FeCl_3 + 3/2\ Ca \rightarrow Fe + 3/2\ CaCl_2$ 8 h, steel balls ($d = 9.5$ mm)	
	50–60	$FeCl_3 + Al \rightarrow Fe + AlCl_3$ 90 h, steel balls ($d = 6.4$ mm) Control agent: NaCl load/agent mass: 1	

Ni	10–20	$NiCl_2 + 2\,Na \rightarrow Ni + 2\,NaCl$ Control agent: NaCl load/agent mass: 0.5 SPEX 8000, B/P load ratio: 3/1 Atmosphere: argon, 16 h, steel balls (d = 4.8 mm) Washing with deionized water, then methanol under ultrasounds	[DIN 96b, DIN 99]
	10–500	$NiCl_2 + Mg \rightarrow Ni + MgCl_2$ SPEX 8000, B/P load ratio: 10/1 Atmosphere: argon 3 h, steel balls (d = 4 mm) Washing with deionized water, then methanol under ultrasounds	[BAB 97]
Si	20	$SiBr_4 + 2\,Mg \rightarrow Si + 2\,MgBr_2$ SPEX 8000, B/P load ratio: 10/1 Atmosphere: argon 20 h, steel balls (2 of d = 11.1 mm and one of 14.3 mm) washing 40 mL of HCl (37%) and heated at 70°C for 30 min, then ethanol.	[SAN 07]
	    51 13	$SiCl_4 + 4\,Li \rightarrow Si + 4\,LiCl$ Glen Mills Turbula T2C, B/P load ratio: 90/1 Atmosphere: argon, 24 h, steel balls (d = 7.9 mm and 12.7 mm), washing, tetrahydrofuran, Without control agent With control agent: LiCl load/agent mass: 10	[CHA 14b]
	 2.5	$3\,SiO_2 + 4\,Al \rightarrow 3\,Si + 2\,Al_2O_3$ SPEX 8000, B/P load ratio: 23/1 Atmosphere: argon 15 h, stainless steel balls (total weight 24.87 g)	[ARA 03]
	 5	$SiO_2 + C \rightarrow Si + CO_2$ Planetary mill, B/P load ratio: 10/1 Atmosphere: argon, 240 h, stainless steel balls (d = 7.5 mm)	[LAM 00]
Sb	19	$Sb_2S_3 + 3\,Fe \rightarrow 2\,Sb + 3\,FeS$ Fritsch Pulverisette 6, B/P load ratio: 120/1 Atmosphere: argon, 3 h, 50 balls WC (d = 10 mm)	[BAL 07]

Table 1.2. *Mechanosynthesis of nanosized metal particles*

Oxide	d (nm)	Conditions	Reference
Al_2O_3	10–20	$2\ AlCl_3 + 3\ CaO \rightarrow Al_2O_3 + 3\ CaCl_2$ SPEX 8000, B/P load ratio: 8/1 Atmosphere: argon, 24 h, balls (d = 9.5 mm) Annealed at 400°C, 1 h Washing with deionized water under ultrasounds	[DIN 96d]
CeO_2	10	$CeCl_3 + 3\ NaOH \rightarrow Ce(OH)_3 + 3\ NaCl$ Control agent: NaCl 12 equivalents SPEX 8000, B/P load ratio: 10/1 Atmosphere: argon, 4 h, steel balls (d = 6.4 mm), annealed at 500°C, 1 h Washing with deionized water under ultrasounds	[TSU 01c]
	19	$CeCl_3 + 3/2\ CaO + 1/4\ O_2 \rightarrow CeO_2 + 3/2\ CaCl_2$ SPEX 8000, 24 h, annealed at 400°C, 6 h	[GOP 00]
	40–70	$2\ CeCl_3{\cdot}6H_2O + 3\ Na_2CO_3{\cdot}10H_2O \rightarrow Ce_2(CO)_3{\cdot}n\ H_2O + 6\ NaCl + (42\text{-}n)\ H_2O$ SPEX 8000, B/P load ratio: 5/1 24 h, annealed at 900°C, 1 h Washing with deionized water under ultrasounds	[LI 05]
Cr_2O_3	 3 50 10–100 10–80	$Na_2Cr_2O_7 + S \rightarrow Cr_2O_3 + Na_2SO_4$ Control agent: NaCl 187% SPEX 8000, B/P load ratio: 10/1 Atmosphere: argon, 6 h, steel balls (d= 12.7 mm), annealed at 450°C, 1h or Annealed at 500°C, 1 h or Annealed at 600°C, 1 h Washing with deionized water under ultrasounds With Na_2SO_4 as control agent and annealed at 800°C, 1 h	[TSU 00]
Fe_2O_3	10–30	$2\ FeCl_3 + 3\ CaO \rightarrow Fe_2O_3 + 3\ CaCl_2$ Control agent: $CaCl_2$: load/agent mass: 1 SPEX 8000, B/P load ratio: 5/1 Atmosphere: argon, 24 h, steel balls (d = 4.8 mm), annealed at 150°C, 1 h Washing with methanol under ultrasounds	[DIN 97b]
	20–50	$2\ FeCl_3 + 3\ Ca(OH)_2 \rightarrow Fe_2O_3 + 3\ CaCl_2 + 3\ H_2O$ Same conditions except annealed at 200°C, 1 h	
	  5 50–100 Hexagonal platelets D: 50–200 E: 20–40	$Fe_2(SO_4)_3 + 3\ Na_2CO_3 \rightarrow Fe_2(CO_3)_3 + 3\ Na_2SO_4 \rightarrow$ $Fe_2(CO_3)_3 \rightarrow Fe_2O_3 + 3Na_2SO_4 + 3\ CO_2$ SPEX 8000, B/P load ratio: 10/1 Atmosphere: air, 4 h, 6 steel balls (d = 12.7 mm), annealed at 350°C, 1 h or Annealed at 500°C, 1 h or Annealed at 700°C, 1 h Washing with deionized water under ultrasounds	[TSU 11]

Gd_2O_3	5–50	$GdCl_3 + 3\ NaOH \rightarrow Gd(OH)_3 + 3\ NaCl$ Control agent: NaCl load/agent mass: 0.1 SPEX 8000, B/P load ratio: 10/1 Atmosphere: argon, 24 h, balls (d = 6.4 mm), annealed at 500°C, 1/2 h Washing with deionized water under ultrasounds	[TSU 99]
	Platelets length 20–150 E: 5–20	$2\ GdCl_3 + 3\ CaO \rightarrow Gd_2O_3 + 3\ CaCl_2$ SPEX 8000, B/P load ratio: 40/1 Atmosphere: argon, 8 h, balls (d = 12.7 mm) Washing and then annealed at 800°C, 1 h	[TSU 98]
Nb_2O_5	10–100	$2\ NbCl_5 + 5\ Na_2CO_3 \rightarrow Nb_2O_5 + 10\ NaCl + 5\ CO_2$ Control agent: NaCl 8 equivalent SPEX 8000, B/P load ratio: 10/1 Atmosphere: argon, 6 h, steel balls (d = 12.7 mm), annealed at 550°C, 1/2 h Washing with deionized water under ultrasounds	[TSU 01a]
	200–1,000	$2\ NbCl_5 + 5\ MgO \rightarrow Nb_2O_5 + 5\ MgCl_2$ Same condition, milling duration 8h Annealed at 400°C	
SnO_2	5–30	$2\ SnCl_2 + 2\ Na_2CO_3 \rightarrow 2\ SnCO_3 + 4\ NaCl$ $SnCO_3 \rightarrow SnO + CO_2$ $SnO + ½\ O_2 \rightarrow SnO_2$ Control agent: NaCl: 100% equivalents B/P load ratio: 10/1 Atmosphere: argon, 6 h, chromed steel balls (d = 6.4 mm), annealed at 400°C, under air Washing with deionized water under ultrasounds	[CUK 01]
TiO_2	37	$TiOSO_4\ xH_2O + Na_2CO_3 \rightarrow TiO_2 + Na_2SO_4 + xH_2O + CO_2$ TB-1 Kadan ltd, B/P load ratio: 20/1 Atmosphere: argon, 6 h, corundum balls (d = 11 mm), annealed at 700°C, 1 h under air Washing with deionized water under ultrasounds	[BIL 07]
Y_2O_3	28	$2\ YCl_3 + 6\ LiOH \rightarrow Y_2O_3 + 6\ LiCl + 3\ H_2O$ SPEX 8000, Atmosphere: argon, 4 h, 20 steel balls (d = 9.5 mm), annealed at 500°C, 1 h Washing with deionized water then methanol under ultrasounds	[DOD 01a]
ZnO	10–40	$ZnCl_2 + NaCO_3 \rightarrow ZnCO_3 + 2\ NaCl$ $ZnCO_3 \rightarrow ZnO + CO_2$ Control agent: NaCl 8.6 equivalents SPEX 8000, B/P load ratio: 10/1 Atmosphere: argon, 6 h, steel balls (d = 6.4 mm), annealed at 400°C, 1/2 h under air Washing with deionized water under ultrasounds	[TSU 01b]

	5–10	$ZrCl_4 + 2\ CaO \rightarrow ZrO_2 + 2\ CaCl_2$ SPEX 8000, B/P load ratio: 5/1 Atmosphere: argon, 20 h, steel balls ($d = 9.5$ mm), annealed at 350°C, 1/2 h Washing with methanol under ultrasounds	[DIN 97a]
	14	$ZrCl_4 + 4\ LiOH \rightarrow ZrO_2 + 4\ LiCl + 2\ H_2O$ SPEX 8000, Atmosphere: Argon, 4 h, 20 steel balls ($d = 9.5$ mm), annealed at 500°C, 1 h Washing with deionized water then methanol under ultrasounds	[DOD 01a]
ZrO_2	7	$ZrCl_4 + 2\ MgO \rightarrow ZrO_2 + 2\ MgCl_2$ SPEX 8000, Atmosphere: argon, 12 h Annealed at 500°C, 1 h Washing with deionized water under ultrasounds	[DOD 01b]
	32	$ZrCl_4 + 2\ Li_2O \rightarrow ZrO_2 + 4\ LiCl$ Control agent: LiCl, 4 equivalents SPEX 8000, B/P load ratio: 10/1 Atmosphere: argon, 6 h, 20 steel balls ($d = 9.5$ mm), annealed at 400°C, 1 h Washing with deionized water then methanol under ultrasounds	[DOD 01c]

Table 1.3. *Mecanosynthesis of nanosized metal oxide particles: D: diameter, E: thickness, L: length*

1.1.1.5. *Conclusion*

It is possible to obtain nanosized particles through the direct milling of metals of oxides by mechanical milling of brittle materials, and as well of ductile materials such as copper or aluminum, provided that a surfactant is employed during the milling.

The same applies to reactive milling, which offers the possibility to obtain metals or metal oxides in the form of crystallites or nanosized amorphous particles. This often requires the use of a control agent, which will act as a diluent by preventing or delaying autoignition reactions, and due to its dilution effect, it will also strongly reduce the effects of agglomeration between nanosized grains. The drawbacks of the mechanosynthesis technique are, on the one hand, the low quantities produced and, on the other hand, the duration of the process, which is often of several hours. However, the most important problem is the contamination of powders by the milling media.

It is worth noting that oxides manufactured by this method have low "reactivity" when they are used for the preparation of nanothermites. Examples of the most reactive oxides for this application include WO_3, MoO_3, CuO, Bi_2O_3, I_2O_5 or Ag_2O.

1.2. Liquid-phase elaboration

1.2.1. *Sonochemistry*

1.2.1.1. *Principle*

Sonochemistry is a technique that uses ultrasonic waves to activate or accelerate chemical reactions. This simple method is recommended in particular for organizing nanoscale structures. The use of this method allows us to obtain amorphous or crystalline particles or even metal or metal oxide colloidal suspensions. The ultrasounds used in sonochemistry are essentially in the range of low frequencies (between 16 and 100 kHz) [PIN 08, KOL 96, KOL 12] and high frequencies (between 100 kHz and 1 MHz) [ABU 14, REY 15, FUJ 14]. Very high frequency ultrasounds (above 1 MHz) are generally used for diagnostic purposes, even though frequencies above 1 MHz may sometimes be used in sonochemistry [REY 15].

1.2.1.2. *Effects of implementation parameters*

1.2.1.2.1. Power of emission

A second criterion for differentiating ultrasonic waves is the power of emission. When this is below 1 W, ultrasounds induce no modification of the medium. In this case, there is only vibratory interaction between the wave and the matter it passes through. Such low powers are employed in the non-destructive testing of materials or structures, or in medical diagnosis. When the wave has higher levels of power, its passing through matter is accompanied by nonlinear physics phenomena and associated chemical reactions. In contrast with photochemical phenomena, the chemistry associated with acoustic waves does not depend on the wave–matter direct interaction. It is essentially due to acoustic cavitation. A characteristic of the cavitation phenomenon is the formation of vapor bubbles in a liquid. This phenomenon emerges when hydrostatic pressure decreases below the value of the liquid vapor tension for a given temperature. A cavitation bubble's evolution depends on its initial volume, the field of pressure it is subjected to and the liquid surface tension. Small size bubbles that are subjected to low pressure fields disappear by rapidly dissolving in the surrounding liquid.

At high values of the pressure field, the bubbles' volume fluctuates. During depression, the bubble's volume increases under the influence of vaporization and gas diffusion. When the bubble is compressed, its gas content is dissolved in the liquid and vapor condenses at the liquid/gas interface. The implosion of the cavitation bubble results in a very sharp increase in temperature, which can reach values above 5,000°C, pressures above 500 atmospheres, local accelerations above 10 g, photon emissions and shockwaves over a duration of less than a microsecond [SUS 89]. These extreme conditions enable the development of chemical reactions that occur only at high energies [SUS 91]. Moreover, due to the rapid cooling of vapors in contact with the liquid interface, metallic glasses can be obtained, such as amorphous iron powder [SUS 91].

Volatile precursors, such as metal carbonyls, are to a large extent thermally decomposed inside the cavitation bubble and lead to the formation of metal nanoparticles [SUS 96, KOL 96], oxides [KOL 12, ABU 05], carbides [SUS 96] or sulfides [LOP 14].

As far as non-volatile precursors are concerned, chemical reactions take place in the vicinity of the liquid/gas interface of the cavitation bubble. On the one hand, they take place by the interaction of radicals formed during solvent pyrolysis in the cavitation bubble with ions in solution, and on the other hand, as a result of temperature, which, despite being lower in the vicinity of the interface, it is however high enough to trigger thermal decomposition of the present solutes and to generate radicals that will also interact with the species that are present in solution. Several other experimental parameters have an influence on the nature and size of the particles produced.

1.2.1.2.2. Frequency of emission

Reyman *et al.* have shown that the rate of Fe^{3+} sonochemical production from Fe^{2+} decreased linearly as a function of frequency. By contrast, the size of iron oxide particles formed does not seem to depend on frequency [REY 15].

1.2.1.2.3. Amplitude of emission

In their works on the sonochemical synthesis of lead dioxide, Ghasemi *et al.* have shown that the size of the particles formed depended on the amplitude of the wave. For an amplitude of 36 μm, lead oxide particles take the form of aggregates of 300–350 nm. For an amplitude of 60 μm, the size of aggregates ranges between 50 and 100 nm. Thus, an increase in the amplitude of ultrasonic waves reduces the particles' growth rate [GHA 08].

1.2.1.2.4. Duration of emission

The impact of the duration of ultrasonic emission on the reaction medium was most notably analyzed by Aslani and colleagues, in relation with the production of nickel oxide (NiO) nanoparticles. For a power in the range from 12 to 18 W, the mean size of particles goes from 60 nm, for a 2 h emission time, to a mean size of 20 nm, for a 3 h emission time [ASL 11].

1.2.1.2.5. Impact of solvent

There are other solution-dependent parameters that can influence wave propagation, such as the viscosity of solution, which depends on the solvent's nature and temperature. The increase in the viscosity of a liquid increases the power threshold for reaching the cavity. Compared to low viscosity liquids, viscous liquids require higher variations of acoustic pressure in order to allow the cavitation phenomenon.

The energy provided by acoustic emission is partly dissipated as heat, which leads to an increase in the medium's temperature. Therefore, it is sometimes necessary to exert thermal control of the medium by a heat transfer fluid to ensure constant temperature during work. The impact of temperature increase during reaction is a complex issue, as temperature influences the viscosity of the reaction medium, vapor pressure and concentrations of dissolved gases, but also the reaction kinetics involved [GHA 08].

Reports of several examples of sonochemical production of metal oxide and metal nanoparticles are provided by Tables 1.4 and 1.5.

Oxide	*d* (nm)	Conditions	Reference
Bi_2O_3	40–100	$Bi(NO_3)_3$, PVP, H_2O, pH 11 US: 600 W, 20 kHz, 75 min, 60°C	[ZHA 06a]
	<23	$Bi(NO_3)_3$, PVP, H_2O, pH 1 US: 200 W, 24 kHz, 1 h, 80°C Annealed at 300°C, under air, 1 h	[AZI 15]
Cr_2O_3	200	$Cr_2O_7(NH_4)_2$, H_2O US: 100W, 20 kHz, 3 h	[DHA 97]
CuO	15–40	$Cu(NO_3)_2$, NaOH aq. (0.1–0.3M) US: 20 kHz, 600 W, 90°C, 0.5–2 h, Annealed at 100–300°C	[PEN 14]
	Nanorods 70 × 350	$CuSO_4$, NaOH aq. (0.05 M), polyvinyl alcohol US: 80°C, 30 min, then maturing at 25°C, 12 h	[SHU 13]
	50–70	$Cu(NO_3)_2$, H_2O, NaOH aq. (0.05 M) US: 20 kHz, 750 W, 1/2 h Annealed at 700°C	[WON 11]
Fe_2O_3	20 nm	$Fe(acac)_3$, tetraglyme, 2% vol. H_2O	[PIN 08]
	5–10	$Fe(acac)_3$, tetraglyme, 6% vol. H_2O US: 20 kHz, 13 W, 8 h	[BAN 07]
	6 <20	$Fe(CO)_5$, hexadecane C (4–12 nm), $Fe(CO)_5$, hexadecane and annealed at 450°C, 2 h US: 20 kHz, 50 W, 20°C, 3 h, atm. Ar.	[REY 15]
Amorphous hollow spheres	21–23	$FeSO_4$, NaOH aq. (0.2 M), ethylene glycol US: 581 kHz, 29W, 20°C, 35 min	

Fe_3O_4	3–14	$Fe(CO)_5$, SDS (0.02M), H_2O US: 20 kHz, 25°C, 3 h, atm. Air.	[ABU 05]
HfO_2	14–25	$HfCl_4$, NaOH aq. US: 500W, 1 h Annealed at 500°C, 2 h	[AMA 12]
In_2O_3	15–30	$In(acac)_2$, H_2O, pH : 9–10 US: 20 kHz, 750 W, 1 h, 80–90°C	[HAF 14]
MnO_2	 50–150 200–1,000	$KMnO_4$, H_2O, US: 200 kHz, 13 W, 20°C, pH 6.0, 9 min US: 200 kHz, 13W, 20°C, pH 9.3, 8 min	[ABU 14]
Mn_2O_3	50	$KMnO_4$, H_2O US: 100 W, 20 kHz, 3 h	[DHA 97]
Mn_3O_4	60 10–40 30–40 20–70	$MnSO_4$ $Mn(NO_3)_2$ $Mn(CH_3COO)_2$ Mn(acetylacetonate) US: 180 W, 12 h	[KUM 03]
NiO	20–150	$Ni(CH_3COO)_2$, NaOH aq. (0.1 M), H_2O Ethanol, polyethylene glycol US: 6–18 W, 30 kHz, 0.5–3 h, then calcination 500°C, 1/2 h	[ASL 11]
PbO_2	50–150	PbO, $(NH_4)_2S_2O_8$, H_2O US: 20 kHz, 600 W, 75°C, 1 h	[GHA 08]
SiO_2	60	$Si(OC_2H_5)_4$, methanol, H_2O, BS, US: 20 kHz, 100 W, 1/2 h, 25°C, atm. Air then ethylene diamine, washing with methanol, calcination 600°C, 2 h.	[MAS 14, MAS 16]
SnO_2	23–30	$SnCl_2$, PEG, H_2O, 1,2-ethylene diamine US: 10 min Then calcination, under air, 500°C, 3 h	[WAN 15a]
TiO_2	17	$Ti(OC_2H_5)_4$, ethanol, H_2O, acetic acid US: 38 kHz, 30 W Calcination 530 K	[AWA 03]
$WO_3 \cdot H_2O$	Nanoplates L:250 E: several dozens	WCl_6, H_2O US : 20 kHz, 1,000 W, 200°C	[CHA 11a]
WO_3	50–70	$W(CO)_6$, Diphenylmethane US: 20 kHz, 100 W, 90°C, 3 h, atm Ar/O_2 (8/2), then annealed at 1,000°C, under air	[KOL 12]
ZnO	4–7	$Zn(CH_3COO)_2$, LiOH, Ethanol, US: 20 kHz, 900 W, 1–120 min	[QIA 03]

Table 1.4. *Sonochemical synthesis of metal oxide nanoparticles acac, acethylacetonate; BS, Schiff base; bis(acethylacetonato) propylene-1,3 diimine or bis (acethylacetonato) butylene-1,4 diimine; SDS, sodium dodecylsulfate; PVP, polyvinylpyrrolidone*

Element	d (nm)	Conditions	Reference
Fe	10	$Fe(CO)_5$, decane US: 20 kHz, 100 W, 0°C, 3 h, atm. Ar	[SUS 91]
	8	$Fe(CO)_5$, octanol, PVP (M_w 40,000), US, 20°C, 1 h, atm. Ar	[SUS 96]
	8	$Fe(CO)_5$, octanol, oleic acid US: 30°C, 3 h, atm. Ar	
Cu	50–70	$Cu(N_2H_3CO_2)_2$, H_2O US : 20 kHz, 100 W, 2–3 h, 80°C, atm. Ar/H_2 (95/5)	[DHA 98]
Ni	10	$Ni(CO)_4$, decane US: 20 kHz, 100 W, 3 h, 0°C, atm. Ar	[KOL 96]
C	150–400	Toluene / H_2O (1/100 in volume) US: 480 kHz, 2.5 W, 6 h, 20°C	[FUJ 14]

Table 1.5. *Sonochemical synthesis of metal and carbon nanoparticles*

1.2.1.3. *Conclusion*

Sonochemical synthesis proves to be a technique that is easy to implement. Nevertheless, the quantities manufactured are low, of the order of several grams per experiment. Sonochemistry therefore remains a laboratory technique. It is worth noting that pollution problems may occur due to the erosion of metals in the probe alloy when waves are generated by an emission probe immersed in the solution.

1.2.2. *Microemulsion synthesis*

1.2.2.1. *Definition*

The term microemulsion was introduced by Schulmann *et al.* [SCH 59] and it denotes a thermodynamically stable dispersion of two immiscible liquids. Adding a surfactant stabilizes the system. At low concentration, the surfactant is dissolved and is present as a monomer. When its concentration in solution reaches a critical value, the spontaneous self-assembly of monomers into various structures and aggregates takes place.

The phase diagrams of microemulsions are very rich, exhibiting a very wide variety of structures. These result from geometrical constraints, curvature energy, interface tension and solubility of surfactant in the continuous phase. Thus, it is possible to obtain oil-in-water or water-in-oil dispersions and also bicontinuous structures. The form taken by dispersed structures, called micelles, is generally spherical. This form is the most frequent because, due to its surface/volume ratio, it is the most advantageous for minimizing the contribution of interface tension to the system's overall energy. Cylindrical forms have nevertheless been experimentally observed [SVE 00].

Micelles emerge when surfactant concentration exceeds critical micelle concentration. "Micelles" are referenced when the polar side of the surfactant is at the outer surface of the micelle in contact with the continuous polar phase and the lipophilic organic phase inside the micelle. On the contrary, a "water in oil" microemulsion is formed when water is dispersed in a continuous organic phase. In this case, the surfactant molecules self-assemble to form aggregates called "reverse micelles", the polar parts of the surfactant being directed toward the inner part of the micelle.

These emulsions can be used to carry out chemical reactions and they most notably permit the synthesis of metal or metal oxide nanoparticles.

In 1982, Boutonnet *et al.* obtained noble metal (Pt, Pd, Rh and Ir) nanoparticles of very small sizes (from 2 to 5 nm) by reduction of chloride salts with either hydrazine or a hydrogen bubbling in an emulsion consisting of water, octanol and cetyltrimethylammonium bromide (CTAB), or by using a microemulsion composed of water, hexane and pentaethyleneglycol dodecylether (PEDGE) [BOU 82].

1.2.2.2. *Preparation of nanoparticles*

The most widespread methodology for the synthesis of nanoparticles consists of mixing two identical inverted microemulsions (water: oil: surfactant) each containing a different reagent dissolved in the aqueous phase confined in the micelles [KIM 01]. The second method consists of adding a reagent in pure state [BOU 82] or dissolved in the aqueous phase [PAN 02, QIU 99] directly to a solution of inverted micelles containing the other reagent. There are numerous syntheses in micellar phase, and only those reactions leading to the production of solid nanoparticles will be developed here.

1.2.2.3. *Mechanisms involved*

There are several works that attempt to explain the mechanisms of formation of nanoparticles in microemulsions. Destrée and Nagy [DES 06] propose a mechanism that is based on LaMer diagram [LAM 50] representing the concentration of dissolved species as a function of time during crystal formation. In the case of emulsion synthesis, nucleation occurs at the beginning of the reaction in micelles that, on the one hand, have sufficient ion concentration, allowing for the nucleation to generate a stable nucleus and, on the other hand, have been in contact with the second reagent. According to this model, the first stage in nanoparticle formation is the nucleation, followed by the growth of particles, in the second stage. Growth kinetics being faster than nucleation kinetics, the latter can be neglected and only growth of nanoparticles takes place. When the reaction is completed, a population of quasi-monodisperse particles will be obtained.

According to a second model, the reagent will thermodynamically stabilize the formed nanoparticles. The size of particles will thus remain constant, although nucleation goes on simultaneously with the particle growth stage. The particles having reached a certain size cease to grow due to the surfactant stabilizing effect, irrespective of the micelles' concentration in metal salts or size.

These mechanisms seem to be supported by the works of Concha. Tojo *et al.* conducted a simulation-based study of nucleation and particle growth processes [TOJ 97]. Their works show that separation in time of the nucleation and growth stages is clearly facilitated by a high concentration of reagent in the droplets. When concentration is low, nucleation and growth phenomena take place simultaneously.

These works also show that the growth of particles may be due to the "Ostwald ripening" process, which consists of the solubilization of a small particle that afterward redeposits on a larger particle. Because of the difference in free energy between small and large particles, they cannot be in equilibrium at the same concentration in solution. Due to Gibbs–Thomson effect, smaller particles having higher solubility than the large ones, they are slowly dissolved. Thus, larger particles end up within a slightly supersaturated solution and the solute precipitates on their surface. In the case of microemulsions, this mechanism develops inside the micelles and by exchange among micelles when they interact. The second process of growth of particles is the autocatalytic process [BES 05].

For the classical synthesis path involving the mixture of two emulsions, Fletcher has proposed two limit cases describing the exchanges of compounds solubilized in water inside micelles [FLE 87]. First of all, the reactive species should remain confined inside the droplets, which excludes the possibility to diffuse in another droplet by passing through the continuous phase formed by the organic solvent. In effect, given the low solubility of salts in organic media, a dynamic exchange of reagents among the droplets through the continuous organic phase is not favorable. Second, the exchange of solutes inside the micelle has to be the key stage in the reaction kinetics.

The first limit case involves the two droplets getting closer and in contact, but without coalescing. The solubilized species diffuse through the surfactant bilayer at the point of contact between the droplets.

The second limit case involves the fusion of the two droplets and formation of one "dimer" droplet, a structure whose lifetime is below one microsecond, and which afterward breaks up into two new droplets that contain a mixture of solutes of the two initial droplets.

The simulation works of Bandyopadhyaya *et al.* referring to the method of classical synthesis with two emulsions corroborate previous works. The mechanism proposed for obtaining nanoparticles requires that micelles exchange by coalescence the reagents under soluble form, thus triggering the formation of nuclei, followed by the growth of nanoparticles, supplied by the reagents in solution in the micelles when they come into contact [ETH 06a, ETH 06b, BAN 97, BAN 00]. In order to be able to predict the size of particles, Bandyopadhyaya *et al.* have added a stage of nuclei coagulation during the coalescence of two micelles, each containing a nucleus. The threshold for taking this mechanism into account becomes important when the resulting particle is larger than the micelle issued from the two micelles' fusion [ETH 07].

1.2.2.4. *Influence of implementation parameters*

The diameter of the synthesized particles depends on several implementation parameters. Correlations have been observed between the form of the micelles present in the microemulsion and the form of the particles produced [BUR 05, PIL 03].

1.2.2.4.1. Concentration of surfactant

When water and oil concentrations are constant, an increase in the concentration of surfactant leads to an increase in the number of micelles, but their mean diameter will be smaller. This results in a reduction of the number of metal ions present in a micelle, a fact less favorable to nucleation. By contrast, when there are more micelles, they will collide more frequently, and this can favor the growth process by Ostwald ripening. On the contrary, the quantity of ions in aqueous solution provided at each micelle's contact is lower than it would have been if its size were higher, and thus growth by autocatalysis slows down. This suggests that the growth of particles is subjected to conflicting effects. The work carried out by Lopez-Quintela and Rivas on the control of iron particles' size shows that the size of particles is higher when the concentration of surfactant is low, compared to high surfactant concentration [LOP 93].

1.2.2.4.2. Nature of surfactant

Surfactants are amphiphilic molecules whose hydrophilic part can consist of ions (anions or cations). Their solubilization in water leads to the release of a counterion and the presence of a charge on the polar head: CTAB [CHE 11], sodium dodecylsulfate (SDS) [QIU 99] and even didodecyldimethylammonium bromide [BUM 11] are part of it. There are amphoteric surfactants that, depending on the pH, simultaneously act as anionic and cationic molecules.

Some nonionic surfactants that are commonly used in the microemulsion synthesis of nanoparticles are sorbitan esters (sorbitan monolaurate; SPAN®) or polyethylene glycol esters (TWEEN 80®) [DIC 15], ethers such as IGEPAL co-520® (polyoxomethylene nonylphenylether) [ARR 99],Triton X-100® (4-(1,1,3,3 tetramethylbutyl) phenyl-polyethylene glycol) [WAN 14a, PAN 02] or the PEDGE [BOU 82].

There are also surfactants that have a double hydrophobic chain such as AOT (bis (2-ethylhexyl) sulfosuccinate sodium salt) [LU 15, OH 01] or ammonium carboxylates perfluoroether [MEZ 05] or even perfluorothiols [FER 07]. These are used for the synthesis of metal particles, especially when supercritical carbon dioxide is employed as solvent.

The use of a surfactant for a particular water–oil mixture is not trivial, as besides the resulting phase diagram, micellar exchanges are affected by the very nature of the surfactant. Lopez-Quintea *et al.* have proposed an explanation that involves the flexibility of the surfactant film. This parameter influences the curvature of the film that is at the origin of the creation of a channel between two micelles that are in contact. The more rigid the film, the more reduced the channel, which will induce a decrease in intermicellar exchanges, whereas a more flexible film will facilitate the transfer of reactants through the interdroplet exchange zone [LOP 03, DE 05].

Finally, the charges carried by the surfactant and also by the counterion can trigger modifications of local concentration of reactants, inducing a modification of the nucleation rate.

1.2.2.4.3. Reaction rate

Depending on the nature and concentration of the reduction agent used, the reduction process can be more or less rapid. The higher the reaction rate, the sharper the decrease in particle size. It is possible to prepare metal nanoparticles by the reduction of metal salts, using agents such as sodium borohydride [QI 97] or potassium borohydride [CHE 11], hydrazine [BOU 82] or dihydrogen [LU 15]. Thus, in the case of platinum salts, the reduction with hydrazine is faster than the reduction with bubbled hydrogen [BOU 82]. In order to decrease the time of reaction, the reducing agent is often added in excess relative to the concentration of metal salt [WAN 01]. The simulation works conducted by Lopez-Quintea *et al.* indicate that the final size of particles is inversely proportional to the reaction rate. They also show that a high reaction rate results in the distribution of particle size being narrower than the one for lower rates [DE 05, BAR 05].

1.2.2.5. *Conclusion*

This method offers a simple way to obtain metal or metal oxide particles of very small size (several nanometers) while allowing for very good control of the particle size and narrow size distribution, and even getting near-monodisperse particles. A weak distribution of particle size is needed for obtaining uniform properties. This is all the more so in the case of mixtures intended for the manufacturing of nanothermites. This will enable us to obtain a more homogeneous mixture of constituents. This control is readily achieved by modifying experimental parameters such as the reactant concentrations and the water:surfactant concentration ratio. In some cases, it is also possible to control the form of particles depending on the ternary system employed (water–oil–surfactant).

One drawback of this technique is the low quantity of product obtained, in comparison with the reaction volume. This is due to the low concentrations of dissolved reactants and to the fact that a significant part of the reaction volume (the organic phase) is reduced to the role of liquid that separates micelles. This synthesis method is not adapted for the production of significant quantities of metal or metal oxide particles and in particular for producing nanothermites, therefore it is recommended for laboratory use.

1.2.3. *Solvothermal syntheses*

The first hydrothermal synthesis related works were conducted by Schafhäult in 1845 starting from silicic acid present in the water heated for 8 days in Papin's steam digester in order to obtain quartz crystals [SCH 45]. Starting from the 1890s, this methodology was developed on corundum. A significant number of metal oxides are today obtained by this method.

1.2.3.1. *Principle*

The term solvothermal is used to describe the technologies in which a heterogeneous chemical reaction occurs under conditions of pressure above one atmosphere and temperature above ambient temperature, irrespective of the solvent used. When water is used as solvent, then the synthesis is "hydrothermal".

These methods can be divided into two different categories, depending on the temperatures and pressures involved:

– the classical or subcritical solvothermal synthesis corresponds to reactions conducted at temperatures above the solvent's boiling point, but below its critical temperature, and within a range of pressures below the critical pressure of the solvent [YAN 07c, QIN 11];

– the supercritical solvothermal route corresponds to reactions conducted at pressures and temperatures above the critical values of pressure and temperature of the solvent.

Critical temperatures and pressures of the main solvents used in the synthesis of metal oxide and metal nanoparticles are provided in Table 1.6.

Solvent	T_c (°C)	P_c (MPa)
Water	374	22.10
Carbon dioxide	31.2	7.38
Methanol	240	8
Ethanol	243	6.39

Table 1.6. *Common solvents used in the solvothermal synthesis of metal and metal oxide nanoparticles (Tc: critical temperature, Pc: critical pressure)*

Solvothermal reactions can be conducted in discontinuous mode in a closed reactor or in continuous mode. The first works related to the continuous synthesis of nanoparticles in supercritical medium were conducted in the 1990s by Adchiri and Arai [ADS 92b]. These works refer to the hydrolysis of a dozen metal salts in order to obtain the corresponding oxides by continuous injection of aqueous solutions of salts in a flow of water in the supercritical phase. The time of reaction is of the order of 2 min and the size of the particles obtained ranges between 20 and 600 nm, depending on the nature and concentration of salts [ADS 92a].

The main advantage of carrying out reactions in subcritical or supercritical aqueous medium is that the properties of the reaction medium can vary within a wide range by modifying pressure and/or temperature. Thus, subcritical and supercritical water were used as reaction medium for the synthesis of crystalline particles, in particular of metal and metal oxide nanoparticles. This method also allows us to control morphology, size and size distribution of particles by relatively slight variations of temperature and pressure [JUN 01].

In the case of metal oxide formation, a hydrothermal synthesis method can be described according to two chemical reactions: hydrolysis of salts in solution, followed by dehydration as follows:

$$M(L)_x + xH_2O \rightarrow M(OH)_x + xHL \quad \text{Hydrolysis} \qquad [1.6]$$

$$\mathrm{M(OH)}_x \rightarrow \mathrm{MO}_{x/2} + \frac{x}{2}\mathrm{H_2O} \quad \text{Dehydration} \qquad [1.7]$$

When these reactions take place in supercritical water, it is possible to obtain metal particles by adding a reducing agent such as formic acid [ARI 11], formaldehyde [ZHO 15a] or glycerol [KIM 14].

The following mechanisms come into play: in supercritical medium, the formic acid decomposes [YU 98, YOS 04] either by decarbonylation (or dehydration):

$$\mathrm{HCO_2H} \rightarrow \mathrm{H_2O} + \mathrm{CO} \qquad [1.8]$$

or by decarboxylation (or dehydrogenation):

$$\mathrm{HCO_2H} \rightarrow \mathrm{H_2} + \mathrm{CO_2} \qquad [1.9]$$

Formaldehyde decomposes into:

$$\mathrm{HCOH} \rightarrow \mathrm{H_2} + \mathrm{CO} \qquad [1.10]$$

and glycerol decomposes into:

$$\mathrm{C_3H_5(OH)_3} \rightarrow 4\mathrm{H_2} + 3\mathrm{CO} \qquad [1.11]$$

In turn, carbon monoxide can react according to the gas reaction to water:

$$\mathrm{H_2O} + \mathrm{CO} \rightarrow \mathrm{H_2} + \mathrm{CO_2} \qquad [1.12]$$

Then, hydrogen reacts with the metal oxide resulting from the hydrolysis and dehydration reactions [ARI 11, SEO 11, KUB 14].

$$\mathrm{MO}_x + x\,\mathrm{H_2} \rightarrow \mathrm{M} + x\,\mathrm{H_2O} \qquad [1.13]$$

While the vast majority of solvothermal reactions are conducted in water for the fabrication of metal particles, these can also be obtained in other media.

Titanium oxide nanoparticles were obtained either by the hydrolysis of titanium (IV) isopropoxide [STA 03] or by the hydrolysis of titanium (IV) diisopropoxide bis(acetylacetonate) [ALO 07] solubilized in an aqueous phase and injected in supercritical carbon dioxide. The particles obtained are spherical, but their size distribution is wide, ranging between 20 and 800 nm in the first case, and somewhat narrower in the second case (between 100 and 600 nm).

Alumina nanoparticles have been obtained by oxidation of aluminum particles in a medium of water/supercritical carbon dioxide. This leads especially to producing particles with sizes ranging between 300 and 500 nm, composed of crystallized alumina in its alpha form, and smaller particles of crystallized alumina in the gamma phase, with sizes between 20 and 70 nm [VOS 10].

Various works show that it is possible to use supercritical methanol or ethanol for the continuous or discontinuous synthesis of metal oxide nanoparticles and influence the form of the particles obtained. For example, the use of supercritical methanol allows us to elaborate spherical ZnO nanoparticles [VER 10b, HAN 11], while particles obtained with supercritical water take the form of nanorods [KIM 15c].

Depending on the implementation conditions, the use of supercritical alcohol allows us to obtain also metal particles [KIM 09, CHO 10].

In Choi *et al.* [CHO 10, KIM 09], it has been shown that it was possible to form nickel, silver and copper nanoparticles with supercritical methanol without reduction agent above a certain temperature threshold, which was specific for each metal. Below this threshold oxide and hydroxide formation can be observed. The presence of oxides and hydroxides was equally noticed during the continuous synthesis, when the residence time is too short. According to the explanation advanced, at high temperature, supercritical methanol acts as a hydroxylation agent. In effect, the hydroxyl ions are dissociated from the methanol molecules, when in the supercritical phase [BUL 08, AND 06]. The "free" hydroxyl ion can thus participate in the reduction by transfer of electrons to the metal precursor or to a metal intermediary [ZHO 15a].

The examples reported hereafter refer predominantly to the synthesis in water in subcritical or supercritical conditions. Several examples using an alcohol [KIM 09] or carbon dioxide [VOS 10] are reported in the tables below.

Oxide	d (nm)	Conditions	Reference
Bi_2O_3 nanowires	*D*: 40 *L*: several μm	$Bi(NO_3)_2$, Na_2SO_4, NaOH aq., H_2O, 120°C, 12 h	[WU 14b]
CeO_2 particles	13–17	$Ce(NO_3)_2$, hexamethylene tetramine, H_2O, PVP, 180°C, 100 min	[XIE 15]
Co_3O_4 nanocubes	20	$Co(NO_3)_2$, H_2O_2 aq., pH 8–9, PEG (M_w 20 000), H_2O n-butanol, 160°C, 10 h	[YAN 07c]
Cr_2O_3	29–60	CrO_3, ethanol, H_2O, 190°C, 1 h, calcination 500°C 1 h	[PEI 09]
CuO platelets	*L*: 250–300 *l*: 50-100	$Cu(NO_3)_2$, NaOH AQ., sodium citrate, H_2O, 160°C, 12 h	[SON 15]
CuO nanotubes	*L*: 20–120 *D*: 7	$Cu(NO_3)_2$, ethylene diamine, NaOH aq., H_2O, 100°C, 24 h, calcination 400°C, 5 h	[IPE 14]
α-Fe_2O_3 nanocubes	100–200	$Fe(NO_3)_3$, triethylamine, H_2O, 160°C, 24 h	[QIN 11]
α-Fe_2O_3 polyhedrons	90–110	$FeCl_3$, NH_3 aq., H_20, 180°C, 8 h	[MA 10]
α-Fe_2O_3 polyhedrons	20–50	$FeCl_3$, CH_3CO_2Na aq., PVP, H_20, 200°C, 18 h	[ZHU 12]
In_2O_3 platelets	15–35	$In(NO_3)_3$, Triton X-100, hexamethylene tetramine, H_2O, 180°C, 8 h, calcination 400°C, 3 h	[SEL 14]
α-MnO_2 nanowires	*D*: 15–20 *L*: several micrometer	$KMnO_4$, $(NH_4)_2S_2O_8$, H_2O, HNO_3, 180°C, 15 h	[TOU 14]
MoO3 nanoribbons	*l*: 100–300 *L*: several μm	H_2MoO_4, oxalic acid, H_2O, HNO_3, 180°C, 24 h	[ZEN 13]
α-MoO_3 nanorods	*L*: 3 μm *D*: 200	Na_2MoO_4, HCl, H_2O, 120°C, 18 h	[WAN 13c]
MoO_3	<100	Na_2MoO_4, cyclohexane, sc CO_2, 35°C, 150 bars	[REV 10]
NiO particles	100–3,000	$Ni(NO_3)_2$, H_2O, 400°C, 300 bars 10 min	[KIM 14]
Sb_2O_3 polyhedrons Sb_2O_3 nanoribbons	100 *L*: 4 μm, *l*: 400, E : 20	$SbCl_3$, pH 7–9, 150°C, 16 h, CTAB butanol-1/ethanol Water/butanol-1/ethanol	[LIU 10]
Sb_2O_3 nanorods	*L*: 1–2 μm, *l*: 50–150	$SbCl_3$, H_2O, pH 8–9, 120–180°C, 12 h	[CHE 05b]
SiO_2 spheres	2–14	Na_2SiO_3, H_2O, AOT, sc CO_2, 5.60 MPa, 308.2 K	[ZHA 06b]
SnO_2 nanorods spheres	7–27 10–15	$SnCl_4$, NH_3 aq., H_2O, pH 10.0, 180–250°C, 40 atm. without surfactant, calcination 600°C, 4 h with CTAB	[JAI 06]

SnO_2 nanorods nanoflower aggregate	270 for the nanorods	$SnCl_4$, polyacrylamide, NaOH aq., H_2O, 200°C, 12 h	[TIA 14]
SnO2	19	$Sn(Cl)_4$ CTBA, urea, pH 8, H_2O, 100°C, 24 h, calcination 600°C, 2 h	[BLE 14]
SnO_2 platelets	2–19	$SnCl_4$, CTAB, urea, H_2O, 100°C, 24 h	[FAR 10]
WO_3 hexagonal nanorods	*L*: 500–1,000 *D*: 50–60	$(NH_4)_6H_2W_{12}O_{40}.nH_2O$, Na_2SO_4, citric acid, H_2O, pH 1.5, 180°C, 24 h	[WAN 14b]
WO_3	*D*: 80 *L*: >2.5 μm	Na_2WO_4, Na_2SO_4, K_2SO_4, H_2O, pH 1.5–2, 180°C, 24 h	[MIA 15]
WO_3 nanowires	*D*:15 *L*: >5μm	Na_2WO_4, H_2O, glycine, pH 1.3, 180°C, 12 h	[LIN 15]
WO_3 Particles agglomerate	500 20–50	Ammonium tungstate, HNO_3, pH 3, 180°C, 30–60 min, assisted by microwaves (200W), annealed at 250°C, 10 min under air	[HER 14]
WO_3 nanorods	<50	Na_2WO_4, H_2O, pH 2.0, 180°C, 0.75–3 h, assisted by microwaves (200 W)	[CHA 11b]
h-WO_3	*D*: 100–150 *L*: 1–2 μm	Na_2WO_4, H_2O, pH 1–1.2, 180°C, 4–12 h,	[HA 09]
ZnO particles	<300	$Zn(CH_3COO)_2$, NaOH aq., pH 9, H_2O, 120°C, 12 h	[BAM 14]

Table 1.7. *Hydrothermal synthesis of metal oxide nanoparticles in closed reactor and in subcritical medium; L, length; l, width; D, diameter; E, thickness; CTAB, cetyltrimethylammonium bromide; PVP, polyvinylpyrrolidone; sc, supercritical*

Materials	*d* (nm)	Conditions	Reference
Al_2O_3	20–400	Al, H_2O, sc CO_2, 446–723 K, 0.85–38.03 MPa	[VOS 10]
WO_3	25–100	Na_2WO_4 cyclohexane, sc CO_2, 35°C, 150 bars	[REV 10]
TiO_2 spheres	20–800	TTIP, Zonyl® FSP fluoro surfactant–2-propanol–water (35–20–45), sc CO_2, 20–30 s	[STA 03]

Table 1.8. *Hydrothermal synthesis of metal oxide nanoparticles in closed reactor and in supercritical carbon dioxide medium; TTIP, titanium tetraisopropoxide*

Materials	*d* (nm)	Conditions	Reference
α-Fe_2O_3 particles	30–60	$Fe(NO_3)_3$, H_2O, pH 1.8, 250–350°C, FW/FS: 3/2, flow rate 50–100 mL/min	[KRI 15]
ZnO particles	20–25	$Zn(NO_3)_2$, NaOH aq., H_2O, 210°C, 260 bars, flow rate 10 mL/min	[SON 11]
TiO_2 spheres	200 ± 100 (a) 270 ± 125 (b)	DIPBAT (a) or TTIP (b), ethanol, sc CO_2, 300°C, 20 MPa, residence time: 2 min	[ALO 07]

Table 1.9. *Continuous hydrothermal synthesis of metal oxide nanoparticles in subcritical medium: FW, subcritical fluid flow; FS, solute solution flow; DIPBAT, diisopropoxititanium bis (acetylacetonate); TTIP, titanium tetraisopropoxide*

Materials	*d* (nm)	Conditions	Reference
γ-Al_2O_3 spheres		$Al(NO_3)_3$, H_2O, 585°C, FW: 24-43.5 g/min, FS: 3.5–20 g/min, 25–35 MPa	[NOG 08]
CeO_2	30–50	$Ce(NO_3)_3$, sc methanol, 400°C, 30 MPa, residence time 40 s	[VER 09]
Co_3O_4 nanocubes	9–90	$Co(CH_3COO)_2$, H_2O_2 aq., 430°C, 240 bars, residence time 2s	[LES 12]
CuO	8–14.8	$Cu(NO_3)_2$, H_2O, 400°C, 30 MPa, residence time: 0.129–2.08 s	[SUE 11]
Fe_3O_4 particles aggregate	16–25 90–150	$Fe(NO_3)_3$, methanol, 400°C, 30 MPa, residence time: 38 s	[VER 10a]
α-Fe_2O_3 spheres	37 ± 6	$Fe(NO_3)_3$, H_2O, 400°C, 30 MPa, residence time: 38 s	[VER 10a]
Fe_2O_3	4–6.7	$Fe(NO_3)_3$, H_2O, 400°C, 30 MPa, residence time: 0.129–2.08 s	[SUE 11]
Fe_3O_4 sphere	21 ± 2	$Fe(NO_3)_3$, sc methanol, 400°C, 30 MPa, residence time 38 s	[VER 10a]
NiO particles	20–55	$Ni(NO_3)_2$, H_2O, 400°C, 30 MPa, total flow rate: 30–60 g/min	[KAW 10]
NiO	6.6–9.7	$Ni(NO_3)_2$, H_2O, 400°C, 30 MPa, residence time: 0.129–2.08 s	[SUE 11]
TiO_2 particles	13–30	$Ti(SO_4)_2$, KOH, H_2O, 400°C, 30 MPa, residence time 1.7 s	[KAW 09]
ZnO nanorods agglomerates particles	 *D*: 200–500 *L*: 1–5 μm 200–1000 5–10	$Zn(NO_3)_2$, FW: 9 g/min, FS: 2 g/min, sc H_2O, 400°C, 30 MPa sc methanol, 270°C, 20 MPa	[KIM 15c]
ZnO particles	20–25	$Zn(NO_3)_2$, NaOH aq., H_2O, 210°C, 260 bars, flow rate 10 mL/min	[SON 11]
ZnO nanorods	*L*: 22–166 *D*: 20–331	$Zn(NO_3)_2$, KOH, H_2O, 410°C, 305 bars, FW = FS = 20 mL/min	[DEM 14]

Table 1.10. *Continuous hydrothermal synthesis of metal oxide nanoparticles in supercritical medium: L, length; D, diameter; FW, flow of supercritical fluid; FS, flow of solute solution*

Metals	*d* (nm)	Conditions	Reference
Ag particles	390 ± 114 320 ± 73	$AgNO_3$, CH_3OH, 400°C, 300 bars, 5 min $AgNO_3$, C_2H_5OH, 400°C, 300 bars, 5 min	[KIM 09]
Co cubes	30	$Co(NO_3)_2$, CH_3OH, oleic acid, 400°C, 300 bars, 15 min	[SHI 10]
Co agglomerates	200-400	$Co(CH_3COO)_2$, H_2O, CH_3OH, HCHO, 340–420°C, 22.1 MPa, 10 min	[SEO 11]
Cu particles	420 ± 119	$Cu(NO_3)_2$, CH_3OH, 400°C, 300 bars, 5 min	[KIM 09]
Cu particles	14-,85	$CuSO_4$, EDTA$(Na)_2$, NaOH aq., H_2O, HCHO, 400°C, 25 MPa, 3 min	[ZHO 15a]
Cu particles	100-1,000	$Cu(NO_3)_2$, glycerol, H_2O, 400°C, 300 bars, 10 min	[KIM 14]
Ni particles	50 ± 10	$Ni(NO_3)_2$, CH_3OH, 400°C, 300 bars, 5 min	[KIM 09]
Ni particles	100–1,000	$Ni(NO_3)_2$, glycerol, H_2O, 400°C, 300 bars, 10 min	[KIM 14]

Table 1.11. *Discontinuous hydrothermal synthesis of metal nanoparticles in supercritical medium (EDTA(Na)$_2$, disodium ethylenediamine tetraacetate dihydrate)*

Materials	*d* (nm)	Conditions	Reference
Ag spheres	148 ± 32	$AgNO_3$, 150°C, 30 MPa, FS 6 mL/min, sc methanol	[CHO 10]
Cu particles	14–85	$Cu(SO_4)$, EDTA$(Na)_2$, NaOH aq., sc H_2O, HCHO, 400°C, 25 MPa, residence time: 3 min	[ZHO 15a]
Cu spheres	240 ± 44	$Cu(NO_3)_2$, 250°C, 30 MPa, FS 6 mL/min, sc methanol	[CHO 10]
Ni spheres	119 ± 19	$Ni(NO_3)_2$, 400°C, 30 MPa, FS 6 mL/min, sc methanol	[CHO 10]

Table 1.12. *Continuous hydrothermal synthesis of metal nanoparticles sc, supercritical; FS, flow of solute solution*

The characteristics of the particles (mean size, size distribution, forms, etc.) are governed by numerous parameters, such as temperature, pressure, concentration of precursor solution, its nature, the nature of the solvent, purity of the raw materials, design of the reactor, duration of reaction in the case of continuous reactions, time of residence in the flow, etc.

1.2.3.2. *Effect of temperature*

Several works have shown that an increase in temperature leads to an increase in the size of the resulting particles, when hydrothermal reactions are conducted at various temperatures [LES 12].

To give an example, Patzke *et al.* use temperature as a parameter when adjusting the diameter of the molybdenum oxide nanorods that they manufacture. For a temperature between 90 and 12°C, they obtain thin nanorods, with a mean diameter of 40 nm. For a temperature of 150°C, nanorods nearly reach their maximal diameter of 150 nm [PAT 04]. Kolen'ko and colleagues obtain anatase nanoparticles of 8 and 26 nm at 423 and 523 K, respectively [KOL 03].

A further effect of temperature was observed by Choi *et al.* during the continuous hydrothermal reactions of metal nitrates in supercritical methanol. When temperature increases above a certain value that depends on the metal ion, the particles obtained are metals and not oxides. Reduction takes place starting from 150°C for silver, 300°C for copper and 400°C for nickel. In this case, an increase in the reaction temperature activates a mechanism of metal oxide reduction by supercritical methanol [CHO 10].

1.2.3.3. *Effect of precursor concentration*

Several authors have shown that the size of particles formed was directly influenced by the concentration of precursor salt. The higher the precursor concentration, the larger the particles. Furthermore, strong concentrations facilitate the formation of aggregates [LES 12, ZHU 12, JAI 06].

For example, Søndergaard and colleagues obtained an increase in the size of zinc oxide particles from 29 to 34 nm for a precursor concentration of 0.005 mol/L and from 49 to 59 nm for a concentration of 0.5 mol/L [SON 11]. Similarly, Zhou *et al.* have varied the size of copper particles from 14 to 50 nm by varying the concentration of copper sulfate from 0.05 to 0.5 mol/L. Above this value, the size of particles slightly changes. However, their size distribution becomes more heterogeneous. At low concentration, nuclei are rapidly generated, and the low concentration of ions still present does not allow for further growth. The low density of nuclei generated prevents or in any case significantly slows down aggregation-induced growth. This results in small particles with low polydispersity. On the contrary, at high concentration, nucleation can extend over a longer period, which results in larger particles with higher polydispersity [ZHO 15a].

1.2.3.4. *Effect of surfactant presence*

The effect of surfactant on the solvothermal synthesis of nanoparticles is manifested along two axes. The first consists of preventing or reducing aggregate formation [XIE 15]. The use of non-ionic surfactant such as polyethylene glycol facilitates proper dispersion of Co_3O_4, while the use of an ionic surfactant such as sodium dodecylbenzenesulfonate does not allow us to prevent the formation of aggregates [YAN 07c].

The second is to induce morphological modifications of the particles. These are due to a preferential mechanism of growth of particles alongside privileged crystallographic directions that depend on the material's structure. As crystallographic planes have different surface energies, surfactant molecules show preference for the interaction with the surface of certain planes. The surfactant molecules that are thus interacting prevent the deposition of solute molecules on these planes, which prevents their growth. Besides the absence of particles' aggregation in the presence of surfactant, Jain *et al.* obtained a change in the morphology of tin oxide particles generated, going from nanorods to spheres with a diameter ranging from 10 to 15 nm [JAI 06].

1.2.3.5. *Effect of pH*

The influence of base concentration, which is linked to the pH of a solution, has also been highlighted. Yang *et al.* have shown that pH influences the morphology of the particles obtained during cobalt oxide formation. For a pH of 8–9, regular nanocubes of 20 nm are obtained, while for a pH of 11–12, irregular nanocubes are observed [YAN 07c].

In the case of CeO_2 particle elaboration, Xie *et al.* have used hexamethylenetetramine to facilitate the growth of a structure according to a particular crystalline plane for particles' size ranging between 13 and 17 nm. When the base is absent, the particles have sizes between 200 and 300 nm and a spherical form [XIE 15].

Finally, the works of Søndergaard *et al.* relating to the zinc oxide synthesis, in which morphology variations are obtained depending on pH, should be cited. Large nanorods are obtained in acid medium, small particles are obtained in neutral medium, while in basic medium platelets and nanorods are formed [SON 11].

1.2.3.6. *Effect of solvent*

Depending on the solvent nature, it is possible to influence the morphology of the final structures. Liu *et al.* modify the form of antimony oxide particles by using mixtures of alcohol and water. The use of a butanol-1/ethanol mixture results in polyhedral oxide particles with cubic crystalline structure. When water is added to the alcohol mixture, nanoribbons with a mean length of 4 μm, width of 400 nm and thickness of 20 nm are obtained. These nanoribbons have an orthorhombic crystalline structure [LIU 10]. Using subcritical (120–180°C) hydrothermal synthesis, Chen *et al.* obtain orthorhombic Sb_2O_3 nanorods with a length ranging from 1 to 2 µm and width ranging from 50 to 150 nm [CHE 05b].

1.2.3.7. *Effect of anion*

The modification of counterions (nitrate, sulfate, acetate) in the precursor solutions can lead to modifications in the morphology of particles. Yang *et al.* have shown during the formation of Co_3O_4, the existence of irregular nanocubic forms, quasi-spherical forms or regular nanocubes when nitrate, sulfate or acetate was respectively employed [YAN 07c].

1.2.3.8. *Effect of duration*

It has repeatedly been shown in the literature that when solvothermal reactions are carried out in closed reactor, the increase in the reaction duration typically facilitates the increase in size of formed particles [QIN 11, MA 10]. Similarly, in continuous synthesis, the increase in residence time leads to an increase in the size of formed particles [LES 12].

Duration influences first of all the rates of conversions of solubilized salts into particles. For example, the works of Lester *et al.* on cobalt oxide show that when residence time (T_s) varies from 0.5 to 2 sec for a reaction conducted at 430°C, conversion rate goes from 89% to 99%. At 200°C, due to slower kinetics, conversion rates go from several percent to less than 30% after 20-sec residence time. As for the size of particles, it goes from 28 to 47 nm for residence times of 1 and 7 sec, respectively [LES 12].

1.2.3.9. *Microwave-assisted synthesis*

The literature mentions the use of microwave sources to assist hydrothermal reactions, as a means to accelerate reactions by efficiently providing radiation energy. This allows us to take into consideration larger-scale syntheses due to more rapid and uniform heating. Furthermore, radiation effects of microwaves allows uniform seeding of the solution. This assistance may also be useful in certain cases when there is an interest not to use additives [HER 14], surfactants or model agents, and it has allowed Hu *et al.* to obtain hematite nanorings with outer diameter of the order of 100 nm and inner diameter between 20 and 60 nm depending on the duration of irradiation [HU 07].

1.2.3.10. *Conclusion*

Hydrothermal synthesis, either classical or in supercritical medium, has several advantages in terms of control of size, distribution and morphology of the formed particles. The vast majority of solvothermal reactions are conducted in water. A supercritical medium is preferred for the reactions, as in this state the decrease in the dielectric constant improves the hydrolysis and dehydration stages. The most important drawback resides in the discontinuous fabrication. In effect, it is important to have a good management of the reaction time, which should be relatively short in order to obtain nanosized particles. Very rapid increases in temperature are therefore needed and can be obtained easily in smaller reactors, and this, combined with low concentrations, which are also required when nanometric sizes are aimed at, leads to obtaining low quantities of product. The continuous method allows us to avoid this obstacle, provided that the injection of the solute phase in the sub- or supercritical fluid stream is brought under control. The mixture should be carried out very rapidly, as residence times range from several seconds to several dozen seconds. But the injection nozzles should not be clogged by a sudden deposit of particles formed when the reaction takes place in the mixture area due to rapid precipitation of metal species. Geometries of continuous reactors have been modified in order to mitigate these problems [RUI 13]. This allows us to consider the synthesis of quasi-monodisperse small size nanoparticles of metal oxides that are of interest for nanothermites.

1.2.4. *Sol-gel syntheses*

The first works in the field of sol-gel synthesis were carried out by the French chemist Jacques-Joseph Ebelmen in 1846. He brought tetraethyl orthosilicate to react in the presence of water and form gelatin-silica [EBE 46]. The first patents related to sol-gels were issued at the end of the 1930s for surface treatments of rearview mirrors [GEF 39]. In the 1950s, colloidal silica spheres obtained by the sol-gel method were commercialized under the trade name Ludox® by Dupont de Nemours [ILE 52, BEC 51]. A very large number of works on this subject have been published since then.

The sol-gel method allows us to elaborate a wide variety of oxides in various forms (dense monolithic structures or, on the contrary, extremely porous structures, thin films, fibers, powders). Due to the wide diversity both in terms of materials and implementation technique, this method has become very attractive for a large number of industries.

1.2.4.1. *Principle*

Sol-gel methods rely on the solution-gelation method as their basic principle. This consists of preparing a stable solution (sol) that contains the molecular precursors and initiating hydrolysis-condensation reactions to obtain a gel.

The sol-gel method can be approached based on two synthesis paths. The inorganic or colloidal path resorts to metal salts (chlorides, nitrates, sulfates, etc.) in aqueous solution. It has the advantage of being less expensive, but controlling it may prove a complex task, due to the high number of reaction parameters (pH, temperature, mixture method, rate of hydrolysis, condensation, oxidation, etc.). When adopting this path, the counterion may influence the morphology, structure and even chemical composition of the chemical phase obtained. Finally, eliminating the counterion may prove problematic. This is why the organometallic method is preferred. The most frequently used precursor in the organometallic method is a metal alcoholate that can be easily converted into oxide. The sol-gel method is based on two chemical reactions: hydrolysis and condensation of precursors.

The hydrolysis stage consists of the following reactions:

$$[M(H_2O)_x]^{y+} + H_2O \rightarrow [(HO)M(H_2O)_{x-1}]^{(y-1)+} + H_3O^+ \xrightarrow{H_2O} M(OH)_y \quad [1.14]$$

$$M(OR)_x + H_2O \rightarrow HO - M - (OR)_{x-1} + ROH \xrightarrow{H_2O} M(OH)_x \quad [1.15]$$

The condensation stage consists of the following reactions:

– olation, in which a hydroxy bridge is formed, and a metal hydroxide is generated:

$$M - OH + H_2O - M \rightarrow M - OH - M + H_2O \quad [1.16]$$

– oxolation in which an oxo bridge is formed and a metal oxide is generated:

$$M - OH + HO - M \rightarrow M - O - M + H_2O \quad [1.17]$$

1.2.4.2. *Influence of operating conditions*

The synthesis of metal oxide nanoparticles can be influenced by many parameters, among which the most important are the pH and ratio of the water to metal precursor concentration.

1.2.4.2.1. Effect of temperature

Temperature usually accelerates the reaction kinetics, and this influences the reaction duration. On the contrary, the effect on the particle size seems much more complex. In effect, Kojima and Sugimoto have not observed a significant effect of temperature on the size of titanium oxide particles synthesized from titanium butoxide in the presence ammonia in aqueous medium, for temperature variations between 25 and 80°C. The size of the particles obtained is of the order of 450 nm [KOJ 08]. However, Qiu *et al.* have observed a complex effect of temperature during the synthesis of silica particles. There is an increase in particle size when temperature goes from 26 to 28°C and then a decrease from 28 to 32°C [QUI 13].

Finally, during the zirconium oxide synthesis Santos *et al.* have observed an increase in the size of particles, which in their view is caused by an increase in temperature. Zeta potential may vary when temperature increases, and this may affect the stability of particles in solution and facilitate size increase [DOS 14].

1.2.4.2.2. Effect of solvent

Water is commonly used as solvent when inorganic salts are used as precursors [WAH 07], but if the salt is soluble, it is also possible to employ alcohols [MBA 15, CAM 08]. Alcohol or a water–alcohol mixture is generally employed when using alcoholate-type precursors, as they only slightly miscible or non-miscible with water [SON 15, DEM 15, DOS 14, QUI 13].

When water concentration increases compared with metal precursor concentration, a decrease in the size of the particles obtained is observed [WAN 14c, KOJ 08].

1.2.4.2.3. Effect of pH

During sol-gel reactions under acid catalysis, hydrolysis kinetics is favored over condensation reactions, which start in general when hydrolysis is completed. Condensation reactions in basic medium are more rapid than hydrolysis reactions, which results in the precipitation of agglomerated thin particles.

The influence exerted by pH on the size of particles has been particularly studied by Pottier *et al.* in relation to the formation of anatase nanoparticles produced by hydroxylation of titanium tetrachloride. The size of particles was measured for a pH range between 2 and 6. The results indicate that the more acidic a solution is, the smaller the size of particles will be. Their mean diameter varies in the range between 5 nm, for pH of 1, and 8.6 nm, for pH of 5. This size variation is explained by a decrease in interfacial tension due to the protonation of groups at the surface of particles [POT 03]. Works on the synthesis of magnetite particles have also shown, though for a basic medium, that the higher the pH, the smaller the size of particles [VAY 98]. This indicates that the further the pH gets from the zero charge point, the smaller the size of particles [JOL 04].

1.2.4.2.4. Effect of salt addition

Several teams have shown that size and polydispersity of metal oxide particles obtained by sol-gel methods were highly sensitive to the presence of salt being added during reaction. Using electron microscopy, Eiden-Assmann *et al.* have observed perfectly spherical TiO_2 colloids obtained during the synthesis in the presence of alkaline chloride or nitrate (LiCl, NaCl, KCl, CsCl, KNO_3). When using alkaline chlorides, the size of particles decreases when the zeta potential and the ionic radius of cation increase. The size of particles goes from 2,500 nm in the presence of LiCl (zeta potential of 9 mV) to approximately 200 nm in the presence of CsCl (zeta potential of 25 mV). The same behavior is observed when the chloride anion is replaced by a bromide or iodide one. It appears that the size of particles is also correlated to the ionic force of the solution. When KCl concentration is 4×10^{-4} mol/L, the size of particles is in the range between 500 and 900 nm. For a concentration of 8×10^{-4} mol/L, the size drops down to nearly 50 nm and when concentration doubles, no particles are observed [EID 04]. The conclusions of these works are similar to those of Vayssiere *et al.* on the formation of magnetite nanoparticles in basic medium, when the ionic force of the solution evolves. When it increases, the size of particles decreases [VAY 98]. It had already been shown by the works conducted in 1991 by Bogush and Zukoski that the ionic force had an influence on the size of silica particles formed, but in this case the increase by two orders of magnitude of the concentration of electrolyte resulted in a doubling of particle diameter [BOG 91].

Cai *et al.* observed an increase in the polydispersity of silica particles synthesized when the concentration of electrolyte increases to a certain value. According to the explanation proposed, colloidal silica particles are negatively charged due to the presence on their surface of silanol groups. However, given that the surface charge of particles is weak, the forces of repulsion between colloidal silica particles are generally weak, which facilitates particle coagulation and consequently an increase in polydispersity. When interactions are very weak, especially when salts are absent or their concentration is low, the reaction medium generates very few free silica particles, as it instead produces agglomerated spheroids. The increase in zeta potential of the solution would lead to an increase in the force of repulsion between silica colloidal particles, which might be beneficial to monodispersity in a certain range, as an effect of competing forces of attraction and repulsion [CAI 12].

1.2.4.2.5. Effect of surfactant

An increase in the concentration of surfactant generally leads to a decrease in the size of particles and less agglomerates are formed [MIR 10, EID 04].

Adding surfactant during sol-gel synthesis can also influence the crystalline structure of the particles obtained. For example, in their works on the preparation of alumina nanoparticles, Mirjalili *et al.* and Park *et al.* have shown that the addition of surfactants resulted in the formation of an amorphous phase of boehmite coupled with surfactant hydrates. This further generated amorphous alumina, while in the absence of surfactant, crystalline alumina particles had been obtained [MIR 10, PAR 05b].

1.2.4.3. *Conclusion*

The sol-gel method allows the elaboration of materials as nanoparticles at temperatures close to ambient value and at atmospheric pressure. The fact that many synthesis parameters can be influenced allows us to control the size and form of oxide and metal hydroxide particles. As for the latter, a stage of calcination will be required to obtain the desired oxide. The same is applicable for obtaining particular crystalline phases. This methodology also allows us to obtain very intimate mixtures of mixed oxides that are definitely more homogeneous than those resulting from other techniques [TOB 94, AND 97a, AND 97b]. It is also possible to carry out continuous synthesis of nanoparticles [GIE 94]. The main drawbacks of this technique are the costs of alcoholate precursors and the significant volumes of solvent required for all the operations (synthesis, washing, etc.).

1.3. Gas-phase elaboration

1.3.1. *Condensation in inert gas*

1.3.1.1. *Principle*

This method consists of setting a high purity (>99%) metal sheet, wire or pellet either in a pot made up of ceramic material such as alumina [REC 94] or metal such as tungsten [SAN 98] or on a metal wire [FUK 74]. The assembly is then placed in a vacuum chamber, in which an atmosphere of low pressure neutral gas is generated (helium or argon [BOW 81, WAD 67, KIM 63, YAT 73], xenon [FUK 74, WAD 68] or nitrogen [SAN 98]). The metal is then resistively heated to a temperature that is specific to the metal to be evaporated. The atoms of the evaporated material lose kinetic energy due to their collisions with the gas molecules. They condense in small particles that grow both by the addition of vapor atoms to particular clusters and by aggregation due to collisions between clusters. These are afterward collected on a cold finger. For metals that are highly sensitive to oxidation, such as aluminum, the nanoparticles thus formed are subsequently passivized by small quantities of oxygen diluted in a neutral gas before being exposed to air [SAN 98].

1.3.1.2. *Influence of operating conditions*

The modification of various experimental parameters allows for the variation in the size of metal nanoparticles synthesized:

– the increase in the evaporation temperature leads to an increase in the size of particles [PAN 95];

– the size of particles increases with the distance between the target and the recovery device [PAN 95, GLE 89];

– the increase in the pressure of inert gas generates an increase in the size of nanoparticles formed [PAN 95, KIM 63];

– the increase in molar mass of the inert gas leads to an increase in the size of nanoparticles [WAD 67, WAD 68].

Many metal nanoparticles can be obtained by this method (see Table 1.13).

Metal	Mean diameter or size range (nm)	Experimental conditions	Reference
Ag	1–15 5–20, 20–60, 70–300 100–300	He, $P = 30$ kPa, $V = 25$ m/s $T = 1000$°C Ar, $P = 0.13, 0.67, 2.0, 4.0$ kPa Xe, $P = 0.27, 2$ kPa	[IWA 92] [KIM 63] [WAD 68]
Al	10–40 24–100 5–20, 20–60, 70–300 100–400 100–2,300	N_2, $P = 0.13$ kPa, $T = 1350$°C Ar, 1.3 kPa $< P <$ 6.7 kPa, 1,400°C $< T <$ 1600°C Ar, $P = 0.13, 0.67, 2.0, 4.0$ kPa Xe, $P = 0.27, 2$ kPa He, $P = 6.67$ kPa, 1,100°C $< T <$ 1,500°C	[SAN 98] [PAN 95] [KIM 63] [WAD 68] [YAT 73]
Au	5–20, 20–60, 70–300 30–200	Ar, $P = 0.13, 0.67, 2.0, 4.0$ kPa Xe, $P = 0.27, 2$ kPa	[KIM 63] [WAD 68]
Be	100–400	Xe, $P = 0.27, 2$ kPa	[WAD 68]
Bi	5–20, 20–60, 70–300	Ar, $P = 0.13, 0.67, 2.0, 4.0$ kPa	[KIM 63]
Cd	5–20, 20–60, 70–300 100–400	Ar, $P = 0.13, 0.67, 2.0, 4.0$ kPa Xe, $P = 0.27, 2$ kPa	[KIM 63] [WAD 68]
Co	5–20, 20–60, 70–300 10–20	Ar, $P = 0.13, 0.67, 2.0, 4.0$ kPa Xe, $P = 0.27, 2$ kPa	[KIM 63] [WAD 68]
Cr	5–20, 20–60, 70–300 >20 50–100	Ar, $P = 0.13, 0.67, 2.0, 4.0$ kPa Ar, $P = 0.67$ kPa Xe, $P = 0.27, 2$ kPa	[KIM 63] [NIS 74] [WAD 68]
Cu	5–20, 20–60, 70–300 10–200 2–5	Ar, $P = 0.13, 0.67, 2.0, 4.0$ kPa Xe, $P = 0.27, 2$ kPa He or Ar, 26.66 kPa $< P <$ 53.33 kPa, 1,727°C $<$ $T <$ 2127°C	[KIM 63] [WAD 68] [BOW 81]
Fe	30 5–20, 20–60, 70–300 50–200 6	Xe or Ar, 1;33 kPa $< P <$ 2.67 kPa Ar, $P = 0.13, 0.67, 2.0, 4.0$ kPa Xe, $P = 0.27, 2$ kPa He, $P = 2$ kPa	[FUK 74] [KIM 63] [WAD 68] [BIR 84]
Mg	5–20, 20–60, 70–300 500–3,000 2–6	Ar, $P = 0.13, 0.67, 2.0, 4.0$ kPa Xe, $P = 0.27, 2$ kPa He, $P = 1.3$ kPa	[KIM 63] [WAD 68] [KIM 87]

Mn	5–20, 20–60, 70–300 20–300	Ar, P = 0.13, 0.67, 2.0, 4.0 kPa Ar, P = 2.0 kPa, T = 1,500°C, V = 6 m/s	[KIM 63] [IWA 95]
Nb	10	Ar, 2.0 < P < 6.67 kPa, 2,500°C < T < 2,600°C	[SAIT 80]
Ni	5–20, 20–60, 70–300 5–60	Ar, P = 0.13, 0.67, 2.0, 4.0 kPa Xe, P = 0.27, 2 kPa	[KIM 63] [WAD 68]
Pb	5–20, 20–60, 70–300	Ar, P = 0.13, 0.67, 2.0, 4.0 kPa	[KIM 63]
Sn	5–20, 20–60, 70–300	Ar, P = 0.13, 0.67, 2.0, 4.0 kPa	[KIM 63]
Sb	20	Ar, P = 0.8 kPa, T = 750°C, V = 4.2 m/s	[IWA 95]
V	10–100	Ar, 0.03 kPa < P < 5.33 kPa, 2,000°C < T < 2,300°C	[SAIT 80]
Zn	5–20, 20–60, 70–300 50–200 10–20	Ar, P = 0.13, 0.67, 2.0, 4.0 kPa Xe, P = 0.27, 2 kPa Ar, 13.33 < P < 26.66 kPa, T = 900°C	[KIM 63] [WAD 68] [REC 94]

Table 1.13. *Examples of sizes for various metal particles obtained by thermal evaporation techniques as a function of experimental conditions: pressure (P), temperature (T), gas velocity (V)*

There is a variant of this method that consists of heating the metal by radiofrequency (RF) induction in a cryogenic fluid in the liquid phase (argon, nitrogen or helium). This method allows for a significant nucleation rate and rapid cooling, which results in limiting the particle growth. Finally, the high vapor pressure contributes to a significant yield [CHA 98].

Nanoparticles are afterward collected by bubbling the cryogenic gas that contains them in suspension through an organic solvent such as hexane. The metal nanoparticles obtained have a granulometry between 20 and 200 nm [CHA 96a]. Starting from an aluminum block (30 g) subjected to RF induction heating (150 KHz) under argon atmosphere (used instead of liquid nitrogen, to avoid formation of aluminum nitride (AlN), Champion and Bigot [CHA 98] produce approximately 150 mg aluminum nanoparticles per minute. These are spherical particles, with a diameter of the order of 100 nm. They are passivized by an oxide layer with a thickness of 3 nm. The cryogenic fluid used for iron and copper nanoparticles is nitrogen, and size dispersion is weaker [CHA 96a, GOU 93].

1.3.1.3. *Conclusion*

This is a simple method, which can be readily implemented, though with low yields [REC 94].

This technique can also be used for manufacturing nanosized metal oxides, by evaporating metal in the presence of oxygen or air diluted in the flow of inert gas.

Zeng *et al.* [ZEN 04] have thus obtained Sb_2O_3 nanoparticles with a diameter between 50 and 180 nm.

1.3.2. *Explosion of metal wires*

1.3.2.1. *Principle*

Electric explosion of wire is one of the vapor-phase methods for the production of particles by evaporation of a thin metal wire through which a strong current flows.

Nairne was the first to describe this phenomenon in 1774 [NAI 74], when he carried out the discharge of a battery of interconnected Leyden jars through copper and silver wires.

It is in a US Atomic Energy Commission report from 1946, drafted by Abrams *et al.* [ABR 46], that the production of ultra-thin powders was for the first time described, together with the fabrication of nanoparticles of apparently non-crystalline plutonium, uranium and aluminum, with a diameter of the order of 200 nm, and whose particles form aggregates at least 1 µm in length.

In 1962, Karioris and Fish described a metal particle aerosol generator based on explosion of metal wire, and applied to about 15 metals, including aluminum and copper [KAR 62].

This method is still being developed throughout the world [KOT 03, IVA 03]. Many other metal and alloy nanoparticles have also been obtained in a gas atmosphere [NAZ 06, LEE 10, JAN 11]. The implementation parameters are the following:

– 5–30 KV high voltage source [DRE 09];

– current density above 10^{10} A/m^2 [KWO 01];

– pulse duration that can vary from several nanoseconds [ROM 06] to several microseconds [JIA 98];

– metal wire diameter that can vary from a dozen to hundred micrometers [DAS 12, BAG 11, SUE 07].

1.3.2.2. *Influence of operating conditions*

1.3.2.2.1. Effect of pressure

The influence that inert gas pressure exerts on the formation of particles has been studied by Liu *et al.* [LIU 13]. Results show that aluminum nanoparticles produced under high pressure argon are more spherical in form and furthermore the mean diameter of aluminum nanopowders increases with argon pressure. This effect is accompanied by an increase in the size distribution of aluminum nanoparticles and in yield [SAR 07a].

1.3.2.2.2. Effect of gas nature

Sarathi *et al.* [SAR 07b, SAR 10] have shown that the nature of the inert gas (helium or argon) employed does not seem to influence the size of particles formed or their composition (aluminum and oxygen rates) for an equivalent implementation pressure. On the contrary, the use of nitrogen leads to the formation of AlN.

The aluminum nanoparticles obtained by explosion under inert gas have very high surface reactivity. Before being subjected to other treatments, these particles have to be passivated to avoid autoignition when they are exposed to air.

1.3.2.3. *Passivation*

Most of the aluminum nanoparticle powders for commercial use are passivated by inert oxide layers composed of amorphous or crystalline alumina. This air passivation can be carried out after obtaining the particles or during the fabrication process.

The works of Kwon [KWO 03] present several passivation methods for thin aluminum particles (between 200 and 300 nm) either by using argon (at a pressure of 1.1 atm) with a very low air concentration (below 0.1% in volume), or by employing mixtures based on argon and nitrogen or argon and hydrogen (10% in volume) with air traces. The table below includes several examples. The passivations carried out during particle preparation in the presence of reactive gases (oxygen or nitrogen) diluted by argon lead to an increase in the size dispersion of particles obtained [KOT 99]. Passivation by nitrogen is not stable, as AlN [KIN 03] becomes oxidized and hydrolyzed during its air storage and alumina is formed in the end. Ilyin and colleagues have in turn developed a method of passivation by boron during the fabrication process. The aluminum particles are enveloped by a compound that approximately consists of AlB_2 [KWO 02].

Passivation can also be conductive starting from organic coatings such as stearic or oleic acids or fluoro polymers [KWO 03, GRO 06].

1.3.2.4. *Conclusion*

This method is currently used for the fabrication of aluminum nanopowder for commercial use under brand names such as Elex or Alex.

It is worth noting that metal nanopowders can be produced by explosion of metal wire in liquids such as deionized water, ethanol, isopropanol or acetone [CHO 10, HA 10, GOO 09, YUN 11] when this method is used for producing non-oxidized metal powders. Cho [CHO 07] has described the process for the elaboration of silver nanoparticles in deionized water. This work notes the formation of a plasma sphere composed of atoms of vaporized material. This plasma bubble is contained in the water under pressure generated by the shockwave. By cooling, vapor condensation generates particles inside this sphere before its collapse and subsequent dispersion of particles in the liquid. There is high dispersion of the size of particles, which varies from 10 to 500 nm depending on the solvents and type of metal used.

1.3.3. *Thermal plasma synthesis*

There is a broad spectrum of plasma generation techniques, which can be classified depending on plasma temperature, gas pressure, frequency or electrode presence. The latter will be described here taking into account their excitation mode. A special section will be dedicated to plasmas generated by pulsed laser (see section 1.3.4). Plasmas generated in the gas phase can be classified into three groups:

– plasmas initiated by direct current (DC) and low-frequency (AC) discharges;

– plasmas initiated by RF;

– plasmas initiated by microwave sources.

1.3.3.1. *Direct current (DC) and low frequencies (AC) discharges*

These plasmas are obtained starting from an arc plasma torch. These can be divided into two categories depending on their design geometry: the so-called blown arc plasma and the transferred arc plasma.

1.3.3.1.1. Blown arc plasma in direct current

In 1957, Gage [GAG 57] designed the first coaxial configuration, in which the electric arc develops between a central cathode and a concentric nozzle that plays the role of the anode. The cathode is generally made up of 2% thoriated tungsten [FAU 96a]. The electrodes are powered by a direct current source and the arc is confined by the flow of cold gas established around the plasma column. Plasma jet temperatures vary between 6,000 and 15,000 K [FAU 96b]. Cold gas also prevents electrodes from overheating. A cooling system using heat transfer fluid may be needed for high power systems.

This type of system sustains injection of precursor in the form of powder [LEE 07], liquid (aqueous or organic solutions) [SIV 13, FAU 11] or gas [RAO 95] [HUA 10a]. Powder can be injected along a direction that is parallel or perpendicular to the plasma flow. It is worth noting that when injecting solvent, its nature strongly influences the temperature and the nature of species present in the plasma.

The use of this method often leads to strongly agglomerated powders [ZHO 09a]. However, by optimizing the flow rate parameters of the carrier gas and plasma gas, Im *et al.* have obtained tin oxide particles with diameters varying between 3 and 21 nm and few aggregates formed. The size of these aggregates is approximately 20 nm. They have also evidenced that a decrease in the flow rate of gas carrying the tin tetrachloride used as precursor led to a decrease in the quantity of oxide formed, and notably to a decrease in the size of the particles formed. Furthermore, by increasing the flow rate of plasma gas, the particle time of residence in plasma diminishes, which considerably reduces the particle growth time and their coalescence, and this also facilitates the formation of small particles with narrow size distribution and few aggregations of particles [IM 08].

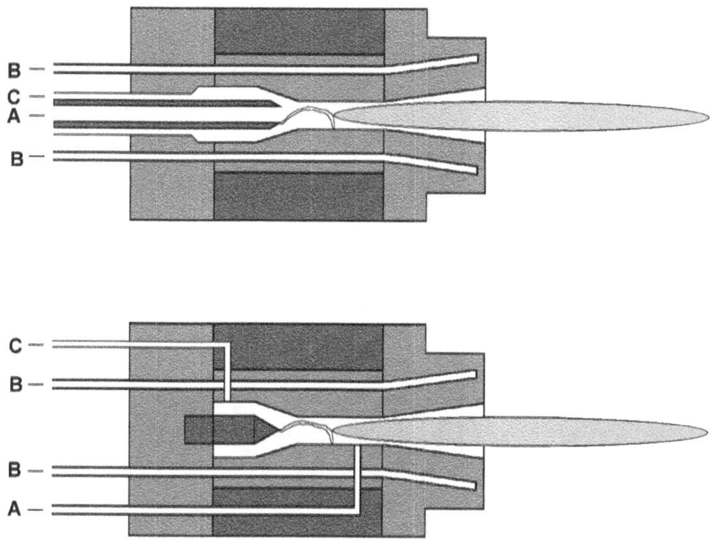

Figure 1.2. *Principle of blown arc plasma torches: top: axial injection, hollow cathode (A); bottom: perpendicular injection of precursor (A), cooling fluid (B) and plasma gas (C)*

1.3.3.1.2. Transferred arc plasma

In the case of transferred arc plasmas, the arc is initiated between two coaxial though non-concentric electrodes powered by direct current. The cathode is made up of tungsten, while the anode may consist of a metal to be vaporized in order to obtain metal or oxide nanoparticles (aluminum [CHA 06], copper [RYU 09], iron [BAR 06]), or it may be made up of tungsten or copper and sustain a pot that allows the fusion of metal to be transformed (aluminum [PAR 05a], silver [ZHO 09a], tungsten [SU 08, SU 05], bismuth [CHE 05a] and nickel [WEI 05]). The temperatures of plasmas generated by transferred arc vary between 9,000 and 25,000 K.

The arc is stable for a voltage between 15 and 20 V, and for a current between 25 and 50 A [PAR 05a]. The plasma chamber is swept by a plasma flow, generally consisting of argon. A reducing reactive gas such as hydrogen may be added during the synthesis of metal particles in order to prevent the formation of oxide [ZHO 09a]. If, on the contrary, obtaining oxides is desired, diluted oxygen or air can be added [CHA 06, BAR 06].

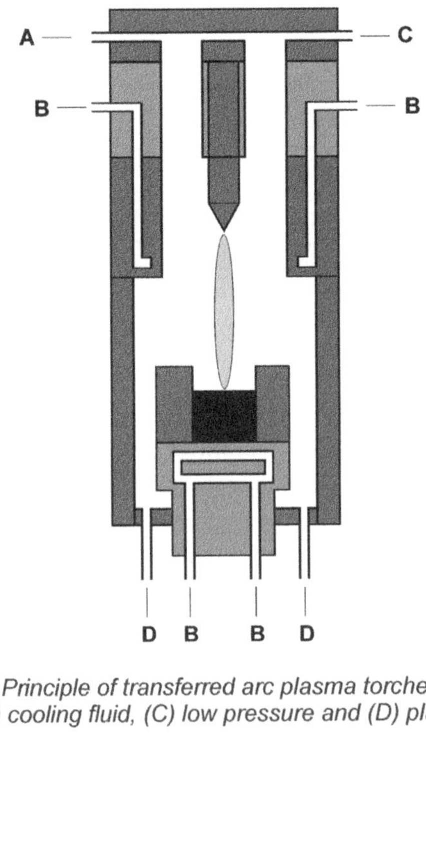

Figure 1.3. *Principle of transferred arc plasma torches: (A) quench gas, (B) cooling fluid, (C) low pressure and (D) plasma gas*

Nanopowders thus produced are highly agglomerated [VOL 07], which is not advantageous for the elaboration of particles used in the formulation of nanothermites.

1.3.3.2. *RF plasma*

Numerous studies were dedicated to RF plasma torches. The vast majority of these torches are based on inductive coupling, but there are also some torches that use capacitive coupling to generate nanopowders [MAT 06, BUS 97].

1.3.3.2.1. RF inductively coupled plasma

The first experiments on low pressure discharge without electrodes were conducted by the end of the 19th Century, but it was not until the 1940s that atmospheric pressure inductive discharge was discovered. The Russian physicist George Babat [BAB 47] was the first to report, in 1947, that once established at low pressure, a discharge can be maintained, while the pressure is brought to the level of atmospheric pressure.

The key development that led to the currently used inductive plasma originated in the works of Reed [REE 61], who in 1961 showed that an inductively coupled plasma discharge can be maintained in an open tube in the presence of a streaming gas. When exiting the discharge area, the partially ionized gas forms a low velocity plasma jet with an average temperature in the range between 6,000 and 11,000 K [BOU 95].

The design of the RF torch is relatively simple. It consists of two concentric tubes placed at a small distance from each other. These tubes are made up of quartz for low powers and ceramics for higher powers. The plasma confinement tube is cooled by compressed air when powers are below 10 kW or by circulating water when powers are up to 150 kW. The induction coil has several turns (three to five), depending on the power source [FAU 96a, FAU 96b]. The range of frequency used by this technology for generating metal or oxide nanopowder goes from 50 kHz to over 10 MHz [VOL 08].

Plasma is initiated and maintained by the RF powered helicoidal winding. As energy is transferred from the source to plasma by inductive coupling, neither the electrons nor the ions can reach the torch walls at the applied frequencies. For this reason, and provided that the wall being in contact with plasma is appropriately cooled, this method allows us to obtain powders that are free from contamination resulted from erosion of electrodes, by contrast with the production methods using direct current arc. On the other hand, this technique does not allow the use of a sacrificial anode, it requires precursor injection. This injection is generally axial [YE 07, SAT 03, ZHA 12a], and in the direction of plasma gas propagation. Schulz and Hausner have used a plasma torch whose geometry allows for reactive precursors being injected tangentially to the plasma gases (argon) and cooling gases (air) [SCH 92]. The rate of this precursor injection varies from 10 to 30 m/s. Particles are obtained with sizes ranging between 100 and 400 nm, and in quantities of the order of 100 g/h. Optimized industrial systems produce good quality nanopowders with particles sized below 100 nm. The range of products includes metals, oxides, carbides and nitrides.

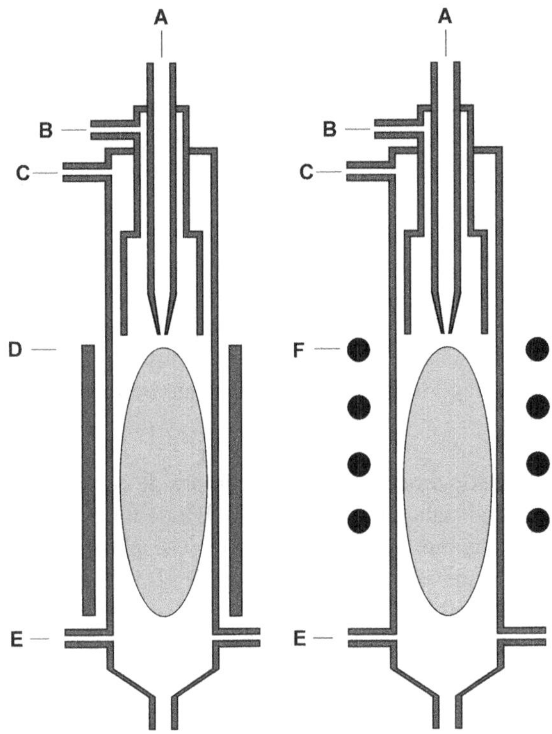

Figure 1.4. *Principle of radiofrequency (RF) plasma torches: left: capacitively coupled torch; right: inductively coupled torch. (A) Precursor and carrier gas, (B) central plasma gas, (C) sheath gas, (D) capacitive plate, (E) quench gas and (F) inductive coil*

1.3.3.2.2. RF capacitively coupled plasma

Several reactors generating capacitively coupled plasmas have been developed for the synthesis of nanopowders [MAT 06, BUS 97]. A capacitively coupled plasma source is composed of two parallel conductive electrodes placed in a vacuum chamber. The discharge is confined by the reactor walls, which can be either conductive or insulating. The electric field that accelerates electrons is generated between the two electrodes. The alternating electric field allows the electrons present in the respective area to gain the energy required for atom ionization. The electrons thus freed contribute in turn to gas ionization. Due to the avalanche effect, the whole chamber will be filled with plasma whose density and temperature will depend on the applied RF power and also on the gas pressure. Gas sheaths separate plasma from electrodes and walls [BUS 97].

Matsui and colleagues [MAT 06] have developed a reactor with perforated stainless steel electrodes placed inside, perpendicular to the flow of plasma gas (argon) and precursor. To control the size of particles produced, a discontinuous plasma discharge cycle is chosen. The cycle duration of the plasma discharge is between 0.5 and 30 s, with a 4s duration of plasma extinction. The size of the particles thus obtained varies depending on pulse duration. The longer the pulse duration, the longer the particle residence time in the plasma, which increases their growth and coalescence time, and therefore their mean diameter increases. According to the works of Matsui and colleagues, this diameter varies between 10 and 120 nm.

1.3.3.3. *Microwave discharge plasmas*

All microwave systems operate according to the same principle. They have no electrodes: microwaves are generated by various sources such as magnetrons or klystrons. The frequencies used are within the range from 200 to 3,000 MHz [FAU 80]. In industry, the most commonly used frequencies are 0.915 GHz and 2.45 GHz. Microwaves are then guided through waveguides and transfer their energy to the electrons in the plasma gas.

Microwaves allow generation of plasmas with a variety of characteristics [TEN 06]. The gas temperature of microwave-generated plasmas can vary from 300 to several thousand kelvin [VOL 93a, PET 11] for several eV electron energy (or an equivalent electronic temperature of several tens of thousands Kelvin).

The first works on the nanopowder synthesis using microwave-induced plasma were carried out in the beginning of the 1990s by Mehta and colleagues [MEH 91], Chou [CHO 92] and Vollath [VOL 93a].

The reactor basically consists of a microwave resonant cavity connected to a waveguide supplied by a microwave generator and an impedance matching system. Plasma is confined in a quartz tube, with axial injection of plasma gas and precursor. Sheath gases are tangentially injected to trigger a swirling type of displacement.

There are many parameters, often interdependent, that may influence the quantity and quality of the nanopowders generated.

First of all, several works have evidenced strong dependency of the size of particles obtained on the precursor concentration. This can be readily explained by the fact that, when concentration is high, there is a stronger probability that a heavy atom meets another one, and this facilitates nucleation, and then coalescence phenomena. This has most notably been studied by Szabo in relation with the size of SnO_2 particles [SZA 12] and by Schumacher [SCH 07].

For their part, Vennekamp *et al.* [VEN 11] have shown that a too high flow rate of central plasma gases leads to reaction quenching due to the rapid transfer of matter out of the plasma area. On the contrary, an increase in the microwave power leads to increase in the length of plasma jet, which results in a longer residence time and allows for growth and coalescence reactions to develop for a longer period, which in the end increases the size of particles. Finally, they have shown that the size of particles is a linear function of the logarithm of overall gas pressure, and that when precursor (tetramethylsilane) concentration is constant, the growth of SiC particles is a linear function of time.

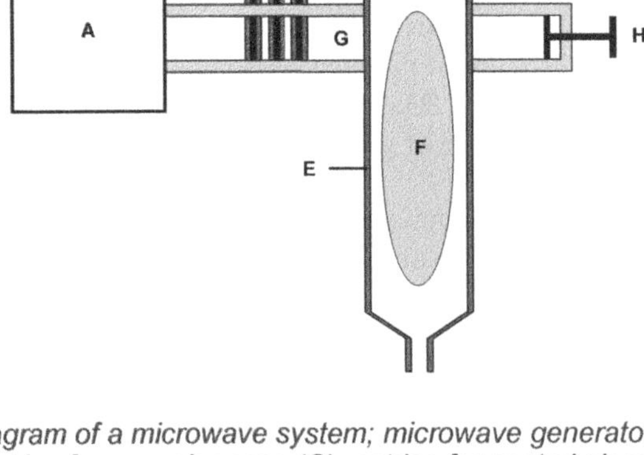

Figure 1.5. *Diagram of a microwave system; microwave generator (A), tuners (B), tangential entries for quench gases (C), entries for central plasma gases and precursors (D), discharge tube (E), plasma ((F), waveguide (G) and mobile piston (H)*

M	Precursor	Gas	Type	Generator	Diameter (nm)	Reference
Al	Al micron	Ar	MW	1.5 kW, 2.54 GHz	1–100	[PHIL 04]
	Al 50 µm	Ar	MW	1 kW	7.4–34.2	[WEI 04]
	Al pellets	Ar	DC	375–1,000 W	19	[PAR 05a]
	Al (anode)	Ar	DC	750 W	200–600	[CHA 06]
Ag	Ag_2CO_3	N_2	MW	2.45 GHz	9–10	[CHA 05]
	Ag micron	Ar, Ar/H_2	DC	7–9 kW	<100	[LEE 07]
	Ag bulk	Ar, N_2, He	DC	1.2–4.8 kW	20-100	[ZHO 09a]
B	BCl_3	$Ar/H_2/CH_4$	RF	45 kW	2–50	[MAR 05]
	BCl_3	Ar/H_2	DC	23–25 kW	50	[HUA 10a]
Co	$C_5H_5Co(CO)_2$	Ar	MW	0–6 kW, 2.45 MHz	12	[POD 04]
	$CoCl_2$	N_2/H_2	MW	2.45 GHz	41.4–21.6	[CHA 06]
Cu	Cu 40 µm	Ar/H_2	RF	28 kW, 2 MHz	50–280	[KOB 08]
	Cu bulk	Ar	DC	Unspecified	30–90	[WEI 06]
	$CuCO_3$	N_2, N_2/H_2	MW	5 kW, 2.45 GHz	60	[CHA 14a]
Fe	Fe 5–9 µm	Ar	RF	20 kW, 2.9 MHz	10–100	[GIR 93]
	$Fe(CO)_5$	Ar	MW	0–6 kW, 2.45 MHz	10	[POD 04]
	$Fe(CO)_5$	Ar	MW	0–6 kW, 2.45 MHz	10–15	[KAL 98]
	Fe bulk	Ar	MW	1.2 kW, 2.45 GHz	3–5	[HAY 90]
Mo	$Mo(CO)_6$	N_2	MW	5 kW, 2.45 GHz	20	[CHA 14a]
Mn	Mn bulk	He/H_2	DC	Unspecified	60	[PAN 07]
Ni	Ni Bulk	Ar	DC	Unspecified	20–70	[WEI 05]
	$NiCl_2$	N_2/H_2	MW	2.45 GHz	35–57.5	[CHA 07]
	$Ni(OH)_2$	Ar/H_2	RF	30 kW, 4 MHz	50–500	[BAI 09]
	$NiCO_3$	Ar/H_2	RF	30 kW, 4 MHz	50–500	[BAI 09]
Si	$SiCl_4$	N_2/H_2	MW	5 kW, 2.45 GHz	26	[KUM 12]
	SiH_4	Ar/H_2	MW	6 kW	4–50	[PET 11]
	Si 30–40 mm	Ar/H_2	RF	10–15 kW	13–37	[LEP 08]
	$SiCl_4$	Ar/H_2	RF	20–100 W, 70 MHz	2–10	[DIN 14]
	$SiCl_4$	H_2	MW	0.8 kW, 2.45 GHz	54	[WU 10]
	$SiCl_4$	Ar/H_2	DC	7.6–13 kW	<100	[RAO 95]
	SiH_4	Ar	RF	0.2 kW, 27.12 MHz	2–8	[MAN 05]
	$SiCl_4$	Ar/H_2	RF	15 W, 144 MHz	3–15	[NOZ 07]
	SiH_4	Ar	RF	200 W, 13.56 MHz	35 ± 4.7	[BAP 07]
W	APT	Ar/H_2	RF	30 kW, 4 MHz	6–48	[PLA 13]
	$W(CO)_6$	N_2	MW	5 kW, 2.45 GHz	30	[CHA 14a]
	APT	Ar/H_2	DC	13 kW	88.5	[RYU 09]

Table 1.14. *Examples of metal nanoparticles synthesized by thermal plasma, according to the literature; APT, ammonium paratungstate*

Oxides	Precursor	Gas	Type	Generator	Diameter (nm)	Reference
Al_2O_3	Al 3 μm	Ar/O_2	RF	30 kW, 4 MHz	65.6–126	[YE 07]
	$Al(NO_3)_3$	Air	MW	0.915 GHz	100–2,000	[VOL 93a]
	Al (anode)	Ar/air	DC	750 W	<40	[CHA 06]
Bi_2O_3	Bi bulk	O_2	DC	Unspecified	<1000	[CHE 05a]
Cr_2O_3	$Cr(CO)_6$	Ar/O_2	MW	0.915 GHz	7–15	[VOL 96]
Fe_2O_3	$Fe(CO)_5$	Ar/O_2	MW	0.7 kW, 2.45 GHz	4.5 ± 1.3	[BAU 06]
	$Fe(CO)_5$	$Ar/H_2/O_2$	MW	2.45 GHz	25	[DAV 14]
	$Fe(CO)_5$	$Ar/H_2/O_2$	MW	2.45 GHz	<30	[DAV 13]
	$Fe(CO)_5$	Ar	MW	180 W, 2.45 GHz	5.5–22	[SYN 11]
	$Fe(Cl)_3$	n.s.	MW	0.915 GHz	4–7	[VOL 95]
	Fe (anode)	Ar/O_2	DC	8.4 kW	10–100	[BAR 06]
	Ferrocene	Ar/O_2	DC	6.7–7 kW	8–9	[LEI 12]
HfO_2	$Hf(OC_4H_9)_4$ or $HfCl_4$	Ar/O_2	MW	2.45 GHz	<5	[FOR 08]
MgO	Mg granule	Ar/O_2	MW	2.45 GHz	10–50	[HON 06]
MoO_3	$(NH_4)_6Mo_7O_{24}\cdot 4H_2O$	n. s.	MW	700 W	100–1,000	[KLI 12]
NiO	Ni bulk	Ar/O_2	DC	Unspecified	25	[WEI 09]
SiO_2	SiO_2 μm	Ar/O_2	RF	40 kW	10–300	[GOO 06]
SnO_2	$Sn(C_4H_9)_4$	Ar/O_2	MW	900 W, 2.45 GHz	10	[SZA 12]
	$SnCl_4$	Ar/O_2	MW	2.45 GHz	2–5	[SZA 07]
	$SnCl_4$	Ar/O_2	MW	2 kW, 2.45 GHz	<10	[SCH 07]
	$SnCl_4$	$Ar/N_2/O_2$	DC	6.6–10.2 kW	3–20	[IM 08]
TiO_2	$TiCl_4$	$Ar/O_2/SO_2$	MW	1.2 kW, 2.45 GHz	24–101	[HON 11]
	$Ti(OC_4H_9)_4$	Ar/O_2	MW	2.45 GHz	<5	[SCH 07]
	TiN 28 μm	Ar/O_2	RF	25 kW, 2 MHz	50–70	[OH 05]
V_2O_5	$VOCl_3$	$Ar/O_2/H_2$	MW	2.45 GHz	50–60	[KIM 07]
WO_3	ATP	Ar	RF	30 kW, 4 MHz	100	[PLA 13]
	W bulk	Ar/O_2	DC	2.34 kW	20–30	[SU 08]
	W bulk	Ar/O_2	DC	2.34 kW	6–18	[SU 05]
	W wire	H_2O	MW	200 W, 2.45 GHz	7–13	[HAT 13]
	$W(CO)_6$	Ar/O_2	MW	300 W, 2.45 GHz	3–4	[SAG 03]
ZnO	$Zn(CH_3)_2$	Ar/O_2	MW	Unspecified	3–7	[JAN 02]
	$Zn(CH_3)_2$	Ar/O_2	MW	2.45 MHz	4.9–10.3	[KLE 02]
	Zn powder	N_2/Ar or N_2	DC	70 kW	30	[KO 06]
	Zn powder	Ar/O_2	RF	300 W, 13.56 MHz	30	[SAT 03]
	Zn powder	Ar/O_2	MW	2.45 GHz	100–200	[KIM 07]
ZrO_2	$Zr(NO_3)_4$	Ar	MW	0.915 GHz	100–500	[VOL 93a]
	$ZrCl_4$	Ar/O_2	MW	0.915 GHz	5	[VOL 93b]
	$Zr(OC_4H_9)_4$	Ar/O_2	MW	2.45 GHz	<5	[SCH 07]
	$Zr(OC_4H_9)_4$ or $ZrCl_4$	Ar/O_2	MW	2.45 GHz	<5	[FOR 08]

Table 1.15. *Metal oxide nanoparticles synthesized by thermal plasma; ATP, ammonium paratungstate*

1.3.3.4. *Thermal plasma in solution*

The techniques that use discharges in liquids to obtain nanopowders can be differentiated depending on whether they are based on the conversion of liquid or components in solution or on electrode erosion. Several examples of metal or oxide nanoparticles obtained by these two options are reported in Tables 1.16 and 1.17.

In the case of the method based on erosion of metal electrodes, the latter are used as source material, and are immersed in an inert liquid, which means that its decomposition in the plasma phase does not generate solid particles. Deionized water is the most frequently used solvent, in which various salts can be dissolved in order to control ionic conductivity, pH or the oxidoreduction capacities of the medium in order to obtain metal or oxide particles. Either direct or alternating current can be used, and it can be applied in continuous, sequential or high frequency pulsed modes. To obtain the discharge arc, electrodes can be both made of the same metal or, if direct current is applied, the cathode only can be made up of a metal to be eroded. The anode is then made up of platinum. Finally, the electrodes can be placed facing each other or in parallel (Figure 1.6). Discharge-based erosion of electrodes creates a vapor starting from which nanoparticles are generated, and then condensates are formed.

A difficulty of this process relates to the control of the size of particles produced. The smaller ones, of the order of several nanometers, result from the condensation of metal vapor and are cooled by the liquid before being able to grow or coalesce with other particles. Those of intermediary size (between 20 and 30 nm) have already stayed for some time in the intermediary region between plasma discharge and the liquid. Finally, the larger particles (of the order of 100 nm) are generated by the metal droplets resulted from the fusion of the electrode during the erosion process. The nanoparticle size distribution is largely determined by how the electrical discharge is delivered [CHE 04].

Works conducted by Ashkarran *et al.* have evidenced that the size of particles produced is correlated with the intensity of the discharge current used during the production of TiO_2 [ASH 10d] and WO_3 [ASH 08] particles.

Still, some other important parameters should be taken into account. Some of these are, for example, the melting temperature of the electrode material, the time and amount of energy deposited, the distance between electrodes or the solvent nature, dielectric character, viscosity, etc. [BEL 14].

Oxide can be obtained directly by metal oxidation in the liquid phase or from the species resulted during the liquid decomposition. At breakdown voltage, besides the evaporation of the electrode metal (copper or zinc, for example) [SAI 11, HU 14a], there is water decomposition into reactive ionized radical species that oxidize the metal atoms. A second method consists of bubbling gaseous oxygen into the liquid phase [ASH 10d].

As for metals that are sensitive to strong oxidation of the aqueous medium due to the presence of hydroxide radicals and atomic oxygen resulted from water decomposition in the plasma phase, in their case it is possible to use another solvent, such as styrene [OMU 07]. The resulting metal particle is coated with a thin layer of polystyrene [SUL 09]. As described by Yao *et al.* a further method consists of modifying the aqueous medium by adding a reducing agent such as hydrazine hydrate [YAO 05] or by using a citric acid buffer solution (pH 4.8) as per Saito *et al.* study [SAI 11]. In the example mentioned, these researchers obtained metallic copper nanoparticles in porous spherical form. The proposed mechanism involves the transformation of molten copper nanoparticles containing oxygen resulted from water decomposition, via the formation of hydroxylated species, followed by a phase of solidification of a Cu-Cu_2O eutectic. The oxide is dissolved by the buffer solution, and non-oxidized copper emerges in porous spherical form.

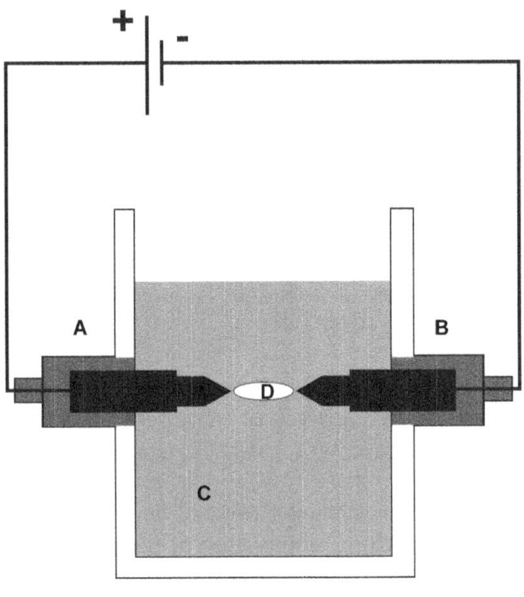

Figure 1.6. *Diagram of a plasma arc system in liquid medium: anode (a), cathode (b), liquid (c) and plasma (d)*

A second method is to transform metal ions in solution in the liquid by generating plasma between two tungsten or platinum electrodes [LEE 14a, BRA 11]. When nanoparticles are fabricated by plasma in liquid medium, a surfactant such as SDS or CTAB is often added. The surfactant generally facilitates the dispersion of particles and eliminates or strongly reduces particle aggregation. This results in slower growth of particles [LEE 14b]. Moreover, surfactant molecules tend to preferentially bond to certain crystal faces, if they are characterized by significant differences in the surface energy, which results in anisotropic growth of the particle. Surfactants are thus perceived as contributors to the generation of non-spherical particles [KIM 13a].

NP	Liquid	Precursor	Electric conditions	Diameter (nm)	Reference
Ag	H_2O H_2O	$AgNO_3$ Ag electrodes	DC 15A DC 4 A, 135 V	27 ± 14 5–55	[ASH 10a] [TIE 09]
Al	H_2O, CTAB	$AlCl_3{\cdot}6H_2O$	DC 250 V, 5 μs, 30 kHz, W electrodes	10–100	[KIM 15a]
Au	H_2O, CTAC H_2O C_2H_5OH	$HAuCl_4$ Au electrodes Au electrodes	DC, 250 ns, 10 kHz, Pt electrodes DC, 70–100 V, 2–3 μs DC, 135V, 6.4A, 50 μs	1–10 15–30 4–15	[BRA 11] [LUN 07] [TSE 09]
Co	H_2O, SDS	$CoCl_2$	DC 250V, 5 μs, 30 kHz, W electrodes	10–30	[KIM 13a]
Cu	styrene H_2O, CTAB H_2O, citric buffer H_2O Hydrazine	Cu electrode $CuCl_2$ Cu cathode Cu electrodes	AC 200 V, 5A DC 250 V, 5 μs, 30 kHz, W electrodes DC 105–130 V Pt anode AC 150 V	3 1–5 100 30–50	[OMU 07] [LEE 14b] [SAI 11] [YAO 05]
Fe	H_2O CTAB	$FeCl_2$	DC 250 V, 5 μs, 30 kHz, W electrodes	5–20	[HEO 15]
Mn	H_2O, CTAB	$MnCl_2$	DC 250 V, 5 μs, 30 kHz, W electrodes	50–150	[KIM 13b]
Ni	H_2O, SDS	$NiCl_2$	DC 250 V, 5 μs, 30 kHz, W electrodes	20	[KIM 15b]

Table 1.16. *Examples of metal nanoparticles (NP) synthesized by plasma in liquids; SDS, sodium dodecylsulfate; CTAB, cetyltrimethylammonium bromide; CTAC, cetyltrimethyl ammonium chloride*

NP	Liquid	Precursor	Electric conditions	diameter (nm)	Reference
CoO	H_2O	Co acetate	AC 20 kHz, 15 kV, 10 A, 2 µs	2.5	[CHE 12]
Cu_2O	H_2O, ascorbic acid	Cu electrodes	AC 150 V	4–10	[YAO 05]
CuO CuO nanowires	H_2O K_2CO_3 H_2O, $NaNO_3$	Cu cathode Cu electrodes	DC 105–130 V Pt anode AC, 150 V	<100 D = 1– 3 L = 4–10	[SAI 11] [YAO 05]
CuO nanowires	H_2O, NaCl	Cu cathode	DC 10–20 kHz, 1–2 µs,	D = 15–25 L = 50–60	[HU 14a]
TiO_2	H_2O H_2O, O_2	Ti electrodes Ti electrodes	AC 200 V, 5 A DC, 20–40 A, 2–3.5 V	7 5–31	[OMU 07] [ASH 10d]
WO_3	H_2O, SDS	WCl_6 ethanol	DC 250 V, 5 µs, 30 kHz, W electrodes	50–70	[LEE 14a]
	H_2O	W electrodes	DC, 10 V, 25 A	30 ± 12	[ASH 08]
$WO_3.H_2O$	H_2O	W electrodes	DC 100 V, 50 A, 15 µs	5	[CHE 14]
ZnO nanowires	H_2O	Zn electrodes	DC 3.5 kV, 100 Hz, 30 µs	*D* = 30 *L* = 80–100	[TAR 10]
ZnO	H_2O H_2O	Zn electrodes Zn electrodes	DC, 5–15 A, 2-3.5 V DC, 5A, 2–3.5 V	100–260 15–20	[ASH 09] [ASH 10c]
ZrO_2	H_2O	Zr electrodes	DC 10–20 A, 2–3.5V	7–52	[ASH 10b]

Table 1.17. *Examples of metal oxide nanoparticles (NP) synthesized by plasma in liquids; SDS, sodium dodecylsulfate; CTAB, cetyltrimethylammonium bromide; CTAC, cetyltrimethylammonium chloride*

1.3.4. *Laser ablation*

Reports on the use of a laser beam to ablate a solid target date from the 1960s, when ruby lasers were developed [SMI 65].

Nanopowders can be obtained by using various types of laser, which operate in continuous or pulsed mode. Referring to the latter, they differ depending on the pulse duration: there can be nanosecond or ultrashort (pico- or femtosecond) pulse lasers, since ablation mechanisms differ.

In a classical approach, processes resulting from the interaction of a laser beam with the surface of a material are strongly dependent on laser parameters such as pulse duration, wavelength or fluence, on the one hand, and on material's physical properties, especially melting and evaporation temperature, latent heat, thermal diffusivity, etc., on the other hand.

From the perspective of using laser ablation to obtain nanopowders, it is worth taking into account a new parameter related to the nature of the environment in which the plasma plume formed during ablation will expand. The expansion can take place in a "free" medium, which means in the presence of high vacuum or vacuum with low gas concentration or even in a confined medium, such as a liquid. This will significantly influence the formation mechanisms, size, dispersion or the nature of particles obtained, if there is chemical interaction between them and the surrounding medium (gas or liquid).

Laser ablation systems presently feature Nd:YAG or excimer lasers. The first systems are inexpensive and require little maintenance. As for excimer lasers, they are based on gas-filled laser cavities, so their use requires precaution. On the other hand, in comparison with Nd:YAG lasers, these lasers allow us to obtain higher output power and have a more homogeneous beam profile. The target can be a sheet of very high purity metal or metal alloy or a tablet of micron-sized metal or oxide particles. Particles obtained by laser ablation of solid targets result directly from the condensation of the plasma plume generated by the laser pulse irradiating the solid target surface.

1.3.4.1. *Long pulse*

When long pulses (ranging from continuous to nanosecond) are used, particle formation is essentially dominated by thermal transfer. The photons in the laser beam have their energy absorbed quasi-instantaneously by the valence electrons, then electronic thermalization takes place by collisions of electrons with each other over a duration from 10 to 100 fs, depending on the nature of the material and laser fluence [RET 02]. In a next stage, energy is transferred within picoseconds from hot electrons to the ion network in the form of phonons. Throughout this thermalization phase, the system is out of equilibrium, its electronic temperature being higher than its ionic temperature. Complete thermalization of the material is normally reached within several picoseconds from the laser energy deposit.

Since the pulse duration is longer than the duration of the thermal mechanisms involved, the ion network becomes heated during the laser pulse.

The process of ablation taking place during a long pulse irradiation is governed by hydrodynamics and heat conduction, since the pulse duration is longer than both the electron-phonon thermalization time (1–10 ps) and the thermal diffusion time (~10 ns).

When applied fluence exceeds the ablation fluence threshold of the material, matter is ejected in the form of vapor or fragments composed of ions, electrons, neutral atoms, aggregates and droplets of molten material. This fluence threshold depends on the material and laser parameters. The ejected particles remain at first confined near the target surface and form the Knudsen layer [KEL 98]. This layer has the same surface dimensions as the laser spot and its thickness corresponds to the optical penetration length δ, which depends on the laser wavelength and the material's extinction coefficient (n_2) according to the relation: $\delta = \lambda\ 4\pi\ n_2$. The particle density within this layer is close to solid density (10^{19}–10^{21} cm^{-3}) [SIN 90].

When the target stops being able to diffuse the absorbed energy into the network, the Knudsen layer is subjected to vaporization [MIO 95, WIL 00].

Since ablation starts well before the end of the laser pulse, the forming vapor and the laser interact, and plasma will be generated. This will be formed through two different mechanisms [CHA 96, SCH 98]. The first one consists of photoionization, which involves ionization of an atom which is neutral or in an excited state, following the absorption of one or several photons. It is worth noting that this process proves more efficient when lasers operate at short wavelengths [AMO 99, MAO 07] rather than at visible or infrared light wavelengths. The second mechanism involves primary electrons emitted either as result of the absorption of several photons by a neutral atom that thus reaches its ionization potential (multiple photon ionization), or due to electron emission by the heated material (thermionic effect), or finally by ionizing collision between vapor atoms. These free electrons absorb the photons emitted by the laser through inelastic collisions with vapor ions and neutral atoms. Called inverse Bremsstrahlung (or inverse braking radiation), this phenomenon enables electrons to gain kinetic energy that, when reaching sufficiently high values, will allow them to ionize vapor atoms through collisions.

Finally, when electron density increases too much, plasma becomes highly opaque to the laser beam and reflects it [MAO 97, AGU 98]. As a result, the material gains less and less energy, and this slows down ablation phenomena and therefore plasma growth, until the end of the laser pulse.

1.3.4.2. *Ultrashort (picoseconds and femtoseconds) pulses*

In the case of a femtosecond pulse, whose duration is shorter than the time needed for a material to thermalize, energy is transferred to the ion network only after the pulse has ended. Depending on the deposited energy, two major phenomenologies can be observed. If the intensity is strong enough to rapidly excite a significant number of carriers, then non-thermal phenomena will cause ablation. On the other hand, thermal phenomena will predominate: the solid is then composed of a gas of hot electrons surrounding a network of ions that are still cold. Then, electron network thermalization takes place (several picoseconds). Ablation can be produced as follows:

– by evaporation [LOR 06]: when the energy level is below the ablation threshold, only the layers at the surface of the material are ejected through vaporization. Should a thick (several micrometers) material be considered, which is often the case of targets employed for producing nanoparticles, and if this target is irradiated by pulses with fluence above the ablation threshold, this will result in several distinct types of ablation products for one unique irradiation. The energy deposited decreases as it penetrates the solid and thus it is the more in-depth layers that will evaporate;

– by phase explosion [LOR 03]: phase explosion or explosive boiling takes place when the energy deposit allows to obtain at the surface, and in its vicinity, a level of temperature close to the critical value. This triggers homogeneous nucleation of gas bubbles within the overheated liquid layer and the material is subjected to rapid transition from overheated liquid to a mixture of vapor and liquid drops. Contrary to the phase explosion mechanism for long pulses (of the order of 1 ns), material heating induced by a femtosecond pulsed irradiation is carried out at constant volume;

– by spallation fracture [WU 14a, DEM 10, INO 10, ION 11]: an extremely rapid increase in the temperature of ions in the crystal lattice during return to thermal equilibrium (when electronic temperature equals ionic temperature) in a volume that is quasi-constant due to its inertia generates very high local pressure in comparison to the ambient pressure. This results in radial emission of a compression wave. A significant part of it propagates toward the target core, but one part is equally directed toward the surface, and is reflected, being transformed into a stress wave with reversed propagation direction. When propagating through the crystal phase, this stress wave causes fractures and matter ejection. This spallation phenomenon can also occur when the shockwave is reflected and is superimposed on the stress wave that is still propagating toward the back face;

– by fragmentation: separation of a supercritical fluid into droplets by dilution in vacuum above the substrate [LEV 04, WAN 92, ION 13].

When fluence is high, ablation can be caused by non-thermal phenomena. The rapid excitation of the electrons in the conduction band can lead to crystal lattice instability [ROB 86, ROU 01]. Under these circumstances, the direct transition from solid to plasma state is possible by optical breakdown [VON 96]. Coulomb explosion is also possible in the case of dielectrics, and it consists of ablation due to electrostatic repulsion among ionized atoms [STO 02].

Contrary to the expansion of the plasma cloud generated by nanosecond pulses, the plasma cone generated by ultrashort pulse is dilated without further heating. Its temperature and pressure decrease rapidly and therefore its lifetime is short.

1.3.4.3. *Plasma expansion under vacuum or low pressure*

The plume expansion under vacuum is considered adiabatic, particles are subjected to few collisions and the expansion regime is free. Energy is dissipated through various radiation loss processes. At the beginning, free electrons lose a part of their kinetic energy when they move through the Coulomb field of a neutral atom or ion (Bremsstrahlung effect). The same can happen through radiative recombination when the electron is captured by an ion. When temperature and density of free electrons have sufficiently decreased, emissions corresponding to electron transitions from various excited states in the neutral or ionized atoms take place. Thermal exchange processes lead the vapor atoms to condense in small aggregates of several hundred atoms, and then by aggregations and coalescence to form nanoparticles.

When low concentration gas is used, the plume expansion is restricted by the compression of ambient gas, and this expansion in turn compresses the ambient gas [HARI 03]. Plasma propagates according to a high pressure shockwave that attenuates until it reaches the ambient air pressure. Plasma expansion is reduced due to the gas presence. In case the gas presents reactive atoms, they can have chemical interactions with the ionized atoms and form new species.

When plasma is generated by a short pulsed laser, there is no interaction between plasma and laser pulse, as plasma only emerges after the pulse ends. Table 1.18 presents several examples drawn from the literature on the fabrication of metal and oxide nanoparticles by laser ablation in a gas medium.

1.3.4.4. *Laser ablation in liquids*

The first works related to laser ablation of solids being immersed in liquids in order to have their surface altered were reported at the end of the 1980s by Patil [PAT 87] and Ogale [OGA 88]. It was only in 1993 that the first works carried out by the teams of Henglein and Cotton on the synthesis of nanoparticles by laser ablation in liquids, in order to prepare nanoparticles free of surface contamination, were published [HEN 93, NED 93].

Metal	Target	Experimental conditions	Gas	d or d_0 (nm)	Reference
Al	Al	KrF Excimer, λ: 248 nm, pulse: 20 ns, *f*: 20 Hz, *E*: 15 mJ/pulse, 5,000 pulses	$P = 10^{-6}$ Torr	76–2,000	[SIR 11]
Cu	Cu	Ti: Sapphire, λ: 800 nm, f: 1 kHz, pulse 100 fs, *F*: 4 J/cm^2	$P \leq 10^{-4}$ Pa	0.5–12	[NOE 07]
Mo	Mo	KrF excimer, λ : 248 nm, pulse: 20 ns, *f*: 20 Hz, *E*: 15 mJ/pulse, 5000 pulses	$P = 10^{-6}$ Torr	75–2,000	[SIR 11]
Ni	Ni	Nd:Glass, λ: 1,055 nm, pulse: 0.85 ps, *E*: 4 mJ/pulse, *f*: 33 Hz, *T* = 60 min	$P \leq 10^{-5}$ Pa	23–56	[AUS 06]
		Nd:Double glass, λ: 527 nm, pulse: 0.30 ps, *E*: 1.3 mJ/pulse, *f*: 33 Hz, *T* = 60 min	$P \leq 10^{-5}$ Pa	5–65	
	Ni	Nd:Double glass, λ: 527 nm, pulse: 300 fs, *E*: 4 mJ/pulse	$P \leq 10^{-7}$ Pa	37	[AMU 07]
		Nd:Triple glass, λ: 263 nm, pulse: 300 fs, *E*: 4 mJ/pulse	$P \leq 10^{-7}$ Pa	17	
	Ni	Ti:Sapphire, λ: 780 nm, *f*: 1 kHz, pulse 120 fs, *F*: 0.3 J/cm², 200 pulses	$P \approx 10^{-5}$ Pa	40 ± 19	[AMU 05a]
Si	Si	Nd:Glass, λ: 1055 nm, pulse: 0.85 ps, *E*: 4 mJ/pulse, *f*: 33 Hz, *T* = 2–60 min	$P \leq 10^{-5}$ Pa	5–39	[AUS 06]
		Nd:Glass, λ: 527 nm, pulse: 0.30 ps, *E*: 1.3 mJ/pulse, *f*: 33 Hz, *T* = 2–60 min	$P \leq 10^{-5}$ Pa	9–114	
	Si	ArF Excimer, λ: 193 nm, pulse: 12 ns, *F*: 1 J/cm^2	He, P = 333 Pa He, P = 665 Pa	d_0 = 13 d_0 = 17	[YAM 96]
	Si	Nd:YAG double λ: 532 nm, pulse: 7 ns, *F*: 5 J/cm^2	Ar, P = 5 Torr	1.5–3	[MAK 04]
W	W	ArF Excimer, λ: 193 nm, pulse: 15 ns, *f*: 1–25 Hz, *F*: 1.4–8.3 J/cm^2, T=15 min	N_2, P = 1 atm	2–110	[LAN 03]
	W	Nd:Double YAG, λ: 532 nm, pulse: 4.5-5.5 ns, *f*: 20 Hz, *F*: 1.9–3.8 J/cm^2	He, P = 2.7–90 kPa	10–20	[OZA 01]
	W	KrF Excimer, λ: 248 nm, pulse: 20 ns, *f*: 20 Hz, *E*: 15 mJ/pulse, 5,000 pulses	$P = 10^{-6}$ Torr	100–2,000	[SIR 11]
CuO	CuO	Nd:YAG λ: 1064 nm, pulse: 0.5 ms, *f*: 34 Hz, *E*: 5,600 W/mm^2	Air, $P = 9 \times 10^4$ Pa	4–26	[YAN 02]
α-Cr_2O_3	Cr	Nd:YAG λ: 1,064 nm, pulse: 0.24 ms, *f*: 10 Hz, *E*: 1,100 mJ/pulse	Air + O_2 50 L/min	30–100	[LIN 09b]
γ-Fe_2O_3	Fe	Nd:YAG λ: 1,064 nm, pulse: 0.3–20 ns, *f*: 5–150 Hz, *P*: 400W	N_2, P = 0.18 MPa O_2, P = 0.02 MPa	5–90	[WAN 06a]

Table 1.18. *Metal nanoparticles obtained by laser ablation in gas medium d_0, mean diameter; F, fluence, E, energy deposited per pulse; T, ablation duration; P, emitted power*

Many works have meanwhile evidenced the possibility of producing a very wide variety of nanomaterials, depending on a variety of laser parameters or on the nature of the liquids employed.

1.3.4.4.1. Pulse specificity in the liquid phase

In very general terms, when focusing a given wavelength laser beam on a target immersed in a transparent liquid medium, energy is transferred to the target according to the previously described mechanisms, depending on pulse duration [YAN 07a, BER 97]. When the beam irradiates the surface of the immersed target, a plasma is formed at the solid–liquid interface. Confined within the liquid, this plasma undergoes adiabatic dilation at supersonic rate, which triggers a shockwave that, in turn, induces temperature and pressure increase in the plasma [FAB 90] [BER 99, BER 00]. Compared with expansion in air or under vacuum, plasma expansion in a liquid medium is much smaller [WU 08]. This confinement leads to a much higher density of ablated species than in the case of expansion under vacuum or low gas pressure. But on the other hand, plasma lifetime in a liquid medium is also much shorter. These two phenomena have a conflicting effect in terms of size of the particles formed. The first phenomenon tends to facilitate the coalescence of nanoparticles due to the high density of species that are present, while the short lifetime of plasma involves a reduction of the time available, within it, for the reactions of coalescence and agglomeration to last sufficiently and to influence the size of the particles formed [SAI 02].

Finally, the liquid located at the plasma–liquid interface vaporizes and gives rise to bubbles that coalesce and form a cavitation bubble [PAR 96]. The species present in this bubble are dissociated, ionized and mixed with the present species, thus enabling new chemical reactions. Metal ablation in water leads to the formation of metal hydroxides [LIA 04], while if carried out in carbon-rich solvents, it produces carbides [PRO 97]. This bubble will expand its volume and size and will in the end implode. During its implosion, a second shockwave is generated [TSU 07, SAS 05] and all the species present in the plasma will be ejected in the solvent.

As nanoparticles continue to be generated, they accumulate in the liquid phase and interact with the incident beam. Due to absorption and diffusion phenomena, beam transmission in the medium is significantly reduced. This results in an important reduction of efficiency as a function of time [TSU 02]. Several strategies can be deployed as remedy. The first solution consists of using only a small thickness of liquid above the target, but it results in low product quantities, as the liquid volume gets rapidly saturated in particles. The second solution consists of using a steady stream cell that allows us to recover particles as they are produced, while maintaining a flow of liquid free of most of its particles in suspension [BES 10, SAJ 10b]. On the contrary, this reduces the interactions between the beam and the colloids in suspension and consequently the fragmentation and growth phenomena induced by laser pulse irradiation of a colloidal solution [PAN 14, PYA 14].

Thus various types of metal nanoparticles, metal oxides or semiconductor particles such as silicon [SEM 10], have been produced by laser ablation of an immersed massive target [YAN 08] or by fragmentation of particles with sizes of several micrometers placed in suspension in the liquid [PAN 14].

Tables 1.19 and 1.20 report a certain number of examples in the literature referring to metal and oxide nanopowders that have been obtained.

Metal	Target	Experimental conditions	Liquid	d or d_o (nm)	Reference
Al	Al	Nd:YAG, λ: 1.06 μm, *f*: 10 Hz, *T*: 1 h, pulse: 30 ps, *F*: 4 J/cm^2	Ethanol + H_2	30	[VIA 10]
	Al	Ti:Sapphire, λ: 800 nm, pulse: 200 fs, *f*: 1 kHz, *F*: 0.4 J/cm^2, *T*: 10 min Nd:YAG, λ : 1064 nm, pulse: 30 ps, *F*: 8 J/cm^2, *f*: 10 Hz, *T*: 60 min Nd:YAG, λ: 1064 nm, pulse: 150 ps, *F*: 1.5 J/cm^2, *f*: 10 Hz, *T*: 60 min	Ethanol Ethanol Ethanol	10–60 10–250 10–50	[STR 09]
Co	Co	Nd:YAG double, λ: 532 nm, pulse: 3 ns, *F*: 6.37 J/cm^2, *f*: 20 Hz, *T* = 60 min	EG, PVP	22	[ZHA 08]
Cu	Cu	Ti: Sapphire, λ: 800 nm, pulse: 120 fs, *f*: 1 kHz, *F*: 0.2 mJ/pulse, *T*: 60 min	H_2O/ethanol 8/2, PVP	10	[FAN 14]
	Cu	Nd:YAG double, λ: 532 nm, pulse: 5 ns, *E*: 0.2 J/pulse, *f*: 10 Hz, *T* = 10 min	Polysiloxane Mw=240 Mw=1200	 5–20 2–5	[SAI 08]
	CuO	Nd:YAG, λ: 1,064 nm, pulse: 5 ns, *F*: 509 mJ/cm^2, *f*: 10 Hz, *T* = 5 min	2-propanol	d_o = 28.9	[YEH 99]
Mg	Mg	Nd:YAG, λ: 1,064 nm, pulse: 5.5 ns, *F*: 0.265J/cm^2, *f*: 10 Hz, *T* = 60 min	2-propanol acetone	15–20 50–100	[PHU 08]
Ni	Ni	Nd:YAG double, λ: 532 nm, pulse: 3 ns, *F*: 6.37 J/cm^2, *f*: 20 Hz, *T* = 60 min	EG, PVP	8	[ZHA 08]
	Ni	Nd:YAG double, λ: 532 nm, pulse: 8 ns, *E*: 40 mJ/pulse, *f*: 10 Hz, *T* = 60 min	H_2O	3–5	[MAH 08]
Si	Si	Nd:YAG, λ: 1,064 nm, pulse: 10 ns, *E*: 160 mJ/pulse, *f*: 10 Hz, *T* = 60 min	H_2O, SDS	10–20	[YAN 08]
	Si	Ti : Sapphire double, λ: 387.5 nm, pulse: 180 fs, *f*:1 kHz, *F*: 0.8 J/cm^2	H_2O	d_o = 20	[SEM 10]
	Si	Nd:YAG, λ: 1064 nm, pulse: 60 ps, *f*: 20 Hz, *F*: 0.9–2.7 J/cm^2, *T* = 60 min, Nd:YAG triple, λ: 355 nm, pulse: 60 ps, *f*: 20 Hz, *F*: 0.4–4.4 J/cm^2, *T* = 60 min,	H_2O H_2O	40 ± 10 3 ± 1	[INT 14]
	Si	Ti : Sapphire, λ: 800 nm, pulse: 120 fs, *f*:1 kHz, *E*: 1 mJ/pulse, *T* = 30 min	H_2O	d_o = 2.4	[RIO 09]
Zn	Zn	Nd:YAG double, λ: 532 nm, pulse: ns area, *E*: 100 mJ/pulse, *f*: 10 Hz, *T* = 60 min	H_2O, SDS	14.7 ± 8.1	[SIN 07]

Table 1.19. *Metal nanoparticles obtained by laser ablation in liquid medium; EG: ethylene glycol, SDS: sodium dodecylsulfate, PVP: polyvinylpyrrolidone, d_0 mean diameter, F: fluence, E: energy deposited per pulse, T: ablation duration, Mw: average molecular weight*

Oxides	Target	Experimental conditions	Liquid	*d* or d_o (nm)	Reference
Al_2O_3	Al_2O_3	Ti:Sapphire, λ: 800 nm *f*: 10 Hz, pulse: ≤ 130 fs, *F*: 0.462 J/cm^2, *T*: 90 min,	H_2O Ethanol	60 80	[ALN 13]
	Al_2O_3	λ: 1,064 nm f: 4 kHz, pulse: 40–55 ns, *F*: 14–50 J/cm², *T*: 5 min	H_2O	29	[SAJ 10a]
Bi_2O_3	Bi_2O_3	Ti/Sapphire, λ: 800 nm, pulse: 120 fs, *f*: 1 kHz, *F*: 0.5 mJ/pulse, *T*: 60 min	Ethanol	10–60	[LIN 10]
	Bi	Nd:YAG, λ: 1.06 μm, pulse: 10 ns, *f*: 1 Hz, *F*: 21 J/cm², *T* = 200 min	H_2O	56.81	[ISM 14]
Co_3O_4	Co	Nd:YAG, λ: 1,064 nm, pulse: 4 ns, *f*: 10 Hz, *E*: 330 mJ/pulse, *T* = 15 min Nd:YAG, double, λ: 532 nm, pulse: 4 ns, *f*: 10 Hz, *E*: 165 mJ/pulse, *T* = 15 min	H_2O H_2O	10–14 13–22	[HU 14b]
CuO	Cu	Nd:YAG double, λ: 532 nm, pulse: 90 fs, *f*: 1 Hz, *T* = 30 min, *F*: 9J/cm²	H_2O	25–200	[NAT 11]
Cu_2O	Cu	Nd:YAG, λ: 1,064 nm, pulse: 10 ns, *f*: 10 Hz, *E*: 80 mJ/pulse, *T* = 60 min	H_2O, PVP	15–60	[LIU 11]
Cr_3O_4	Cr	λ: 800 nm, pulse: 90 fs, *f*: 1 kHz, *T* = 35 min, *F*: 102 J/cm²	H_2O	6.8	[SEM 14]
	Cr	Nd:YAG, λ: 1,064 nm, pulse: 240 μs, *f*: 10 Hz, *E*: 1100 mJ/pulse, *T* = 10 min	H_2O	10–30	[LIN 09a]
Fe_2O_3	Fe_2O_3	Nd:YAG, λ: 1,064 nm, *E*: 40 mJ/pulse, *f*: 10 Hz, *T* = 60 min	H_2O H_2O, CTAB	28–33 10.7–15.6	[PAN 14]
	Fe_2O_3	Nd:YAG triple, λ: 355 nm, pulse: 10 ns, *E*: 40 mJ/pulse, *f*: 30 Hz, *T* = 60 min	Ethanol H_2O Acetone	3–50, 5–30 4–28	[MAN 13]
$Mg(OH)_2$	Mg	Nd:YAG triple, λ: 355 nm, pulse: 7–8 ns, *E*: 100 mJ/pulse, *f*: 10 Hz, *T* = 60 min	H_2O, SDS	*L* > 200 Φ = 10– 30	[LIA 04]
	Mg	Nd:YAG, λ: 1,064 nm, pulse: 5.5 ns, *F*: 0.265J/cm^2, *f*: 10 Hz, *T* = 60 min	H_2O H_2O, SDS	*L* = 150, Φ = 5–10 20–30	[PHU 08]
NiO	Ni	CW, P: 250 W, λ: 1,070 nm, *T*: 1s, Spot: 40 μm	H_2O, SDS	13 ± 9	[LIU 09]
TiO_2	Ti	CW, P: 250 W, λ: 1,070 nm, *T*: 1s, Spot: 40 μm CW, P: 250 W, λ: 1,070 nm, T: 1s, Spot: 40 μm	H_2O, SDS H_2O H_2O, SDS	17 ± 7 27 ± 11 40 ± 18	[LIU 09, ABD 08]
	Ti	Nd:YAG triple, λ: 355 nm, pulse: 7 ns, *E*: 150 mJ/pulse, *f*: 10 Hz, *T* = 60 min	H_2O, SDS	2–6	[SAS 04]
Sb_2O_3 fibers	Sb	Nd:YAG double, λ: 532 nm, pulse: 5 ns, *E*: 140 mJ/pulse, *f*: 10 Hz, *T* = 10 min	H_2O H_2O, SDS	35	[MEN 13]
SnO_2	Sn	Nd:YAG triple, λ: 355 nm, pulse: 7 ns, *E*: 100 mJ/pulse, *f*: 10 Hz, *T* = 60 min	H_2O, SDS	2–6	[SAS 04]

	Sn	Nd:YAG triple, λ: 355 nm, pulse: 7 ns, *E*: 100 mJ/pulse, *f*: 10 Hz, *T* = 60 min	H_2O, SDS	2.5 ± 0.6	[LIA 03]
*H_2WO_4 flakes	W	Nd:YAG double, λ: 532 nm, pulse: 10 ns, *E*: 300 mJ/pulse, *f*: 10 Hz, *T* = 30 min	H_2O	10	[XIA 12]
*H_2WO_4	W	Nd:YAG λ: 1064 nm, pulse: 10 ns, *E*: 80 mJ/pulse, *f*: 10 Hz, *T* = 120 min	H_2O	1	[ZHA 14, [ZHA 12b]
WO_3	W	Nd:YAG double λ: 532 nm, pulse: 5 ns, *F*: 1–7 J/cm^2, *f*: 10 Hz, *T* = 30 min	H_2O	2–6	[BAR 11]
	Zn	Nd:YAG double, λ: 532 nm, pulse: ns, *E*: 60 mJ/pulse, *f*: 10 Hz, *T* = 60 min	H_2O, SDS, O_2	13.7 ± 5.8	[SIN 08]
	Zn	Nd:YAG triple, λ: 355 nm, pulse: 7 ns, *E*: 100 mJ/pulse, *f*: 10 Hz, *T* = 20 min	H_2O, CTAB	10–25	[HE 08]
ZnO	Zn	Nd:YAG triple, λ: 355 nm, pulse: 7 ns, *E*: 100 mJ/pulse, *f*: 10 Hz, *T* = 60 min	H_2O pH : 5.36 pH:11.98 pH: 7.51 NaCl	 15 ± 6 20 ± 8 23 ± 11 26 ± 15	[HE 07]
ZnO nanorods	Zn	Nd:YAG triple, λ: 355 nm, pulse: 7 ns, *F*: 3.2 J/cm^2, *f*: 10 Hz, *T* = 40 min	H_2O, 80°C	*L* = 500–600 Φ = 200	[ISH 06]

*WO_3 precursor before calcination at 800°C, 4 h.

Table 1.20. *Metal nanoparticles obtained by laser ablation in liquid medium. SDS; sodium dodecylsulfate; CTAB, cetyltrimethylammonium bromide; PVP, polyvinylpyrrolidone; d_0, mean diameter; F, fluence; E, energy deposited per pulse; T, ablation duration*

1.3.4.5. *Effect of laser parameters*

1.3.4.5.1. Effect of number of pulses

Several authors have shown that when the number of pulses increased, the mean diameter and the size dispersion decreased. Notable examples in this sense include the works of Mahfouz *et al.* [MAH 08] related to the fabrication of nickel oxide nanoparticles in water or those of Prochazka *et al.* [PRO 97] referring to colloidal silver nanoparticles. These researchers have shown that the size of the particles formed decreased when the number of pulses increased and, moreover, irradiating the colloidal solution with only one laser pulse resulted in a reduction of particle size.

Works conducted by Mafuné *et al.* [MAF 01] on the synthesis of silver nanoparticles have evidenced that the abundance of nanoparticles formed increases linearly as a function of the number of pulses, up to a limit of around 50,000 pulses beyond which the increase slows down. Slope reduction beyond the 50,000 pulses could be explained by weak absorption of the wavelength by the silver nanoparticles that contribute to the attenuation of the laser beam that penetrates the solution above the metal plate. As the energy that reaches the surface of the target is lower, the formation of nanoparticles loses efficiency.

1.3.4.5.2. Effect of pulse duration

Compared with longer pulses, of the nanosecond type [YAN 08], the use of short picosecond [INT 14] or femtosecond [RIO 09] pulses leads to obtaining smaller diameter particles with low polydispersity. This can be due to non-thermal ablation, such as Coulomb explosion for a dielectric, when radiation energy is low. On the other hand, the femtosecond pulse ends well before plasma starts to emerge. This generates a plume of lower intensity and the lifetime of the cavitation bubble is shorter. This tendency increases in the case of ablation in the liquid phase or for adequate pressure of ambient gas. Under femtosecond regime, the growth of nuclei is thus limited, which leads to the production of smaller nanoparticles [SYL 05].

On the other hand, the yield of nanoparticle production can be made up for by using a higher repetition frequency for femtosecond lasers: this is of the order of kilohertz, while for the nanosecond lasers, the repetition frequencies are of the order of hertz or a dozen hertz.

1.3.4.5.3. Effect of wavelength

The rate of metal ablation tends to increase when laser wavelength diminishes [VLA 08, TSU 01]. On the one hand, this is due to higher radiation absorption by the metals when the wavelength is shorter. On the other hand, wavelength reduction generally leads to lower absorption of plasma by inverse Bremsstrahlung [BOG 05].

Furthermore, the secondary interaction between laser and the nanoparticles produced becomes stronger with shorter wavelengths [ZHE 07]. This secondary interaction normally leads to reduced size of the colloids that are fragmented by laser [TSU 02].

The use of shorter wavelengths generally results in a reduction of the mean size of the particles formed [TSU 02, NIC 06a].

1.3.4.5.4. Effect of fluence

Under fixed focus conditions, the mean size increases with pulse energy [SYL 05, KAB 03, AMU 05b]. This is caused by two phenomena. The first arises because of the higher concentration of matter in the plasma plume, which will facilitate coalescence and aggregation during cooling. In particular, Kabashin has shown that during the synthesis of gold colloids in water with a femtosecond laser, he obtained a linear log-normal correlation between the size of nanoparticles and laser fluence [KAB 03]. When fluence exceeds a certain threshold that depends on the nature of the irradiated metal, a second phenomenon emerges, which generates large size particles. These result from the ejection of molten metal drops from the irradiated surface [YAH 10, FIS 06]. In effect, the largest particles formed (several dozen nanometers) during the use of high fluences are due to the phase explosion taking place at the target surface [ACA 10, AME 09, NIC 06b, KAB 03]. An increase in the size of particles is likewise obtained when defocused radiation is used [PET 09, PYAT 04, MAF 00a].

1.3.4.5.5. Effect of gas pressure

During the ablation in gas medium, the increase in ambient pressure limits the expansion of plasma plume and therefore that of ablated species. This results in weaker dilution that depends on time, and facilitates the growth of particles within plasma as the latter gets cooler [SUZ 12, BAR 07]. There are two mechanisms that govern the formation of different nanoparticles. Small nanoparticles mainly result from evaporation followed by condensation. As for larger particles, they result from a phase explosion process at the surface of the target, caused by laser radiation. As described in section 1.3.4.5.4, this last phenomenon depends on the fluence used [OZA 01].

1.3.4.5.6. Effect of solvent nature

Metal powders can be produced by laser ablation in a solvent. For example, ethylene glycol was used as reducing agent for obtaining nickel nanoparticles [ZHA 08]. The use of ethanol alone or in the presence of hydrogen dissolved by bubbling enables the production of titanium [SIM 04] or aluminum [VIA 10] nanopowders. Using water as solvent generally leads to the formation of metal oxides.

1.3.4.5.7. Effect of surfactants

When basic or acidic salts [SYL 04a, HE 07] or surfactants [SAS 04, PAN 14] [LIU 11] are introduced in the liquid medium before laser ablation, dramatic changes may occur in the distribution, nature [MAF 01], form [LIA 04] and/or aggregation density of nanoparticles obtained at given laser parameters.

Mafuné *et al.* have established that the presence of certain surfactants, such as SDS, could slow down or stop the oxidation of zinc particles in water [SIN 07, HAI 05]. Upon plasma extinction, the zinc clusters formed come in contact with the solvent and surfactant molecules present in solution, and this puts in competition the chemical reactions of oxidation and encapsulation of the particles. When the surfactant concentration is low, the first effect is the reduction of the aggregation of zinc clusters; oxidation phenomena are dominant and lead to the formation of zinc oxide nanoparticles. When the surfactant concentration increases, the encapsulation effect becomes dominant and prevents zinc oxidation. This protection against oxidation is a phenomenon that has not been evidenced for other metals.

The use of surfactants influences the morphology of magnesium hydroxide crystallites produced in solutions of deionized water. SDS directs the growth of crystallites either along an axis that privileges fibrous morphologies or along two axes, which results in a platelet-analogous form. The growth mode depends on the surfactant concentration [YAN 07b, LIA 04].

In all cases, the increase in the stabilizer concentration results in decrease in the mean size of nanoparticles [MAF 00b, MAF 03, SYL 04b, SYL 04a].

1.3.4.5.8. Effect on colloids in suspension

When particles, either in suspension in gas or in solution, absorb the beam energy, the result may be fragmentation [TAK 99, INA 05, BES 06, SVR 11, YAM 06], growth [YEH 99, ZHE 07, BES 09], or modifications of the form of particles [AGU 04, PYA 07, WAN 10].

In liquid medium, these phenomena have been studied either by direct exposure to radiation of colloidal solutions of nanoparticles [GIAM 10, WER 11, PYA 09] or during the generation of nanoparticles by laser ablation of a solid target [MUT 08, MAF 01, VAN 13].

These works show that laser fragmentation of microparticles allows us to obtain nanoparticles with production yields that are higher than those obtained by laser ablation of the surface of a massive target [YEH 99, ZHA 03, KAW 06]. The nanoparticles produced by this method generally feature lower size polydispersity. Two possible models that explain these phenomena are reported in the literature. The first one relies on thermal transfer [PYA 13]. The energy deposited at the surface of the particle causes an increase in its temperature followed by melting and surface evaporation, which leads to a decrease in particle size [TAK 99] and also a change in form [PYA 07]. The second model, developed by Kamat *et al.* [KAM 98], proposes the Coulomb explosion phenomenon as cause of the observed size reduction. These two models are not mutually exclusive and enable a description of reality depending on the laser emission conditions. Strong fluence facilitates the fragmentation of nanoparticles by Coulomb explosion, while exposure to lower energy radiation facilitates fragmentation by thermal transfer.

1.3.4.6. *Conclusion*

To conclude, this technique offers the possibility of optimizing favorable effects in order to obtain small size particles, with low levels of polydispersity and aggregation. To this purpose, it is preferable to use low laser fluences, low pulse widths, short wavelength and also long ablation durations. Finally, the use of surfactants facilitates particle size decrease and reduces particle aggregation [YAN 07a]. While small sizes, of the order of several nanometers, and weak polydispersity are obtained, the production yield remains at modest levels.

1.3.5. *Pyrotechnic synthesis*

1.3.5.1. *Detonation synthesis*

The originality of the synthesis by detonation resides in using a phenomenon that is commonly considered destructive to form nanosized grains of matter and "freeze" them while in divided state. Matter dispersion upon detonation allows for the formation of powders, in which elementary nanoparticles are less strongly agglomerated than those prepared in solution or by milling. Due to their small size, but also to the relative independence of the nanoparticles they are made up, the materials synthesized by detonation can be dispersed in liquids for which they have affinity. The formation of colloid suspensions constitutes the most important stage in the formulation of a nanothermite, regardless of whether it has been prepared by simple physical mixture of powders in suspension or by a proper assembly method.

From a historical perspective, nanodiamond was the first detonation synthesized material. Its presence in the secondary explosive detonation products was for the first time observed in July 1963 by Soviet scientists working in the field of nuclear weapons. The first tests conducted in 1962 consisted of the compression of graphite or carbon black contained in spherical or cylindrical storage ampoules due to a detonation shock. Graphite had been afterward incorporated in a cylindrical charge composed of trotyl (TNT) and hexogen. A test conducted in the absence of graphite had at that time shown that nanodiamonds were formed starting from the carbon content of explosive molecules [DAN 04]. The size of nanodiamonds resulted from explosive detonation ranges between 4 and 10 nm. Recent works have shown that the detonation of nanostructured hexolite (TNT/RDX) allowed to produce smaller nanodiamonds and, at the same time, to significantly increase their proportion in the detonation products [PIC 13]. As shown by the carbon phase diagram given by Bundy, at ambient air temperature and atmospheric pressure diamond is a metastable material [BUN 62]. When temperature increases, diamond transforms into graphite or oxidizes depending on the nature of the atmosphere it is exposed to. Diamond is the highest density form of carbon at ambient temperature ($d = 3.52$ g/cm^3) and one of the best heat conductors known [FIE 92]. Moreover, its oxidation energy is close to 33 kJ/g (or 116 kJ/cm^3), which recommends it as a reducing agent of choice for pyrotechnic compositions. The use of nanodiamond as fuel in energetic compositions based on potassium chlorate ($KClO_3$) has been reported by Comet *et al.* [COM 09]. These compressed compositions feature continuous or intermittent combustion, depending on the proportion of nanodiamond they contain and on their porosity. The study of these original energetic materials has allowed us to estimate the rate of oxidation of the nanodiamond resulted from a detonation at 0.8 mm/s and to show that nanodiamond powders have a thermal conductivity that is four orders of magnitude smaller than that of massive diamond [COM 14a]. The small size of diamond nanoparticles and the presence of oxygen functional groups at their surface [SCH 12] render them easier to oxidize than massive diamond. Nanodiamond however remains a much less reactive fuel with respect to other fuels that are for nanothermite formulation, such as aluminum, boron or red phosphorus. This means that the oxidizability of nanodiamond is too low for this material to be used as fuel in nanothermites. However, detonation synthesized nanodiamond was used by Pichot *et al.* as an additive in order to significantly increase the threshold of sensitivity to friction and electrostatic discharge of a nanothermite based on bismuth oxide (Bi_2O_3) and aluminum. For this purpose, nanodiamond is first deposited on the surface of submicron Bi_2O_3 particles starting from a suspension in acetonitrile. The resulting composite material is then physically mixed with nanosized aluminum powder [PIC 15].

The use of detonation synthesis to produce various types of nanomaterials has been proposed by Eidelman and Altshuler. According to these authors, the detonation waves generate plasmas with high energy density, which can be effectively used to produce nanomaterials. The detonation of the most common explosives, which are formed of carbon, hydrogen, oxygen and nitrogen, results in nanostructured carbonaceous phases whose nature depends on the charge composition and on the environment in which it is fired. Depending on the experimental conditions, the detonation of more exotic explosives, containing boron, results in boron oxide (B_2O_3) or nitride (BN). Nanostructured metals are formed by the detonation of metal azides or acetylides. Finally, the incorporation of metal halides into classical explosive charges would be a method to nanostructure some metals such as iridium, plutonium, rhenium, tungsten, vanadium or titanium. Detonation synthesis offers the unique possibility to obtain stable polymorphous products at high pressure due to the drastic cooling triggered by the rapid quenching of the detonation products [EID 93].

There are two approaches to the detonation synthesis of nanostructured powders. The first is the *top-down method*, which consists of the fragmentation of inert matter under the action of the detonation shockwave. The second is the *bottom-up technique*, which consists of building nanoparticles starting from a reactive precursor.

Gibot *et al.* have used the top-down approach for the fragmentation of silicon carbide (SiC) particles that are initially of micrometer size (40–100 μm). To do so, the SiC particles are first coated with polyurethane in order to limit the friction between carbide and the explosive charge. The coated particles are afterwards mixed with trinitrotoluene and a wax; the charge results from the compression of this mixture. SiC particles are miniaturized by detonation, but their size distribution is very wide (0.01–10 μm) [GIB 11c]. This result indicates that the detonation wave impact does not allow to fragment the whole refractory matter contained by the charge. Dividing the small size primary particles into even smaller size secondary particles requires matter to be submitted to high stress. This experimental observation should be considered in connection with the concept of atom, which has for a long time been presumed an indivisible unit of elementary matter, whose fragmentation is only possible by employing significant energies developed by nuclear reactions. Silicon carbide is a ceramic that cannot be used as a reactive component of nanothermites. However, miniaturized particles of silicon carbide could serve as additives in nanothermites, rendering them sensitive to shock, by the conversion of crushing as result of impact into friction force. SiC particles would thus play the original role of pyrotechnic transducer.

Beloshapko *et al.* have subjected aluminum powders composed of petal-shaped particles of submicron thickness (0.25–0.5 μm) and micron planar size (20–30 μm) to the detonation of various explosive compositions. Aluminum is not incorporated in the explosive charge, but set in contact with it. The particles resulted according to this original variant of the top-down approach have spherical geometry and their composition depends on their size. The smaller particles are essentially composed of α alumina (corundum), while the larger particles contain mostly δ alumina and AlN. To explain the transformation of matter, the authors propose an interesting mechanism: due to the compression of the aluminum powder by the shockwave, the metal melts and evaporates. Aluminum then gets oxidized in contact with the oxygen contained in the firing chamber or/and in the explosive charge. The increase in temperature generated by the reaction causes the dispersion of oxide in a liquid state by evaporation or ablation. The growth of particles takes place through a mechanism of coalescence of liquid droplets [BEL 90]. Obviously, the ceramic nanoparticles obtained by Beloshapko cannot be used to formulate nanothermites, but they testify to the effect of the detonation shockwave on micro-sized aluminum. Metal vaporization through shock may allow for explaining the detonating properties of hybrid formulations [THI 11], in which synergy exists between the detonation of an explosive and the aluminothermic reaction.

A method for the detonation-based preparation of chromium oxide (Cr_2O_3) nanoparticles, which is relevant for both top-down and bottom-up approaches, was reported by Comet *et al.* [COM 11c]. This method consists of fragmenting, melting and dispersing a spongy matrix of chromium oxide by the simultaneous action of detonation of an explosive (RDX) solidified in the oxide porosity and a hexolite charge serving as reactive confinement to the nanocomposite material RDX@Cr_2O_3 (see Figure 1.7).

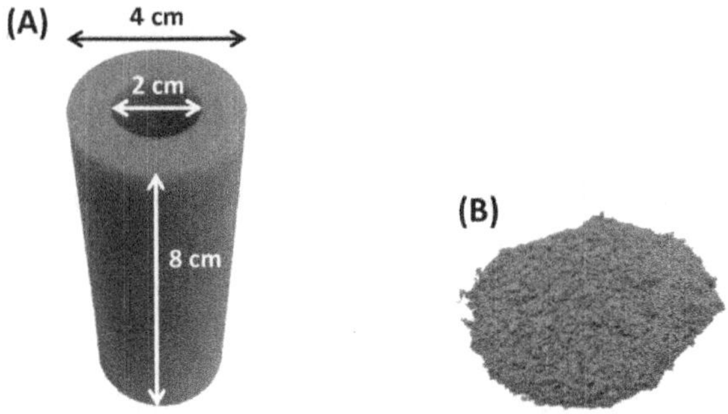

Figure 1.7. *Tubular charge of hexolite (A) used as container for the nanocomposite material RDX@Cr_2O_3 (B) used as precursor for the formation of Cr_2O_3 nanoparticles by detonation [COM 11c]*

Detonation products (see Figure 1.8(a)) contain Cr_2O_3 nanoparticles of spherical or slightly oblong form (5–50 nm) and carbonaceous species, nanodiamonds and graphite particles of crescent form, and also several impurities resulted from the erosion of the walls of the detonation chamber. Carbonaceous phases, which are a major component of detonation soots, are eliminated through an oxidation treatment (air, 500°C), which allows us to recover chromium oxide nanoparticles (see Figure 1.8(b)).

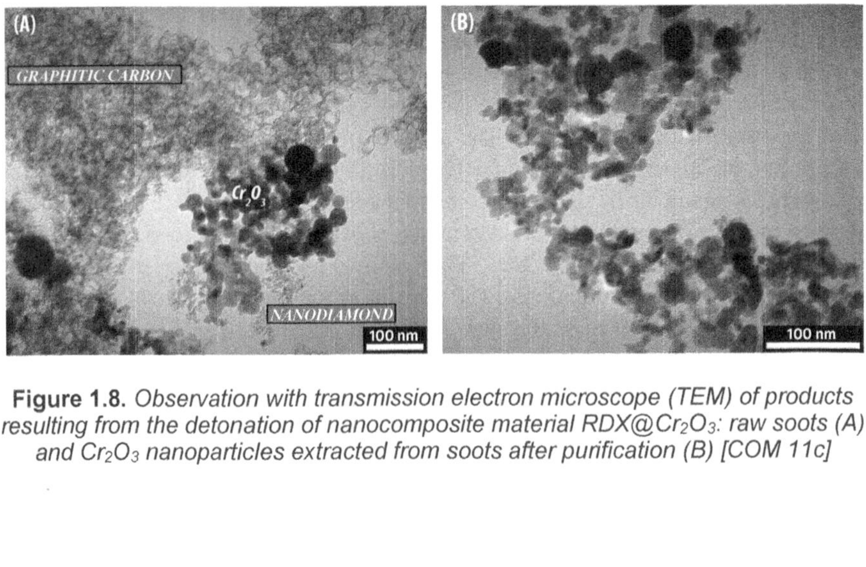

Figure 1.8. *Observation with transmission electron microscope (TEM) of products resulting from the detonation of nanocomposite material RDX@Cr_2O_3: raw soots (A) and Cr_2O_3 nanoparticles extracted from soots after purification (B) [COM 11c]*

Calculation according to Scherrer equation of the mean size of Cr_2O_3 elementary crystallites shows that these are smaller in the Cr_2O_3 of origin (19.4 nm) than in the detonation structured oxide (49.6 nm). This result is confirmed by specific surface area measurements and observations using the scanning electron microscope (see Figure 1.9), which show that Cr_2O_3 primary particles grow larger as a result of detonation. According to Fedoroff *et al.* [FED 60], TNT and RDX detonation temperatures reach 3,450 and 4,500 K, respectively. Therefore, during detonation, Cr_2O_3 is exposed to a temperature between these limits. The detonation temperature is probably closer to that of RDX, insofar as the oxide is in direct contact with this explosive within the nanocomposite material RDX@Cr_2O_3. On the other hand, according to the literature, the melting and boiling temperatures of Cr_2O_3 are 2,539 and 4,273 K. The oxide droplets formed are projected and exposed to rapid quenching in contact with the water in which the charge is fired. This explains why Cr_2O_3 elementary particles produced by detonation are larger than the oxide particles they result from.

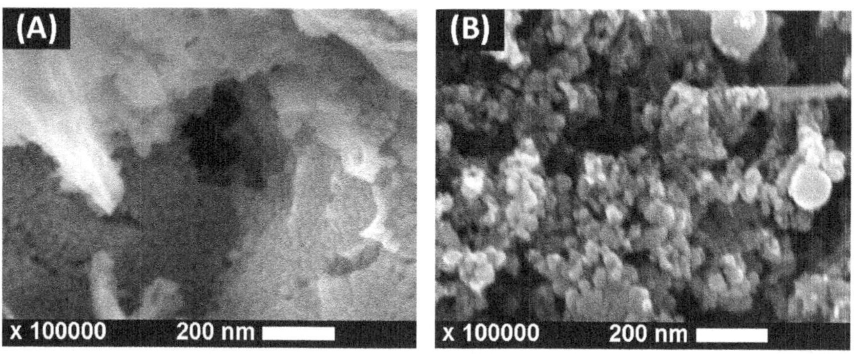

Figure 1.9. *Observation with scanning electron microscope of the morphologies of original Cr_2O_3; a) and detonation produced Cr_2O_3; b); the specific surface areas of these materials are 44.2 and 20.4 m^2/g, respectively [COM 11c]*

Granulometric measurements carried out on aqueous suspensions of detonation structured chromium oxide show that they contain a population of "free" particles of submicron and nanometric size. Moreover, the aggregates of elementary particles are smaller than those observed in the porous chromium oxide suspension (see Figure 1.10). These results show that the Cr_2O_3 particles formed by detonation are more easily dispersible than those of porous chromium oxide, though these are thinner.

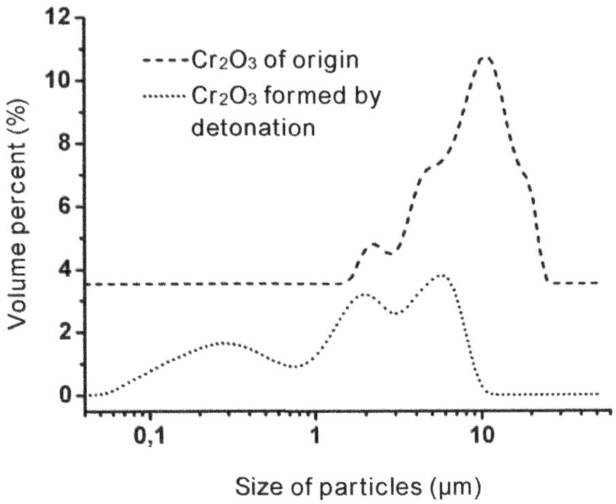

Figure 1.10. *Comparison between the size distribution of Cr_2O_3 of origin and that of Cr_2O_3 formed by detonation [COM 11c]*

Under the same compression conditions, Cr_2O_3/Al nanothermites formulated starting from chromium oxide produced by detonation densify easier than those prepared with porous oxide. This result is an additional illustration of the fact that oxide particles produced by detonation are less agglomerated. The reduction of porosity is also reflected by lower sensitivity to ignition and more regular combustion.

Sphere-shaped (5–25 nm) titanium dioxide (TiO_2) nanoparticles have been produced by the detonation of a charge composed of hexogene and ammonium nitrate containing titanyl hydroxide [QU 11]. In this case, detonation ensures dehydration and dispersion of the precursor. The formed oxide particles contain two allotropic varieties of TiO_2, rutile (65.2%) and anatase (34.8%). The authors use thermodynamic calculation to show that under the given experimental conditions, the reaction temperature obtained (2,577 K) is higher than the melting temperature of rutile (2,153 K). In their opinion, only rutile should form under these circumstances, while the presence of anatase cannot be fully explained by the detonation theory. However, according to Hanaor, rutile is the only thermodynamically stable phase of TiO_2. The formation of anatase, whose structure is less constrained than that of rutile, is facilitated by kinetics [HAN 11]. Thus, it does not seem surprising that anatase forms during the drastic quenching of the molten TiO_2 droplets dispersed by detonation. There is limited interest in the titanium dioxide as far as nanothermite formulation is concerned, since its reaction with nanosized aluminum is one of the weakest known [VAL 03].

Detonation synthesis according to the bottom-up approach consists of the formation of nanoparticles starting from a precursor taking part in the detonation reaction. Thus, nanodiamonds are produced by detonation form starting from the surplus carbon in the explosive. Anisichkin has used isotope marking to show that carbon atoms coming from the TNT molecule are preferentially transformed into diamond [ANI 07]. Frank *et al.* describe the detonation synthesis of the hexagonal variety of gallium nitride (GaN), under the form of small size (2–24 nm) nanoparticles. A primary explosive, triazidogallium, stabilized by various ligands from the trialkylamine family, was used for this purpose [FRA 98]. Many works report on the detonation synthesis of ceramics starting from nitrate-fuel types of compositions. Compositions based on ammonium nitrate have a low temperature of explosion [BÜC 00] and produce a significant quantity of gas. These two effects are particularly important for the production of materials by detonation, since they contribute to limiting the growth of nanoparticles structured by this technique. Zhi-wei Han *et al.* have thus synthesized CeO_2 nanoparticles (55 nm), intended for catalyzing the decomposition of ammonium perchlorate, by using an explosive emulsion in which the dispersed phase consisted of cerium and ammonium nitrates and the continuous phase of paraffin and an emulsifier [HAN 14]. Nitrate-based emulsions have also been used by Neves *et al.* to prepare aluminum-doped zinc oxide (ZnO) nanoparticles (AZO). The observation of formed powders by electron microscope indicates that they are composed of nanoparticles with various morphologies: spheroids, rodlets and platelets. The specific surface areas and the sizes of crystallites do not vary much with aluminum content, which distinguishes detonation synthesis from conventional methods, in which the doping level influences the size of particles. In-depth examination by scanning transmission electron microscopy shows that aluminum is uniformly distributed in the AZO particles formed, both in the ZnO structure where it replaces Zn^{2+} cations, and in the form of a mixed oxide, the gahnite ($ZnAl_2O_4$). Nanoparticles are associated as brittle agglomerates, which are dissociated by milling, using microballs. By this treatment, the specific surface area grows significantly, from 8 to 12 m^2/g [NEV 14]. Zinc oxide nanoparticles (20–50 nm) have been fabricated by detonating a complex nitrate-based emulsion containing 2% epistatic polystyrene microspheres [XIE 06a]. Due to its stability, zinc oxide cannot be used as oxidizer in nanothermites. Nevertheless, an advantage could be drawn from its piezoelectric properties [WAN 06b] in order to increase the shock sensitivity of nanothermites by transforming the energy gained by impact into electrical discharge that can initiate the reaction.

Mixed oxides have been produced by the detonation of compositions containing mixtures of metal nitrates. Thus, Xie *et al.* have prepared primary nanoparticles (20–60 nm) of lithium manganate ($Li_{1+x}Mn_2O_4$) starting from lithium and manganese nitrates. These elementary nanoparticles assemble into spherical secondary particles (1–2 µm) [XIE 06b]. Manganese ferrite ($MnFe_2O_4$) nanoparticles with sizes ranging between 5 and 30 nm have been prepared by the detonation of a complex composition, containing iron (III), manganese (II) and ammonium nitrates. These authors propose that the material forms by a mechanism in two stages, as follows: first, nitrates are decomposed into oxides, and second, oxides associate and form the mixed oxide [WAN 08]. The use of nanostructured mixed oxides as an oxidizing phase in nanothermites seems to have been little studied until present. The association of two reactive oxides ($MnFe_2O_4$) within the same crystal structure or that of a reactive and an inert oxide provides an additional parameter, which could be used for adjusting the performances of nanothermites.

Luo *et al.* have studied an original method, which consists of the introduction of urea as ligand to the metal cation of iron (III), cobalt and nickel nitrates in order to transform these oxidizing salts into intrinsic energetic compositions. These substances are associated with secondary explosives whose role is to reinforce and support detonation. Nanoparticles (5–30 nm) of various natures have thus been produced starting from modified nickel nitrate, hexogen and paraffin. The addition of paraffin allows us to lower the oxygen balance of the explosive charge, and to define the nature of the products formed by detonation. When the oxygen balance is gradually diminished, the phases formed are nickel oxide (NiO), nickel coated with NiO and amorphous carbon, nickel coated with graphite and finally particles that contain a mixture of nickel and nickel carbide (Ni_3C) [LUO 12a]. Similar charges prepared starting from urea complexed ferric nitrate allow us to produce, by progressively lowering the oxygen balance, the following: iron oxides (25–55 nm), a mixture of iron oxide and iron, face-centered cubic iron nanoparticles (6–35 nm), then body-centered cubic iron nanoparticles (6–40 nm) [LUO 12b]. In the absence of paraffin, detonation results in nanoparticles composed of metal cores of cobalt or nickel (10–25 nm) coated with carbon layers of nanosize thickness (3–5 nm) [LUO 10a]. The nanoparticles produced starting from a ferric precursor have a core composed of iron or iron carbides (Fe_2C or Fe_3C) coated with carbon layers [LUO 10b]. Finally, Guilei *et al.* have used the detonation of a mixture of hexogen and cobalt naphthenate to produce nanoparticles of this metal. These nanoparticles have a metal core (22–56 nm) covered by a thick carbon shell (8–41 nm) [SUN 09].

"Pyrohydrolysis" induced by detonation of titanium tetrachloride ($TiCl_4$) has been studied by Ouyang *et al.* It consists of using water resulting from the detonation of a stoichiometric gas mixture of hydrogen and oxygen, initiated by electrical discharge, in order to hydrolyze the droplets of titanium tetrachloride ($TiCl_4$) sprayed in an explosion chamber. The titanium dioxide produced by this method has a mass composition of 12% anatase and 88% rutile. The material contains many nanoparticles (10–30 nm) and also highly agglomerated small particles. SEM observation of this material shows that it is composed of micron and submicron-sized particles that appear to be highly aggregated. The explanation that the authors advance for the presence of large size particles is that they are formed during the transition of the gas mixture from deflagration to detonation [OUY 08].

More recently, Yan *et al.* have used this method to prepare very small nanoparticles (1–10 nm) and non-agglomerated tin dioxide (SnO_2), starting from gaseous tin tetrachloride ($SnCl_4$). To this purpose, the explosion chamber is maintained at a temperature above the boiling temperature of $SnCl_4$ (YAN 13). Tin dioxide is obtained in a pure state, since the second product of the reaction (HCl) is a gas. This preparation mode has many advantages, which could be used in order to prepare oxides of intermediate stability, such as those composing nanothermites, in the form of weakly agglomerated ultimate size nanoparticles.

Nature	Examples	Interest
Reactive oxides	TiO_2; Cr_2O_3; Fe_xO_y; SnO_2	Moderately reactive nanothermites
Mixed oxides	$Li_{1+x}Mn_2O_4$; $MnFe_2O_4$	Adjustment of performances
Ceramics with specific properties	SiC; ZnO CeO_2	Pyrotechnic transduction Catalysis
Oxide@carbon composites	NiO@C	Desensitization
Metals@carbon composites	Fe@C; Co@C; Ni@C	Precursors of free metal oxide nanoparticles
Carbonaceous materials	Nanodiamond	Desensitization

Table 1.21. *Nature of the materials elaborated by detonation and their possible uses for nanothermite formulation*

1.3.5.2. *Deflagration synthesis*

The deflagration of confined explosive compositions, triggered by slow heating (10–20 K/min), is equally used for producing nanomaterials of various natures. This preparation mode, which is often inappropriately termed "detonation synthesis", often leads to nanosized tubular objects, whose formation is catalyzed by metal species.

Lu *et al.* use the decomposition of picric acid for the thermolysis of ferrocene and formulation of iron nanoparticles (5–20 nm) covered by a graphite shell that effectively protects the metal core against oxidation under harsh conditions. Carbon nanotubes have also been observed in the reaction products of the ferrocene-poorest charge [LU 05]. Iron carbide (Fe_7C_3) nanocrystals encapsulated in an amorphous carbon coating have been produced by the reaction of an explosive composition consisting of a nitrated coal pitch and ferric nitrate [WU 03]. Carbon nanotubes can be produced by the explosion of compositions with a nitroaromatic compound associated with an organic metal salt. Multiwall carbon nanotubes are formed, for example, in the explosion of a mixture of picric acid and cobalt (II) acetate. The reaction of picric acid alone gives few solid products; the latter are exclusively composed of amorphous carbon nanoparticles [LU 02]. Nanotubes having a bamboo structure have also been produced by the explosion of this same composition, in which hydrocarbons such as benzene or paraffin had been incorporated [LU 04]. The deflagration of mixtures based on dinitrobenzene (DNB) and cobalt acetate or nickel formate generates multiwall carbon nanotubes and carbon-encapsulated nanoparticles. The highest yields of nanotubes are obtained with high charging densities and DNB/cobalt acetate molar ratios [LU 03]. Finally, carbon nanobulbs have been produced by blowing of carbon nanotubes, resulting from the explosive reaction of a mixture of picric acid and nickel formate [ZHU 04]. Siegert *et al.* have proved that hollow carbon fibers with an inner diameter of about 40 nanometers could serve to desensitize MnO_x/Al ($x \approx 1.9$) nanothermites to friction, by physically separating manganese oxide from aluminum [SIE 10]. The tubular carbon structures with relatively large diameter produced by deflagration could serve for this purpose. More generally, the deflagration of explosive could be used for coating metal oxide nanoparticles with a carbon film in order to formulate nanothermites that are less sensitive to friction and electrostatic discharge.

1.3.5.3. *Combustion synthesis*

For thousands of years, man has been using fire to extract metals from ores and to process metals and ceramics. The use of the "internal fire" produced by the reaction of combustion of a substance to form a material with defined chemical nature and morphology has been studied for the first time by Berzélius in 1825, upon the discovery of the pyrophoric character of zirconium powder. There is however in nature a very common example of this method, namely wood combustion. The pyrolysis of biopolymers, cellulose and lignin produces the porous carbon structure of charcoal. The ashes resulted from complete oxidation of charcoal are nanostructured materials, composed of a cocktail of mineral oxides. Moore and Feng [MOO 95] offer an interesting chronology of the history of combustion synthesis from its start to 1995.

Various modes of combustion synthesis have been reviewed by Patil *et al.* who distinguish four types of methods: (1) self-propagating high temperature synthesis (SHS), (2) solid state metathesis reactions, (3) synthesis by flame pyrolysis and (4) synthesis starting from redox compounds or mixtures [PAT 97].

The conventional SHS method is not adapted to the production of nanomaterials, as it involves the reaction of highly heterogeneous combustible compositions, made up of particles with sizes ranging between 10 and 100 µm. Metathesis reactions serve to prepare ceramics for electronics and have to be carried out under rigorously inert atmosphere.

The synthesis by flame pyrolysis is a method used in industry for producing pyrogenated silicas such as Cab-O-Sil® or Degussa P25 titanium dioxide. This method has been used to produce certain oxides (V_2O_5, Fe_2O_3, SnO_2, WO_2) of interest for the nanothermites [ROT 07]. According to Patil *et al.* flame pyrolysis can be used to prepare boron or refractory metal (Ti, Ta, Zr, Hf, Nb) particles that could serve as fuels for nanothermites. These materials are produced by the displacement of metal halide by an alkaline or alkaline earth metal in gas state [PAT 97].

In an updated literature review, Patil *et al.* mention catalysts and nanomaterials synthesis by solution combustion [PAT 02] and offer examples of refractory oxides that are pure, mixed or doped with metal elements elaborated by this method. Still more recently, Mukasyan *et al.* have described the specific advantages of combustion synthesis for producing nanostructured materials. First, the initial reaction medium is in liquid state, which means that reagents are homogeneously mixed at molecular scale. Second, reaction temperatures are high, which leads to pure and well-crystallized products. This specificity allows to avoid the thermal treatment, which is almost always necessary for purifying or crystallizing nanomaterials elaborated through other methods. Finally, the rapid transformation of matter and the dispersion of products in the condensed phase in the surrounding gas species contribute to limiting the particle growth, stabilizing them in divided state. The nature and morphology of the phases formed in the synthesis by solution combustion depend on the following: (1) fuel/oxidant ratio, (2) nature of the atmosphere in which it is carried out and (3) the chemical nature of the fuel and oxidant [MUK 07]. Today, the synthesis by solution combustion had been used to prepare more than one thousand complex oxides in the form of fine powders. Thin powders of pure oxides are often prepared by combustion of metal nitrates mixed with various fuels. Metal nitrates generally contain a high number of molecules of water of constitution.

The synthesis by solution combustion has been used by Lima *et al.* to produce chromium (III) oxide. Ammonium dichromate is used as precursor; it is associated with urea or glycine to form fuel-rich compositions. The proportion of oxygen in these compositions is increased by adding ammonium nitrate, so as to define fuel-rich, stoichiometric or oxidant-rich compositions. Under optimized conditions, obtained with glycine and oxygen in excess compared to the stoichiometry, Cr_2O_3 nanoparticles with a mean size of around 20 nm have been produced [LIM 06]. Yu *et al.* have nanostructured manganese oxide (ε-MnO_2) by combustion of mixtures of manganese nitrate ($Mn(NO_3)_2$) and probably of glycine ($C_2H_5NO_2$). Combustion of an excessively oxygenated mixture produces platelets with a diameter of 50–150 nm and thickness of 20–25 nm. The reaction of an insufficiently oxygenated mixture leads instead to the formation of spherical particles with a diameter of around 60 nm. In both cases, the electron micrographs presented suggest a quite strong agglomeration of ε-MnO_2 nanoparticles [YU 10]. Nagabhushana *et al.* have prepared α-MoO_3 nanoparticles by the combustion of a solution of peroxopolymolybdic acid and sucrose, and by the combustion of another solution of ammonium heptamolybdate and DL-malic acid [NAG 14]. The observation of these materials using the scanning electron microscope shows that they are both macroporous, but the first appears to be far less agglomerated than the second. The characterization using the transmission electron microscope of the first material shows that it is formed of primary nanoparticles of small diameter (2–10 nm) that seem relatively independent one from another. However, the specific surface area of this material ($\approx 4.97\ m^2/g$) is much lower than it would be expected, given the size of particles it is formed of. Contrary to what the authors state, it therefore seems that α-MoO_3 particles are relatively agglomerated. By using iron nitrate as oxidizer with various fuels, such as glycine, hydrazine or citric acid, Deshpande *et al.* have produced various types of iron oxide nanoparticles: α-Fe_2O_3, γ-Fe_2O_3 and Fe_3O_4, pure or in mixture. The specific surface areas of these materials (50–175 m^2/g) show that they are formed of small size primary particles, but the analysis of their morphology using the scanning electron microscope shows a strong agglomeration of these nanoparticles [DES 04]. One of the most reactive nanothermites was produced starting from bismuth oxide (Bi_2O_3) nanoparticles prepared by combustion. To this purpose, Martirosyan *et al.* have conceived an original variant of synthesis by liquid combustion, which consists of mixing pentahydrated bismuth nitrate and molten glycine. The reaction is initiated by introducing the solution in an oven that was preheated at 250°C. Nanosized (18–95.8 nm) Bi_2O_3 crystallites are formed starting from excessively oxygenated compositions, with a glycine content of only 0.05–0.4 times the stoichiometry. The lower and upper limits of fuel content are defined by non-flammability and explosibility of the solution, respectively. The crystallinity and size of Bi_2O_3 particles increase with fuel content, which is a logical consequence of the reaction temperature increase [MAR 09a].

1.3.5.4. *Conclusion*

A presentation of the characteristics of the three modes of pyrotechnic synthesis of materials of interest for the synthesis of nanothermites is provided in Table 1.22. *Detonation* synthesis allows us to form nanoparticles of refractory materials in a weakly agglomerated state. Chemical and thermal treatment is necessary in order to purify the detonation soots, which contain impurities produced by the erosion of the detonation chamber, and nanostructured carbon species resulting from partial gasification of the explosive charge. The *deflagration* of explosive compositions provides mixtures of carbon materials in various forms, which seem difficult to separate. *Combustion* synthesis allows us to produce a wide range of either pure or mixed nanosized metal oxides, under much suitable conditions than the other two methods. The low temperatures involved in the combustion synthesis seem to be best adapted to the synthesis of thermally vulnerable oxides (Bi_2O_3, CuO, MnO_2, MoO_3).

	Detonation	Deflagration	Combustion
Materials	Little reactive oxides and refractory ceramics	Carbon materials for desensitization	Pure or mixed reactive oxides
Conditions	Very rapid reactions (< 0.1 μs) Detonation chamber	Rapid reactions Manometric chamber	Slow reactions (1–100 s) Glassworks device
Advantages	Independent nanoparticles, brittle agglomerates	Polymorphic materials: metal/carbon tubes, bulbs, onions	Polyvalence, high purity materials
Drawbacks	Presence of various impurities	Mixture of phases, difficult separation	Relatively agglomerated nanoparticles

Table 1.22. *Comparison of pyrotechnic methods for the preparation of precursors that are part of the composition of nanothermites*

2

Methods for Preparing Nanothermites

2.1. Introduction

Nanothermites are obtained by assembling the basic "building blocks" of matter, whose synthesis is described in Chapter 1 of this book. Classical nanothermites consist of a metal oxide and a reducing metal. However, a ceaselessly increasing body of research presently relates to the use of oxygenated metal salts, such as perchlorates, iodates, periodates, sulfates and persulfates, which are better oxidizers than their corresponding oxides. Since the metal content of these compositions remains high, they can be considered thermites. The main methods to prepare nanothermites are the following: physical mixing, coating, sol-gel synthesis, impregnation of porous solids and thin layer deposition. Zhou *et al.* [ZHO 14b] published a literature review of various techniques used in preparing nanostructured energetic composites, with the exception of physical mixing, which is nevertheless the main technique for preparing nanothermites.

Most nanothermites are prepared as loose powders, whose apparent density is low (~0.5–1 g/cm^3). To have their performances enhanced, modify their sensitivity levels, improve their mechanical properties or facilitate their forming, various other substances can be added.

When developing applications that integrate nanothermites, the most important preparation methods are those allowing for the simplest possible ways to fabricate these materials in quantities in the order of kilograms. From this perspective, the random mixing of powders, the assembly of nanoparticles and aerosol synthesis are particularly important.

Evaporation or cathode sputtering is used for the preparation of nanothermites with original compositions and morphologies on the surface of substrates. It is easier to model these "two-dimensional" nanothermites, in which matter is assembled in a compact and ideal manner, than the chemically equivalent mixtures of nanoparticles. The number of applications of these materials outside the field of microsystems seems quite limited, as numerous pyrotechnic effects, such as ignition, transmission or initiation, often require a much higher amount of nanothermite than that which can be deposited on the surface of a substrate in the form of stable coating.

There is a strong connection between the performances of nanothermites and the proportions of fuel and oxidizer they contain. In order to obtain a unique value for describing the composition of a nanothermite, Granier *et al.* have proposed the equivalence ratio (Φ) that can be calculated from equation [2.1], in which the numerator is the ratio between experimentally measured masses (EXP), and the denominator is the ratio of masses corresponding to stoichiometric proportions (ST) of the combustion reaction [GRA 04]:

$$\Phi=\frac{(\text{Fuel/Oxidizer})_{\text{EXP}}}{(\text{Fuel/Oxidizer})_{\text{ST}}} \qquad [2.1]$$

The equivalence ratio value for which nanothermites have optimal performances usually ranges between 1 and 2. It depends on the nature of the composition, the possible presence of impurities, the content of chemical or adsorbed water and the environment in which the nanothermite reacts. The value of Φ should therefore be optimized under experimental conditions that are comparable to those in which the nanothermite is used. The analysis of the research published in this field shows that excess fuel is more often than not required for a nanothermite to reach maximum performances.

Fuel powder particles are coated with an oxide layer that is spontaneously formed in contact with the surrounding gas species. The proportion of oxide in a nanostructured fuel is significant. It increases with the decrease in the size of particles until it reaches a level where it exceeds that of fuel. When formulating a nanothermite, it is therefore essential to determine the exact content of oxide in the nanosized fuel.

There is still a point to be made, which is extremely important for those who are brought to formulate nanothermites as one of their professional tasks. It is recommended that a scientist, or any operator for that matter, who has not yet acquired experimental knowledge of a given nanothermite, should never prepare it in quantities that exceed 0.1 g. Some compositions, such as the mixtures of aluminum and bismuth oxide nanoparticles, which can react when subjected to very low energy stress, would pose a high risk to the uninformed experimenter who prepared them from the start in too high quantities. The two most important causes of accidental ignition of nanothermites are friction and electrostatic discharge.

2.2. Physical mixing

The preparation of nanothermites by physical mixing consists, in most cases, of mixing nanostructured powders by dispersing them in a liquid. This method permits to temporarily dissociate nanoparticles that were aggregated by surface forces, which facilitates the mixing of phases. A review of the literature shows that methods that would qualify as artisanal are often used when evaporating a liquid, such as letting the sample spontaneously evaporate under a laboratory hood or warming it on a hotplate. The most effective way to separate the nanothermite from the liquid in which it is mixed consists of using a rotating evaporator that permits to control temperature, pressure and drying duration, while maintaining the sample under continuous stirring. Nanothermite powders prepared in liquid medium are composed of millimeter-size aggregates. In order to obtain a material with homogeneous macroscopic granulometry, raw powders need to be broken apart by mildly compressing the composition grains with a spatula, and then sifting them.

The preparation of nanothermites by physical mixing of powders without liquid should be reserved for the formulation of complex compositions containing organic species, whose nanosized structure may be impaired by the contact with liquids.

When fuel and reducing agents have very different densities, the compositions prepared by their physical mixing are not stable in time, as the denser phase settles. This behavior can be explained by the loose consistency of nanosized powders, which facilitates the settling of heavier particles. This phenomenon should be taken into account when nanothermites are subjected to vibrations or centrifugation due to which physical separation of phases may be augmented.

2.2.1. *Mixing in hexane*

Hexane is an aliphatic hydrocarbon of a protic nonpolar nature, which is very frequently used to mix the constituents of nanothermites. It has the advantage of being physically and chemically inert in relation with metal oxides, oxidizing salts and fuels. The low boiling temperatures of its isomers facilitate the drying by distillation of the nanothermites prepared in hexane.

Plantier *et al.* [PLA 05] prepared nanothermites by mixing aluminum nanoparticles (52 nm) with sol-gel (xerogel and aerogel) synthesized or commercially available (3 nm) iron oxide powders. Phases are mixed by ultrasonication, either in pure hexane or in hexane containing sodium dioctyl sulfosuccinate as an additive. The compositions prepared with this dispersing agent are much more homogeneously mixed and generate combustion waves with significantly higher velocities (17–53%). It is worth noting that once the solvent evaporated, Plantier *et al.* sifted their compositions through a 200 µm wire mesh in order to disintegrate the macroscopic aggregates. Puszynski [PUS 09] prepared a nanothermite by dispersing titanium dioxide (40 nm) with aluminum (50 nm) in hexane containing sodium dioctyl sulfosuccinate was more homogeneously mixed than the one produced in pure hexane.

Malchi *et al.* prepared alumina enriched nanothermites by dispersing spherical nanoparticles of aluminum (80 nm), cylindrical nanoparticles of cupric oxide (21 nm × 100 nm) and nanoparticles of crystalline alumina (α-Al_2O_3; 40 nm) in hexane. The mixture is homogenized by ultrasonic pulses (0.5 s) applied for 1 min by means of a sonic probe powered at 200 W. Hexane is evaporated on a hotplate (48°C, 10 min). Nanothermites are sifted through a 335 µm mesh. The ratios by mass of nanosized aluminum (22%) and CuO (78%) are close to the stoichiometry of the ideal aluminothermic reaction. The percentage of nanosized alumina (0–20 wt%) in the Al/CuO/Al_2O_3 ternary composition is defined without taking into account the alumina brought by nanosized aluminum [MAL 08].

Prentice *et al.* prepared Al-Fe_2O_3-SiO_2 ternary nanothermites by dispersion in hexane. A first series of materials were produced by mixing aluminum nanoparticles (80 nm) with aerogels Fe_2O_3/SiO_2, in which the two oxides are intertwined at the molecular scale. A second series of materials has been formulated by mixing aluminum with submicron-sized silica and nanosized ferric oxide (3 nm) [PRE 05].

Sullivan *et al.* fabricated ternary mixtures of aluminum (50 nm), boron (62 nm or 700 nm) and cupric oxide (<50 nm) by mixing these powders and submitting them to ultrasonication stirring in hexane. These compositions are dried by allowing hexane to spontaneously evaporate, then placing the composition in a furnace at 100°C for several minutes. The aggregates of the composition are broken apart with a spatula [SUL 09].

Comet *et al.* prepared original compositions by mixing hexane with red phosphorous and nanosized powders of nickel oxide (NiO), iron oxide (Fe_2O_3) and copper oxide (CuO). In these compositions, the micron-sized particles of phosphorous are coated with a layer of oxide nanoparticles [COM 10a].

Martirosyan *et al.* mixed nanostructured aluminum (50–180 nm) and iodine pentoxide (~10 nm) powders in hexane under nitrogen atmosphere. For this purpose, they used a device that permits to mix the sample while milling it [MAR 09b]. Wang *et al.* used this system to mix crystallized Bi_2O_3 nanoparticles (50–150 nm) synthesized by combustion [MAR 09] with aluminum particles of various sizes: 100 nm, 3 µm, 20 µm and 70 µm. The quantity of thermite prepared under these conditions may reach 3 g, where the sample to milling balls mass ratio is 3g/70g.

Attrition duration (0–30 h) very significantly influences the performances of the thermites thus mixed [WAN 11]. An examination of the electron microscope images seems to indicate that this mixing method destroys the spherical morphology of the micron-sized aluminum particles. These results also indicate that the presence of hexane effectively phlegmatizes Bi_2O_3/Al compositions, which are particularly sensitive to friction.

Sullivan *et al.* prepared a nanothermite by mixing aluminum (50 nm) and silver oxide (Ag_2O; < 20 nm) nanoparticles, as well as ternary compositions by adding either cupric oxide (CuO; < 50 nm) or silver iodate ($AgIO_3$) to this basic composition. Powders are mixed in hexane in an ultrasonication bath (~20–30 min). When the solvent has evaporated, nanothermite aggregates are broken up with a spatula in order to obtain a material with the consistency of loose powder. Ag_2O nanoparticles are synthesized by the progressive alkalinization, by the addition of sodium hydroxide (0.025 mol/L), of a low concentration solution (0.005 mol/L) of silver nitrate maintained at 60°C. This reaction generates silver hydroxide (AgOH), which spontaneously transforms into oxide. Nanostructured oxide is then washed through a cycle of dispersion, centrifugation and settling in ethanol, which is applied three to four times, before being dried [SUL 13].

Hexane has been used for mixing aluminum powders (50 nm and 80 nm) with nanostructured silver iodate ($AgIO_3$) by simple agitation in an ultrasonication bath. Silver iodate is prepared by precipitation by adding a solution of silver nitrate (~248 g/L; 125 mL) to a solution of potassium iodate (~42 g/L; 950 mL). After a short agitation period (5 min), silver iodate is filtered, then washed in three stages with water, a mixture of acetone and water, acetone and, in the end, ether. The silver iodate particles thus prepared have platelet morphology and specific surface area of 6.95 m^2/g. The $AgIO_3$ particles fabricated starting from a solution of sodium iodate have a smaller specific surface area (4 m^2/g); their size is thus larger [SUL 10a].

Gibot *et al.* mixed nanosized powders of aluminum (~50 nm) and chromium oxide (10–15 nm) in hexane. The Cr_2O_3/Al nanothermite mass (1.5 g) is small compared to the volume of liquid (30 mL) used for its dispersion. The colloidal medium is agitated for 1 h in an ultrasonication bath. Hexane is finally distilled at 80°C under low pressure (20 kPa). The ultrathin chromium oxide used by Gibot *et al.* is synthesized by an original method, which consists of replicating the interstitial porosity of silica nanoballs (12 nm). Chromium (III) nitrate nonahydrate, used as oxide precursor, is dissolved in a suspension of silica nanoballs in a proportion defined in such a way that silica to chromium salt mass ratio is equal to two. The medium is agitated for 1 h, dried in a furnace (80°C), and then calcined (220°C, 30 min) in order to decompose the chromium nitrate into oxide. The silica matrix is dissolved in a solution of hydrofluoric acid. The oxide is finally washed with water, then with acetone, before being dried at 100°C for one night [GIB 11a]. Gibot *et al.* also used this method to fabricate very small size (13 nm) tungsten trioxide (WO_3) particles, which they subsequently used to formulate WO_3/Al nanothermites. In this case, the oxide precursor is phosphotungstic acid hydrate ($H_3P_4W_{12}O_{40}{\cdot}xH_2O$). The conversion of this heteropolyacid into tungsten trioxide requires quite severe calcination conditions (600°C, 2 h) [GIB 11b].

Foley *et al.* mixed nanostructured powders of copper oxide (33 nm) and aluminum (80 nm) in hexane by ultrasonication. A solution in acetone of a fluorinated polymer ($C = 3.5$ g/L), Viton A®, which precipitates around the nanothermite particles, is added to this mixture [FOL 07].

Even though nanothermites are frequently mixed by dispersion in hexane, this is not an ideal preparation method. First of all, the surface of the metal and metal oxide nanoparticles used for the fabrication of nanothermites contains a significant proportion of hydrophilic oxygenated groups. Given that it is a hydrophobic liquid, the dispersion in hexane is difficult, as particles flee the liquid, instead of dispersing in it. Furthermore, the neurological toxicity of n-hexane is quite high, which leads to loss of peripheral sensitivity and generalized muscle weakening. The n-hexane is unquestionably the most dangerous alkane, due to the specific mechanism the body uses to metabolize it into 2,5-Hexanedione [LOM 09]. It is therefore advisable to use more efficient and less toxic liquids for mixing nanothermites. Another member of the alkane family, heptane, could be used as substitute for n-hexane, even though its higher boiling point may render evaporation more difficult.

2.2.2. *Mixing in isopropanol*

Isopropanol is a protic polar solvent, miscible with water and many organic solvents, which outperforms hexane at dispersing the nanoparticles composing nanothermites. But unlike hexane, isopropanol can interact with fuel nanoparticles by dissolving, for example, the boric acid layer that covers boron nanoparticles or by attacking the metal core of aluminum nanoparticles. Unless traces of water are present, at boiling point alcohols react with aluminum, which leads to alkoxides and hydrogen release [VEI 32]. In an alcoholic medium, water protects aluminum against corrosion, forming a hydroxide layer at the metal surface. Aluminum oxidation by vapors of isopropanol is a method for the preparation of aluminum isopropylate [SMI 60]. Therefore, when mixing nanostructured aluminothermic compositions in isopropanol, any treatment susceptible to heat in suspension, either locally or globally, should be avoided. The application of prolonged ultrasonication to disperse nanoparticles or high temperatures for isopropanol evaporation is not advisable. Nanothermites that are based on oxidizing metal salts, such as, for example, perchlorates, should never be mixed in isopropanol.

Finally, it is worth noting that isopropanol gives rise to organic peroxides, which form within several days in the pure liquid and whose concentration increases over time. These compounds are known to have caused accidental explosions, which occurred during the distillation of this alcohol [BON 84].

Perry *et al.* mixed tungsten oxide nanoparticles, WO_3 and WO_2, with aluminum nanoparticles (44 nm) by dispersion in isopropanol. First, ultrasonication is applied for about 10 min in an ultrasonicating bath in order to disintegrate the oxide (0.3–0.4 g) and obtain its suspension in isopropanol (10 mL). Aluminum powder (0.1–0.2 g) is then added, and the colloidal medium is homogenized by a strong ultrasonic treatment applied by means of a sonic probe (45 s). The tungsten oxide nanoparticles used by Perry *et al.* have platelet morphology (100 nm × 7 nm). Tungsten trioxide is prepared by an original method, which consists of dissolving ammonium paratungstate (35 g/L) in concentrated hydrochloric acid, then rapidly diluting the solution thus formed into distilled water to precipitate the ($WO_3 \cdot H_2O$) hydrated oxide. Nanostructured tungsten trioxide is obtained in cubic or monoclinic form, depending on whether tungstic acid dehydration takes place at 200°C or 400°C. Nanostructured tungsten dioxide is obtained by reducing the trioxide by hydrogen diluted in helium at 650°C [PER 04].

Shende *et al.* also used isopropanol for mixing copper oxide (CuO) nanorods with aluminum powders of various granulometries (18, 80 and 120 nm). A 0.4 g quantity of oxide is dispersed with an appropriate quantity of aluminum in only 1.5 mL of isopropanol. The resulting slurry is homogenized under the action of ultrasounds emitted by a sonic probe (2–10 min), before being dried in a steam room (95°C, 10 min). Shende *et al.* note that sonication time significantly influences the combustion velocity of these nanothermites and define its optimal value (9 min). According to them, a too short sonication results in an improper mixing of nanoparticles, while a too long treatment impairs their chemical and physical properties. The CuO nanorods used for preparing the nanothermites studied by Shende *et al.* are fabricated by the milling of a mixture of cupric chloride dihydrate ($CuCl_2 \cdot 2H_2O$; 5 g) and sodium hydroxide (NaOH; 3 g) in polyethylene glycol (PEG 400; 6 mL). The precipitation of cupric hydroxide ($Cu(OH)_2$) takes place in the surfactant, which orients particle growth along a given crystallographic direction. PEG 400 and sodium chloride resulted from the reaction; they are eliminated by several cycles of washing with water and ethanol. The material is dried (100°C), milled and then finally calcined (450°C, 4 h). The length of CuO nanorods ranges from 100 to 1000 nm, and their diameter varies between 10 and 20 nm. The specific surface area of this material is 46 m^2/g [SHE 08].

Patel *et al.* prepared nanothermites by dispersing aluminum nanoparticles (80 nm) and cupric oxide nanorods in isopropanol. Ultrasonic pulses (10 s) are applied for 5–10 min in order to mix the nanoparticles. Isopropanol is evaporated at 90°C. Copper oxide is synthesized by an original method, which consists of using the gel extracted from aloe leaves (Aloe Vera) instead of PEG 400 to orient the growth of CuO nanoparticles. In a first operating mode, called sonoemulsion biosynthesis, Patel *et al.* dissolve copper chloride dihydrate ($CuCl_2{\cdot}2H_2O$; 1.705 g) in deionized water (10 mL), then add aloe gel (1.5 mL) to this solution, which is maintained under agitation. A solution of potassium hydroxyde (KOH; 1.68 g) is prepared in a similar way, then dropwise added to the copper salt solution that is ultrasonicated at 60°C. The medium is then dispersed in isopropanol, then sonicated for 30–45 min. The precipitate is eventually submitted to several cycles of washing, with water and ethanol. In the end, it is dried at 90°C for 6 h. In a second operating mode, named solidstate biosynthesis, Patel *et al.* use a mortar for milling cupric chloride (3.41 g) with sodium hydroxide (NaOH, 2 g) in the presence of aloe gel (3 mL). The resulting slurry is washed several times with water and ethanol, before being dried (90°C, 6 h). The sonoemulsion synthesized nanorods are less finely nanostructured (10–15 nm × 400 nm) than those fabricated in solid state (5–10 nm × 80 nm) [PAT 13].

Wen *et al.* prepared NiO/Al nanothermites by mixing and milling nanostructured nickel and aluminum oxide powders (~80 nm), then dispersing them in isopropanol by prolonged ultrasound treatment (2 h) in an ultrasound bath. The material was dried on a hotplate, then ground in a mortar. Nickel oxide nanoparticles have filament morphology, with a diameter of 20 nm and a length of approximately 1.5 µm. They are prepared by hydrothermal synthesis in two stages, consisting first of producing nickel hydroxide (NiOH) through controlled precipitation in a weak alkaline medium, and then calcining it in ambient air (500°C, 1 h) [WEN 13].

Thiruvengadathan *et al.* mixed in isopropanol nanosized particles of cupric oxide and aluminum oxide with micro- or nanostructured secondary explosives, such as ammonium nitrate, hexogen or CL-20 [THI 11]. The duration of the ultrasound treatment applied to disperse the particles (4 h) and the drying conditions (70°C, 15 min) of the ternary Al/CuO/explosive nanothermites seem quite severe and probably impair the nanoscale structure of the explosive particles that they contain. Organic nanoparticles should never be mixed with a liquid, not even if only an infinitesimal quantity of their component substance dissolves in the respective liquid. The interaction between liquid and nanostructured matter results in aggregation of particles: this phenomenon is all the more prominent in the case of lengthy contact and high temperature.

2.2.3. *Mixing in water*

From the perspective of pyrotechnic security, it is preferable to mix the constituents of nanothermites in aqueous medium because it prevents the formation of electrostatic charges, which are susceptible to activating the reaction of the prepared composition. Nevertheless, during the rapid oxidation of nanosized aluminum in contact with water, a part of the metal is consumed and hydrogen is released. In order to be able to use water as a liquid for the formulation of nanothermites, Puszynski *et al.* coated aluminum with various substances. The treated aluminum is dispersed in water that is maintained at a temperature of 48°C. Hydrogen release is measured by following the pressure evolution. This technique permits to measure the induction time, which corresponds to hydrogen overpressure being reached, and also the metal consumption rate during hydrolysis. The substances grafted on the oxide layer covering the aluminum nanoparticles are silanes or carboxylic acids, which react with the hydroxyl groups at the surface of the alumina. Hydrophobic substances, such as oleic acid, provide aluminum with long-term protection against hydrolysis, but on the other hand, they render its dispersion in water difficult. The grafting of hydrophilic molecules such as the succinic acid or the ascorbic acid provides aluminum with effective protection while allowing for its dispersion in water [PUS 06]. Puszynski *et al.* also observed that the reaction of aluminum with water can be inhibited by succinic acid, or even more effectively by ammonium dihydrogen phosphate. Aluminum hydrolysis is modified in the presence of bismuth oxide (Bi_2O_3). The oxide dissolves to a small proportion in the form of BiO^+ cations, which can react with aluminum. In a closed reactor, hydrogen resulted from the reaction of aluminum with water is then consumed during the reduction of BiO^+ ions into metallic bismuth. The use of ammonium dihydrogen phosphate in a ratio of 1% by mass of Bi_2O_3/Al nanothermite permits to prepare this type of composition in water. The chemical stability of these dispersions is, nevertheless, limited to several hours, but it can be extended when working at low temperature [PUS 07].

Yang *et al.* prepared ternary MnO_2/SnO_2/Al nanothermites as membranes by using an original method. First, manganese dioxide (α-MnO_2) nanowires are hydrothermally synthesized by the oxidation of manganese (II) by persulfate anion (180°C, 12 h). Then, tin chloride hydrate ($SnCl_4{\cdot}5H_2O$) and sodium hydroxide are added to an aqueous suspension of this oxide to enable the growth of nanosized tin oxide (SnO_2) structures at the surface of MnO_2 wires via hydrothermal route (200–220°C; 3 h). Depending on the reaction temperature, tin oxide grows in the form of particles (220°C) or rodlets (200°C). Nanothermite is then synthesized in an aqueous medium. For this purpose, MnO_2/SnO_2 heterostructures and the aluminum nanoparticles are, respectively, dispersed in an aqueous solution of ammonium dihydrogen phosphate (0.1 wt%) and in isopropanol by ultrasonication. The resulting suspensions are subsequently mixed under continued sonication. The $MnO_2/SnO_2/Al$ membranes are formed by filtering the suspension [YAN 13b].

2.2.4. *Mixing in other solvents*

Puszynski *et al.* used ultrasonication to mix aluminum nanoparticles (41 nm) with ferric oxide (25 nm) or cupric oxide (29 nm) in ethanol [PUS 06].

Bouma *et al.* mixed nanosized aluminum (40 nm) with micro- or nanostructured molybdenum oxide in acetone. The choice of acetone is based on the possibility to incorporate Viton in some of the compositions studied in a ratio of 10% by mass. This fluorinated polymer is a reactive binder, which serves to reinforce the cohesion of the compositions formed by compression. Thermites are prepared in batches of 0.5 g in 5 mL of acetone. Bouma *et al.* used an original method for mixing and drying, which resides in initial agitation of the suspension for 10 min, then carrying on this operation under argon bubbling for 30 min. Once the acetone evaporated, the argon flow rate is increased in order to complete the drying of the composition [BOU 07].

Spitzer *et al.* mixed tungsten trioxide nanoparticles (~30 nm) with micron-sized (~20 µm) or nanosized (~50 nm) aluminum in ethyl ether [SPI 07]. Ether enables the dissolution of water adsorbed at the surface of nanoparticles; nanothermites thus dried are more sensitive to friction and also more reactive. Ether can be used to incorporate organic substances into nanothermites in order to modify their mechanical or pyrotechnic properties. Its rather low boiling point (34.6°C) facilitates its extraction by distillation under mild conditions. But the main drawback of ether is its extreme flammability, particularly when submitted to electrostatic discharges. Moreover, the spontaneous reaction of ether when exposed to light results in highly explosive peroxides whose incorporation in a nanothermite would pose high risks. It is for these reasons that the use of ether when formulating nanothermites should be limited to particular cases in which other solvents cannot be used.

Bach *et al.* prepared ternary thermites by mixing in acetonitrile (CH_3CN) tungsten trioxide with platelet morphology (61.7 wt%) with aluminum nanoparticles (33.3 wt%) and various carbon additives (5 wt%). The nanothermite quantity in suspension (1 g) is small in comparison with the volume of liquid used for the dispersion (60 mL). The carbon substances used for modifying the properties of WO_3/Al nanothermite are a pure carbon black, as well as various products prepared by thermal oxidation or milling of this parent material [BAC 15]. Siegert *et al.* also used acetonitrile as the dispersing liquid for the fabrication of nanostructured aluminothermic compositions, based on manganese oxides and hollow carbon nanofibers [SIE 10]. Comet *et al.* mixed in acetonitrile nanosized amorphous boron with copper oxide (CuO) nanoparticles or submicron bismuth oxide (Bi_2O_3) particles [COM 14b]. Acetonitrile is a polar aprotic liquid with low viscosity and reactivity, which is liquid at atmospheric pressure between –43.8 and 81.6°C, and able to dissolve many organic substances. Due to these physical and chemical properties, it is one of the most appropriate choices when preparing nanothermites by dispersion. Acetonitrile is not in itself toxic, but its metabolization in the body generates hydrocyanic acid, which is a violent poison. More precautions are therefore required when manipulating this liquid than in the case of other organic solvents, such as alkanes or alcohols. It is in particular recommended to avoid any accidental combustion of acetonitrile during the preparation of nanothermites and also to take special precautions when drying powders that are prepared in this liquid.

Petroleum ether is a mixture of hydrocarbons whose composition may vary with low boiling point, which can be used as substitute for hexane to obtain certain formulations in which the solvent is difficult to extract. For example, Comet *et al.* used petroleum ether to incorporate silicone oil in WO_3/Al nanothermites in order to desensitize them to friction [COM 11a].

2.2.5. *Dry mixing*

This is not a common preparation method, as it does not allow for mixing the constituents of nanothermites as homogeneously as the dispersion in a liquid. Moreover, nanothermites prepared by this method need to be insensitive to friction to avoid accidental activation of the reaction.

Comet *et al.* mixed aluminum powder (~100 nm) with various fine micron granulometry metal sulfates by breaking apart powders with spatula and vibration [COM 15a]. Pichot *et al.* used this preparation method to mix nanodiamond (3.7 nm) coated bismuth oxide particles (~200 nm) with nanosized aluminum (~100 nm). For coating the Bi_2O_3 particles with nanodiamond, Pichot *et al.* first dispersed these two phases in acetonitrile, and then they evaporated this liquid while maintaining the suspension under continuous ultrasonication [PIC 15].

Dry mixing by acoustic resonance is a technique implemented by Sullivan *et al.* for preparing nanothermites composed of copper oxide (<50 nm) and aluminum of various granulometries (0.08–108 μm) [SUL 14].

2.2.6. *Aerosol synthesis of the "building blocks" for physical mixing*

Liquids can be sprayed to form of a mist composed of droplets with micron-scale diameters. The rapid drying of an atomized solution can generate submicron- to nanosized particles. Aerosols can be dried by spontaneous evaporation, by molecular diffusion in the presence of a substance that irreversibly fixes the solvent, or by flash evaporation. Aerosolization methods are used more and more to fabricate fine particles of simple or composite oxidizer, as well as organic nanoparticles that enable the enhancement of gas generation by nanothermites. The main advantage of these atomization methods resides in the possibility to continuously synthesize mineral or organic substances that would generally be difficult, if not impossible, to obtain in finely divided state by other methods.

2.2.6.1. *Simple oxidizer particles*

Submicron oxidizer particles can be prepared by drying or pyrolysis of aerosols composed of micron-sized droplets (~1 μm) resulted from the atomization of low concentration aqueous solutions (see Table 2.1). The droplets are first dehydrated in a diffusion dryer and then subjected to heat treatment in tube furnaces.

Produced oxidizer	Method	Precursor	Temperature (°C)	Size (nm)	Reference
CuO	Pyrolysis	$Cu(NO_3)_2$,5%	500	250	[PRA 05a]
KIO_4	Drying	KIO_4	180–200	50–300	[JIA 13a]
KIO_4	Drying	KIO_4,0.4%	180	460±270	[ZHO 15b]
$KMnO_4$	Drying	$KMnO_4$,5%	150	250	[PRA 05a]
K_2SO_4	Drying	K_2SO_4,0.3%	180	420±240	[ZHO 15b]
$K_2S_2O_8$	Drying	$K_2S_2O_8$,0.47%	150	460±190	[ZHO 15b]
$NaIO_4$	Drying	$NaIO_4$	180–200	50–300	[JIA 13a]

Table 2.1. *Examples of oxidizers structured at submicron scale by aerosolization of aqueous solutions*

Jian *et al.* conceived a very interesting variant of the spray pyrolysis method in order to assemble copper oxide nanoparticles (~10 nm) in the form of hollow spheres. For this purpose, they used an ingenious combination of copper nitrate (precursor), sucrose and hydrogen peroxide (H_2O_2) whose reaction generates gases that presumably cause the central cavity of the CuO nanoparticle. The authors observed that the pyrolysis of copper nitrate as such also generates hollow spheres. However, these particles are not composed of pure CuO and their central porosity is less developed. The synthesized materials, with and without gas generating composition, are subjected to a final oxidation treatment (air, 350°C, 1 h) meant to eliminate eventual carbon fractions in the first case, and to ensure full precursor conversion into CuO in the second case [JIA 13b].

2.2.6.2. *Composite oxidizer particles*

Some strong oxidizers, such as potassium permanganate, peroxides or iodic anhydride, when mixed with metals or organic substances, can spontaneously react with them. Coating these super oxidizers with metal oxides permits us to stabilize them in a divided state and to enhance their chemical compatibility at ambient temperatures with the surrounding environment, without significantly impairing their performances.

Prakash *et al.* produced potassium permanganate ($KMnO_4$) nanoparticles (~150 nm) coated with iron oxide (Fe_2O_3). For this purpose, they used an aqueous solution of ferric nitrate nonahydrate and permanganate, with a dissolved salts concentration by mass of 2%. The minimal nitrate/permanganate mass ratio required for the iron oxide shell to be continuous is 3/1. Then nanoparticles contain 86% by volume of permanganate and are coated with an oxide shell whose average thickness is 4 nm. The solution is sprayed with pressurized air (~2.4 bar). Aerosol is dried with a silica gel diffusion dryer before passing through two tube furnaces maintained at 120°C and 240°C. Due to the first heat treatment, the ferric nitrate decomposes into iron oxide. The second treatment, which is applied at a temperature near the melting point of permanganate, enables the separation of phases and their assembly into a core–shell structure. Nanoparticles are finally collected on a membrane with submicron (0.6 µm) porosity. Permanganate nanoparticles that are coated with a sufficient quantity of iron oxide do not dissolve in water [PRA 05b]. Encapsulation of permanganate opens interesting perspectives for the synthesis of new energetic compositions, as it permits not only to stabilize this oxidizer but also to use it in mixtures containing organic substances.

Wu *et al.* used the method developed by Prakash *et al.* to prepare submicron-sized particles of potassium or ammonium perchlorates coated with metal oxides. The temperatures of the two furnaces are set to enable the decomposition of nitrate into oxide (T_1) followed by the formation of the core–shell structure (T_2). The temperature values T_1/T_2 used for the synthesis of $Fe_2O_3/KClO_4$, $CuO/KClO_4$ and Fe_2O_3/NH_4ClO_4 composite oxidizers are 125°C/550°C, 400°C/550°C and 125°C/200°C, respectively [WU 12].

Feng *et al.* prepared particles (~200 nm) consisting of an iodine pentoxide (I_2O_5) core coated with an iron oxide (Fe_2O_3) shell. A significant proportion of the oxide is to be found in the form of nanoparticles aggregated in fractal structures at the surface of I_2O_5@Fe_2O_3 composite particles. The material is synthesized from a starting solution of iodic acid (HIO_3), which is atomized into droplets of submicron size. Aerosol is dried by diffusion through silica gel, then acid particles are calcined by passing through a furnace whose temperature is maintained at 290°C. The pyrolysis temperature should enable to dehydrate the iodic acid ($T > 207$°C) without decomposing the formed anhydride (~390°C). Aerosol is then dried to prevent the reaction-resulted water vapor from interacting with the iodine pentoxide nanoparticles. The iron oxide coating is fabricated by decomposing in a second furnace at ~200°C the iron pentacarbonyl $Fe(CO)_5$ vapor carried by a flow of argon. The thickness of iron oxide coating I_2O_5 is calculated for various argon flow rates. Nanothermites are prepared by mixing I_2O_5@Fe_2O_3 nanocomposite particles with aluminum nanoparticles (~50 nm) in hexane by sonication (30 min). Drying is carried out either by spontaneous evaporation under a laboratory hood for one night or by evaporation under vacuum (3–4 h) [FEN 13]. In their published research, Feng *et al.* do not mention the reduction of iodine pentoxide by carbon monoxide into iodine and carbon dioxide. This classical reaction may consume a significant part of the iodine pentoxide and cause the increase in Fe/I molar ratio in the final product. It could also explain the presence of iron oxide nanoparticles in the form of excrescences with fractal morphology at the surface of I_2O_5 particles by the reaction in homogeneous gas phase between iodine and iron pentacarbonyl.

2.2.6.3. *Organic nanoparticles*

Organic nanoparticles can be incorporated into nanothermites to increase the quantity of gas generated by their combustion or to modify their threshold of sensitivity to a given stress. The flash evaporation drying of aerosol produces a large number of organic compounds in the form of nanoparticles (50–1,000 nm) at a rate of several grams per hour [SPI 13]. This technique consists of atomizing a solution that was preheated at a temperature above its boiling point in a chamber maintained under dynamic vacuum (~5 mbar). The very rapid subsequent evaporation causes the burst of aerosol droplets and the emergence of many crystalline nuclei, which are not given the time to grow and find themselves fixed in nanosized state [RIS 12]. Flash evaporation drying of solutions containing several dissolved organic compounds can lead to various forms of nanomaterials: mixture of nanoparticles, core–shell structures, semicrystallized composites and cocrystals [SPI 14].

2.3. Coating

Coating consists of using a physical or chemical method to get one of the reactive phases of a nanothermite covered with the second one. This method enables an optimal contact between fuel and oxidizer, a decrease in the diffusion distances of reactive species and the conservation of a significant porous volume. Coating modifies the physical and chemical surface properties of nanothermites, such as the friction sensitivity threshold. Spitzer *et al.* have noted that a composition formed of aluminum-coated tungsten trioxide nanoparticles was less sensitive to friction than the corresponding physical mixture [SPI 07].

It is difficult to get nanoparticles coated in liquid medium with a uniformly thick layer of matter, as deposited substances tend to irregularly aggregate on the surface of nanoparticles or to solidify independently of them. The most regular coatings are obtained when dispersing nanoparticles in a fluid bed or in an aerosol, then coating them by the reaction of gas precursors at their surface.

2.3.1. *Coating oxide with fuel*

A chemical method for coating metal oxide nanoparticles with aluminum has been patented by Comet *et al.* [COM 08a]. This technique was developed for formulating low alumina-content nanothermites, which exhibit an optimized contact between metal oxide and aluminum. The first stage of the method consists of preparing an aluminum hydride solution in ether using Schlesinger's double decomposition reaction:

$$AlCl_3 + 3\ LiAlH_4 \rightarrow 4\ AlH_3 + 3\ LiCl$$

The aluminum hydride solution is then separated by decantation from the lithium chloride. The dispersion of WO_3 nanoparticles in this solution is accompanied by a change in oxide color, which goes from pale green to black with bluish undertones. The $WO_3/AlH_3{\cdot}O(C_2H_5)_2$ composite formed by evaporation of ether under low pressure is exposed to heat treatment (T = 120°C) in order to decompose hydride in amorphous aluminum [COM 02]. This stage, which is very difficult to implement, is conducted under dynamic vacuum and in the absence of humidity in order to avoid spontaneous ignition of the material. The WO_3/Al nanothermite is not pyrophoric, but it readily reacts when in contact with a flame. The pellets formed by compressing WO_3/Al composite powders exhibit poor physical stability. When in contact with atmospheric humidity, they gradually loose cohesion. This disintegration is attributed to the presence of residual metal chlorides, which confer a certain hygroscopy to the material. A solution to this problem may be represented by the method of preparing the α variety of aluminum hydride described by Lund *et al.* [LUN 05] using metal oxide nanoparticles as nuclei.

The observation with scanning electron microscopy shows that the original particles of tungsten trioxide are coated with aluminum (see Figure 2.1). The characterization with transmission electron microscope of the smallest particles confirms that they have a WO_3 core coated with a layer of amorphous aluminum.

The submicron-sized particles, with apparently smooth surface, are aggregates of WO_3 nanoparticles coated with aluminum [COM 12].

Figure 2.1. *Scanning electron microscopy images of WO_3 nanoparticles before (a) and after aluminum coating (b); transmission electron microscopy images of a WO_3/Al composite particle (c)*

2.3.2. *Coating fuel with oxide*

Coating fuel nanoparticles with a layer of metal oxide is also described in the literature. This preparation method encapsulates the fuel in the oxide and could be used to enhance the effects of exploding the core–shell structure of certain metal particles. As shown by the works of Kaplowitz *et al.*, the reactive oxide layer does not protect the underlying fuel against oxidation [KAP 14]. Therefore, it is preferable to coat fuel nanoparticles whose surface has already been passivated by spontaneous oxidation.

Kaplowitz *et al.* used an ingenious aerosol pyrolysis method for producing aluminum nanoparticles, and then coating them with iron oxide (Fe_3O_4). Polyhedral aluminum nanoparticles (~90 nm) are first prepared by thermolysis at 350°C of tri-isobutyl aluminum vapor diluted in argon. Aluminum aerosol is then exposed to iron pentacarbonyl ($Fe(CO)_5$) vapor, before being subjected to a treatment at 175°C aimed at decomposing the iron precursor into metal. The metal coating is finally oxidized at 50°C in a mixture of argon and air [KAP 14].

Ferguson *et al.* coated aluminum nanoparticles (~50 nm) with tin oxide (SnO_2) by using an atomic layer deposition technique. Aluminum powder (~0.5 g) is set in suspension in a bed reactor fluidized by argon flow and vibration system. The fluidization chamber consists of a stainless steel tube ($D = 9.52$ mm; $h = 101.6$ mm) in which the powdery sample is confined by means of two metallic filters with average porosity of 10 μm. The tin oxide precursors, tin tetrachloride ($SnCl_4$) and hydrogen peroxide (H_2O_2) are driven by argon bubbling, the same gas being used for purging the reactor. Due to a system of valves, the aluminum powder can be alternatively exposed to a well-defined quantity of each reagent during a treatment of 250 to 1000 cycles. The reactive deposit of oxide is carried out at a temperature of 250 or 350°C. Transmission electron microscopy images of the samples show that aluminum nanoparticles are coated with a layer of SnO_2 with rough morphology. The mass proportion of tin in the samples (1.8–10.4 wt%) is, nevertheless, very low in comparison with the value required by the stoichiometry of the aluminothermic reaction (55 wt%). Fergusson *et al.* attempted to solve this problem by varying the operating conditions, but did not manage to deposit a higher quantity of oxide on the aluminum nanoparticles. As an explanation for the low yields obtained, Fergusson *et al.* proposed a hypothesis according to which exposure to reagents ($SnCl_4$ and H_2O_2) would be insufficient for the completion of surface reactions involved in the formation of SnO_2 [FER 05]. The modest proportion of tin in the Al@SnO_2 nanothermites could also have resulted from the reaction of aluminum with gaseous hydrogen chloride, which was formed by hydrolysis of tin chloride deposited on aluminum. The aluminum oxychloride formed in this case could contaminate the surface of the Al@SnO_2 particles and inhibit subsequent nucleation of tin oxide.

Qin *et al.* used the atomic layer deposition method to coat aluminum nanoparticles with a size distribution between 40 and 180 nm with zinc oxide (ZnO) and tin oxide (SnO_2). Coating is carried out on a low-thickness (<1 mm) fixed bed (~100 mg) of aluminum nanoparticles. The oxide precursors, diethylzinc and tin chloride ($SnCl_4$), are liquid at ambient temperature, which permits to use their vapors, without resorting to vector gas bubbling. Using a valve system, the aluminum powder is alternately exposed to precursor vapor and then to oxygenated water. Nitrogen is used in the purging operations between exposures. The experimental conditions created for coating aluminum with each of the oxides are provided below (see Table 2.2). Exposure duration corresponds to the minimal time required for the exposed surface to reach saturation. It has been optimized by gravimetric follow-up, depending on the number of exposure cycles. Qin *et al.* showed that the particles coated with this method tend to aggregate by forming a matter continuum between particles. Aggregation is all the stronger the thicker the oxide shell is, and the higher the deposit velocity [QIN 13]. The use of a fluidized bed, which maintains nanoparticles in continuous movement during coating, would limit the aggregation phenomena.

Deposited oxides	Precursors	T_R (°C)	D_E (s)	V_D (nm/cycle)	N_C	E (nm)
ZnO	$(C_2H_5)_2Zn$	120	10	0.17	225	~0–39
SnO_2	$SnCl_4$	300	60	0.031	550	~ 0–17

Table 2.2. *Operating conditions used by [QIN 13] for oxide deposit at the surface of aluminum nanoparticles: chemical formulae of precursors, temperature of reactor (T_r), exposure duration (D_E), deposit velocity (V_D), number of cycles (N_C) and range of thickness of oxide layers (E)*

The atomic layer deposition method is not widely used when fabricating nanothermites. Nevertheless, this production method has many advantages, as it enables the production of nanothermites with well-defined core–shell morphology. The thickness of the oxide shell is very precisely set in correlation with the number of cycles of exposure of coated powder to precursor vapors. The contact between fuel and oxidizer is ideal in these materials, and the two phases are as close as possible. The composition can be adjusted to have its stoichiometry coincide with the one of the aluminothermic reaction. However, reaching an optimal fuel/oxidizer ratio at the particle scale requires the use of monodisperse powders.

The atomic layer deposition method could also be employed for a more controlled coating of oxide nanoparticles with metal, using organometallic compounds as precursors. In this case, it would be essential to add a controlled oxidation stage, in which the surface of oxide@metal particles would be passivated before exposing them to air. In these core–shell structures, the direct contact between oxide and metal would probably be accompanied by a partial reaction, consisting of an exchange of oxygen atoms and the formation of a mixed oxide layer at the oxide/metal interface.

2.3.3. *Coating of fuel with a metal layer*

Anecdotal as it may be, this method is worth being noted, as it may serve to modify the surface of metal fuel particles, and consequently to control the combustion of nanothermites.

Wang *et al.* coated micron-sized (5 µm) aluminum particles with a layer of copper nanoparticles. For this purpose, the aluminum particles are dispersed in a solution of copper sulfate and gelatin, whose pH was previously set at 6.4. A solution of ammonium fluoride (NH_4F) is then added in order to dissolve the alumina layer on the aluminum particles and to activate the reduction of Cu^{2+} cations on their surface. Treating coated particles with a diluted solution of hydrochloric acid dissolves the aluminum core. Only the granular layer of deposited copper remains. Al@Cu/WO_3 thermites are prepared by mixing bimetallic Al@Cu particles with tungsten trioxide particles (12 µm) [WAN 07]. It would be interesting to apply this method to reducing metal nanoparticles (Al, Zn, Ti).

2.4. Sol-gel method

Sol-gel synthesis is one of the methods most widely used for producing metal oxides in the form of very small size particles, assembled in three-dimensional porous structures. The first stage of a sol-gel method consists of producing a colloidal suspension starting from a precursor, which is generally a salt or a metal alkoxide. The second stage is the formation of gel, which is carried out by the assembly of colloids and provides the material with a macroscopic form. The third stage consists of drying the liquid-soaked gel, either by simple evaporation or by a specific drying method, such as extraction in supercritical medium. The first drying mode results in xerogels, materials formed by gel contraction under the action of capillary forces resulted from the evaporation of the liquid. The second drying mode, which does not alter gel morphology, leads to materials featuring exceptional physical properties: aerogels. A heat treatment is often necessary in order to dehydrate, purify and crystallize the oxides prepared by the sol-gel method. This last stage damages the fine structure of materials. Moreover, it is very difficult to implement with energetic materials which will react or deteriorate at high temperatures. This is one of the most important limitations of using the sol-gel method for the synthesis of nanothermites.

The vulnerability of the sol-gel method resides in its discontinuous nature. Materials need to be prepared batch by batch, in lengthy stages, and this slows down production. Moreover, sol-gel synthesis is sensitive to many parameters, which introduces the experimental fluctuation of the materials' properties.

Despite its weaknesses, the sol-gel method has many qualities that can be advantageous when preparing energetic materials. In the study of their collective work, Gash *et al.* listed these qualities. The authors consider the sol-gel technique as an ascending synthesis method, which permits to very intimately mix reactive phases, at a scale that would be difficult to reach by other methods. It also enables precise definition of the composition, density, morphology and size of the particles of structured material; in other words, the fabrication of energetic compositions with customized properties. Nanothermites can be prepared by the sol-gel method by forming a metal oxide gel around the fuel nanoparticles [GAS 11] or still by fabricating the oxide before physically mixing it with aluminum.

2.4.1. *Formation of oxide around the metal particles*

Gash *et al.* developed a very original method for sol-gel synthesis, which consists of using an epoxide to induce gelation of solutions of metal salts. This method can be implemented to produce a wide variety of oxides in the form of monolithic aerogel. The synthesis of iron oxide has been studied in detail in order to determine the influence of various experimental parameters on gel formation. An iron (III) salt, nitrate or chloride, is dissolved in various solvents in the presence of a variable quantity of water. Gelation is activated by the addition of propylene oxide to the solution. Gash *et al.* noted that polar protic solvents, such as alcohols with short hydrocarbon chains, were the liquid media most favorable to gel formation by this method. Good results were also obtained with certain polar aprotic solvents, such as acetonitrile, dimethyl formamide or dimetyl sulfoxide. Non-polar solvents, which do not dissolve the ferric salts used as oxide precursors, cannot be used. The minimum quantity of water required for the gelation is that which permits the formation of the hexaaquairon(III) complex. In other terms, water molecules should be at least six times more numerous than Fe^{3+} cations and present in the reaction medium before the addition of epoxide. Propylene oxide binds the hydrogen atoms of complexed water molecules with the iron, and thus activates the polycondensation of these species into oxide. The nucleophilic attack of the protonated epoxide by the anion initially associated with Fe^{3+} cation opens the epoxide cycle. This reaction consumes protons by irreversibly fixing hydrogen within a new organic molecule. Gash *et al.* varied the propylene oxide/Fe^{3+} molar ratio between 3 and 25 and found that for gelation to occur, this ratio had to be above 6. When this condition is satisfied, an increase in the proportion of epoxide shortens the gel time [GAS 01]. Tillotson *et al.* gelated a solution of ferric nitrate nonahydrate (1.0 g) in absolute ethanol (5.0 g) with propylene oxide (1.0 g). Nanothermites have been produced by dispersing in the sol aluminum powders with micrometer (6 µm) or nanometer (30 nm) granulometry, just before gelation takes place. The gels formed are either incubated under vacuum at 70°C for 5–6 days to produce xerogels, or dried by extraction with supercritical carbon dioxide to fabricate aerogels. This drying method consists of replacing the solvent permeating the gel with liquefied CO_2 through several successive washings at 10°C. Once filled with CO_2, the gel is heated to 45°C at a rate of 0.5°C/min, while pressure is maintained at about 100 bars. Carbon dioxide is finally eliminated by slow depressurization at a rate of around 1 bar/min. Tillotson *et al.* noted that aluminum nanoparticles form aggregates in the gel, but they are, nevertheless, well coated with iron oxide. The alumina shell that covers them does not grow thicker during the synthesis. Therefore, the method enables a good contact between the phases, while aluminum is not oxidized [TIL 01].

In subsequent works, Gash *et al.* showed that the nature and morphology of iron oxide depends on the chemical structure of the epoxide used to induce gelation. As epoxides feature a cyclic ether with three links, they react more rapidly than those with four links. Aerogels fabricated with small epoxides, such as propylene oxide, are composed of interconnected spherical amorphous nanoparticles (5–15 nm). Aerogels prepared with larger epoxides, such as oxetane, consist of interconnected fibers of akaganeite (β-FeOOH) whose diameter ranges between 15 and 35 nm. These aerogels are highly porous and exhibit excellent stiffness, which allows us not only to manipulate them but also to process them [GAS 03].

Aerogels with a more complex composition have been prepared by Clapsaddle *et al.* starting from a sol containing ferric chloride and silicon alkoxides, such as tetramethylsiloxane or tetraethylsiloxane. Molecules such as silsesquioxanes or bissilsesquioxanes, which have a fluorinated hydrocarbon chain, have also been used. Spherical aluminum particles with micrometer granulometry (2 µm) are dispersed in the sol before gelation takes place. Silica and fluorine are homogeneously distributed through the iron oxide. In thermite aerogels, aluminum particles are distributed in the oxide matrix, which covers them individually in a very homogeneous manner [CLA 03]. Although nanothermite aerogels have mechanical and pyrotechnic properties of high interest, their fabrication requires lengthy stages, which seem to be incompatible with industrial scale development. According to Gash *et al.*, the typical synthesis of a sample requires gel ageing for 24 h, washing with ethanol for 1 week and finally drying by supercritical extraction for 2–3 days [GAS 04]. Moreover, nanothermite aerogels contain a significant proportion of impurities, such as chlorides, which can have a critical influence on their long-term stability.

Wang *et al.* coated submicron-sized aluminum particles with iron oxide by gelling a solution of ferric nitrate nonahydrate in ethanol by propylene oxide. After a lengthy ageing of 5 days, the gel is dried under vacuum at 55°C to form xerogel, which is then washed at 45°C in ethanol, and finally dried. Oxide nanoparticles (~20 nm) form a compact coating at the surface of aluminum particles [WAN 13b].

Shin *et al.* dispersed aluminum nanoparticles (100 nm) by ultrasonication (30 min) in a solution of ferric nitrate nonahydrate in propan-1-ol. Gelation is induced by the addition of propylene oxide. Nanothermite xerogel is produced by drying the gel in the atmosphere for 2 weeks [SHI 12].

Kuntz *et al.* coated micron-sized tantalum particles (5 µm) with tungsten oxide. Tungsten oxitetrachloride ($WOCl_4$) is used as precursor. This compound is dissolved in a mixture of ethanol and water (95/5 vol%). The solution is mildly turbid; it is clarified by filtration. Gelation is induced by the addition of an epoxide with four links, 3,3-dimethyloxetane. It takes place in less than 5 min. The epoxides with three links react too rapidly and the result is a precipitate rather than a gel. Tantalum powder is dispersed in the sol by magnetic stirring with a teflon-coated magnetic stir bar, which is maintained until gelation occurs. After ageing (24 h), the gel is washed in a bath of absolute ethanol, refilled once or twice a day, for 3 days. The gel is then washed in a similar way in cyclohexane for 2 days. Drying is carried out by spontaneous evaporation at ambient temperature, then it is completed in a steam room at 110°C, under dynamic vacuum for 24 h. The tantalum particles are trapped in the amorphous gel of tungsten oxide. As these compositions are very insensitive to shock, friction and electrostatic discharge, they have been reinforced by flash sintering under high pressure at relative densities of 84–93% [KUN 10].

2.4.2. *Preparation of oxides subsequently mixed with metal*

The epoxide-assisted sol-gel synthesis method developed by Gash *et al.* has been implemented to produce nanostructured oxides, subsequently mixed with nanosized aluminum to formulate nanothermites. The properties of compositions based on iron oxides [PLA 05], tungsten oxides [PRE 06] and nickel oxides [DEA 10] prepared according to this method have been studied. Kim *et al.* conceived an original variant of this method, which consists of the atomization of the sol and triggering its gelation in aerosol in order to form iron oxide nanoparticles [KIM 04]. Bezmelnitsyn *et al.* synthesized ferric oxide by the epoxide method in the presence of a non-ionic surfactant. After having been dried, the gel is submitted to a heat treatment during which temperature is at first increased gradually (1°C/min) up to 500°C, then maintained at this value for 12 h [BEZ 10].

Plantier *et al.* noted that flame propagation velocity in the aluminothermic compositions formulated starting from iron oxide, in the form of aerogel (300 m^2/g) or xerogel (250 m^2/g), was particularly low, considering their fine structure. Calcination under air of these materials permits to eliminate the "chemical water" they contain, to oxidize the residual carbon phases and to induce crystallization of iron oxide in its most stable form (α-Fe_2O_3). These transformations are accompanied by a decrease in specific area and an increase in apparent density. Despite this, the calcined gels mixed with nanosized aluminum (52 nm) prove to be better oxidizers than aerogel and xerogel parents. Hence, Plantier *et al.* concluded that impurities act as flame retardant agents [PLA 05]. This result is not surprising, considering that metal oxide reduction reactions with carbon are much slower than those with aluminum. Carbothermy is indeed used for controlled production of metals in many industrial processes. Moreover, the introduction of carbon in the nanostructured aluminothermic compositions reduces their vivacity due to the formation of metal carbides [PIC 15].

Prentice *et al.* noted that tungsten trioxide nanoparticles prepared in the form of aerogel were not stable in time. The ageing of these materials is caused by the adsorption of water in contact with the atmosphere. Weight gain is higher when an aerogel has simply been dried at 120°C than when it is calcined at 400°C. This evolution of materials proves to be very prejudicial to the performances of nanothermites [PRE 06]. These results show that calcination of oxides prepared by the sol-gel route is essential for improving their stability over time. The temperature employed for this purpose needs to be soundly optimized to avoid destroying the fine structure of the oxide produced by the sol-gel method.

Comet *et al.* described the preparation of $Al_xMo_yO_z$ phases nanostructured by an original sol-gel method consisting of using agar as gelling substance. The first stage of the synthesis consists of dissolving a variable mass of agar in a given volume of aqueous solution of ammonium paramolybdate ($H_{24}Mo_7N_6O_{24}\cdot 4H_2O$,~50 g/L) by heating the medium between 60 and 70°C. Gelation is induced by cooling the solution to ambient temperature. The water contained in the gel is then extracted by around 100 cycles of washing with acetone by means of a Soxhlet apparatus. The composite xerogel consists of very fine filaments, closely interconnected, which form a spongy structure. These filaments have themselves a rosary-like structure, in which the spherical ammonium paramolybdate nanoparticles are associated by the strand of agar containing them (see Figure 2.2).

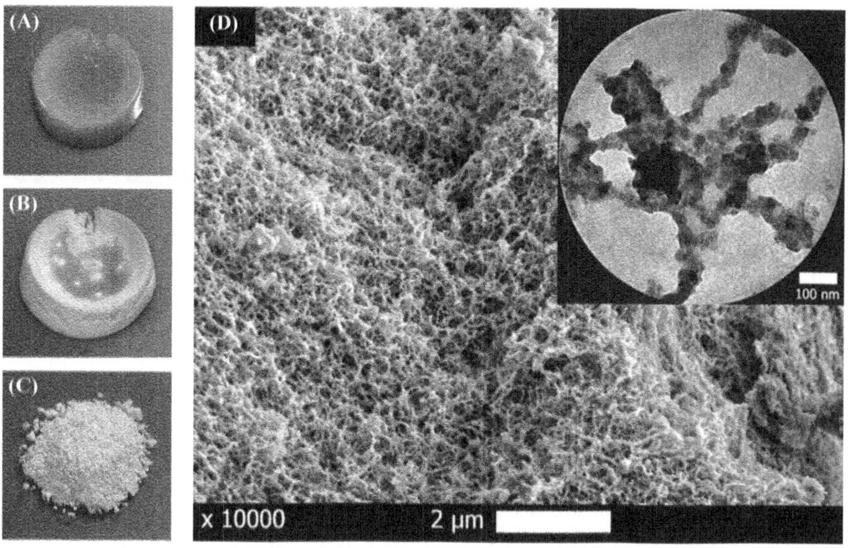

Figure 2.2. *a) Hydrated composite gel; b) monolithic xerogel; and c) xerogel powder; d) observation by scanning electron microscopy and transmission electron microscopy (cartridge) of xerogel morphology [COM 06b]*

The xerogel composite is then transformed into powder by milling in liquid nitrogen, before being soaked in a solution of anhydrous aluminum trichloride ($AlCl_3$) in diethyl ether. After the evaporation of ether, the composite xerogel powder coated with aluminum trichloride is calcined at 550°C for 2 h under continually renewed air atmosphere. The calcination of non-treated composite gel results in micron-sized particles of molybdenum trioxide (MoO_3) in the form of tiles (see Figure 2.3). The aluminum chloride deposited at the surface of the composite xerogel enables to preserve its nanosized morphology during the oxidation heat treatment. In order for this effect to be observed, $Al_xMo_yO_z$ composite oxide needs to contain at least 7.8 wt% in mass of aluminum. As the scanning electron microscopy images show, the higher the mass proportion of aluminum in the $Al_xMo_yO_z$ phase, the more the oxide retains its fine structure (see Figure 2.3). However, this proportion should not exceed 18 wt% to avoid the material densifying due to the filling of its porosity.

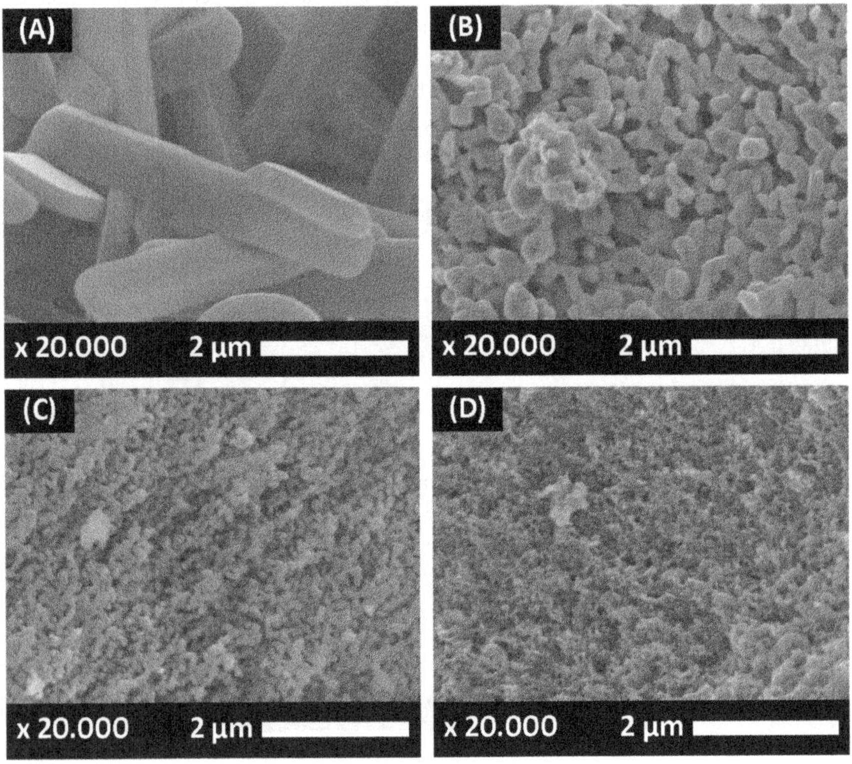

Figure 2.3. *Observation with scanning electron microscope of the morphology of $Al_xMo_yO_z$ phases produced by calcining a composite xerogel depending on their final mass content of aluminum: a) 0%, b) 10.5%, c) 16.8% and d) 24.7% [COM 06b]*

Based on the study of $Al_xMo_yO_z$ phases, Comet *et al.* proposed the following mechanism that explains their formation. The anhydrous aluminum trichloride reacts with the water adsorbed at the surface of xerogel and gives an aluminum oxychloride. During calcination, oxychloride transforms into an alumina layer that covers the structure resulted from the pyrolysis of xerogel filaments. At 550°C, agar is completely consumed, while paramolybdate decomposes into molybdenum oxide, which reacts with the surrounding alumina. Aluminum molybdate, in which molybdenum reaches its highest oxidation degree (Mo^{+VI}), is formed in contact with the alumina layer in the area richest in oxygen. The core of the particles is rather composed of molybdenum dioxide. This metal is at its oxidation degree (+IV) due to the fact that the alumina and aluminum molybdate shell limits the diffusion of oxygen required to transform MoO_2 into MoO_3 [COM 06b].

2.5. Impregnating porous solids

The impregnation of a porous solid is a preparation technique that has been used for the physical separation of the constituents of a nanothermite [SIE 10] in order to assemble the metal used as fuel with the oxide [HU 15] or to incorporate an intrinsic explosive in the nanothermites [SPI 10]. The operation of filling the porosity of a solid material is easy to conceive, but paradoxically, it proves difficult to achieve experimentally. The material to be used as a substrate should exhibit open porosity, the size of its pores being compatible with the size of species, molecules or particles that will impregnate it. The hydrophilic or hydrophobic nature of the liquid used to get the solid matter into the porosity should be defined depending on the nature of the matrix. The substance solidified in a porous matrix can be deposited starting from a solution or suspension by the evaporation of the solvent. In this case, only a part of the porosity is filled, since the volume initially occupied by the solvent remains vacant. Infiltration by molten solid permits a better filling, but several conditions need to be satisfied for this method to be applicable. First, the infiltrating matter should melt, without evaporation or decomposition. Second, it has to be physically and chemically compatible with the substrate it impregnates.

Structuring a compound in nanosized pores permits to maintain it, to a certain extent, in a very divided state. Thus, in black powder, potassium nitrate is stabilized in the form of nanosized crystallites in the porosity of activated charcoal. This energetic material is, nevertheless, very sensitive to air humidity, which impairs its pyrotechnic properties by getting adsorbed in the activated charcoal and by interacting with nitrate. This stability fault is a recurring problem with compositions that have an oxidizing metal salt solidified in the hydrophilic porous matrix. In certain cases, fluctuation in the atmospheric humidity rate can lead to exocrystallization of the matter initially present in the porous matrix. This ageing can be explained by the gradual transformation of the impregnated matter through dissolution/recrystallization cycles. Impregnation of porous matrices with mineral particles or hydrophobic substances does not pose this type of problem.

Siegert *et al.* formed manganese oxide (MnO_x) inside hollow carbon nanofibers (20–40 nm). They then mixed this composite material with aluminum (~50 nm) in order to obtain a ternary nanothermite. The first stage of the synthesis consists of oxidizing carbon tubes in nitric acid (65%) brought to boiling point (120°C, 16 h). This treatment serves to increase the number of oxygenated functional groups at the carbon surface in order to facilitate subsequent impregnation of hollow nanofibers with the precursor of manganese oxide. The filling is carried out by dispersing the activated hollow nanofibers in molten manganese nitrate tetrahydrate in excess (50°C, 3 j). When the treatment is completed, the nitrate-loaded fibers are washed with isopropanol, then calcined at 225°C for 2 h under argon flow, in order to decompose nitrate into oxide [SIE 10].

Hu *et al.* synthesized a Fe_2O_3/Al nanothermite by filling the porosity of iron oxide particles with aluminum nanoparticles. The porous α-Fe_2O_3 particles are first fabricated by replicating pollen grains. For this purpose, pollen is washed with anhydrous ethanol, then dried (air, 60°C), before being impregnated (60°C, 24 h) with a solution of ferric nitrate nonahydrate in ethylene glycol (0.1 mol/L). The impregnation solution moreover contains a cationic surfactant, cetyltrimethylammonium bromide, whose role is to facilitate the dispersion of pollen grains in water. The product is then washed with ethanol, dried (air, 50°C) and then calcined. Calcination is achieved in three stages, at 100°C, 200°C (2 h) and finally at 500°C (2 h). The oxide particles produced by this method have the form of prolate spheroids with a length between 20 and 30 μm. Their surface is covered by micron-sized alveoli. This porosity is filled by drawing up a suspension of aluminum nanoparticles in hexane on the oxide placed under static vacuum. The material is then centrifuged (4,000 rpm, 5 min) to have the aluminum nanoparticles penetrate into the core of the oxide particles, then it is dried (50°C, 4 h). This treatment is repeated three times in order to improve the pore-filling rate [HU 15].

The incorporation of a secondary explosive in the nanothermites facilitates their ignition, augments the proportion of gas in their reaction products and increases their specific heat of combustion. Comet *et al.* solidified hexogen (RDX) in the porosity of chromium oxide (Cr_2O_3, ~46 m^2/g) [COM 08b] and manganese oxide (MnO_2, 300–400 m^2/g) [COM 08c]. The chromium oxide is produced by the combustion of ammonium dichromate in the form of powder composed of millimeter-size particles. Under the action of ultrasounds, in liquid medium, the latter can be partially split up in smaller size porous particles. They have a laced internal structure that forms a very developed open macro porosity. The manganese oxide is composed of nanoparticles with spherical and acicular morphologies, which assemble in submicron-sized aggregates. These latter agglutinate in micron-sized macro porous particles (see Figure 2.4).

The method employed for impregnating oxides consists of dispersing them by ultrasonication in a solution of hexogen in acetone, then submitting this solvent to rapid evaporation under low pressure. Since these porous oxides have high specific areas, the explosive solidifies essentially at their surface, rather than on the walls of the container where the acetone evaporation takes place.

Figure 2.4. *Observation by electron microscopy of the morphology of porous chromium a) and manganese b) oxides, whose porosity was loaded with hexogen by [COM 08c] in order to produce gas generating nanothermites*

Bezmelnitsyn *et al.* impregnated the interparticle porosity of a nanostructured ferric oxide powder produced by the sol-gel method with a non-energetic polymer derived from cellulose, (acrylamidomethyl) cellulose acetate butyrate (AAMCAB). The ferric oxide is first of all dispersed in hexane in order to be covered with a dispersing agent. After drying, the modified oxide is dispersed in a solution of AAMCAB in acetone, in various proportions, so that it forms AAMCAB@Fe_2O_3 composites with a mass content in polymer from 5 to 20 wt%. Nanothermites are formulated by mixing a constant mass of pure or polymer-loaded (200 mg) oxide with a constant mass of aluminum (115 mg). In other terms, those materials that are richest in polymer also contain more excess aluminum [BEZ 10].

Advanced nanothermites can be synthesized by impregnating materials with calibrated porosity, such as porous hybrid solids. These ultraporous materials are composed of metal cations assembled by polydentate organic ligands. They can thus be used to encapsulate oxidizing metal salts whose decomposition would produce the oxygen required for the combustion of the carbon structure. The metal oxides produced by the reaction of the porous hybrid solid and the oxidizing salt could then be reduced by aluminum nanoparticles. It would probably be quite difficult to prepare this type of nanothermites. Nevertheless, the fact that these porous hybrid solids may be prepared in the form of mesoscopic, or even macroscopic [FUR 14] structures, and that a part of their porosity may contain significant quantities of various organic substances, [HOR 10] offers reasons to favorably expect that is it possible to prepare this type of nanothermites.

2.6. Assembly

The assembly consists of producing an ordered mixture of fuel and oxidizer nanoparticles by using the interaction forces that come into play when building supramolecular structures, such as Van der Waals forces, hydrogen bonds or Coulomb attraction. For this purpose, the surface of nanoparticles should be modified either by grafting molecules that can recognize each other or by the accumulation of electric charges. The drawback of chemical or biological methods is that they introduce a non-negligible proportion of carbon in the nanothermite, which degrades its pyrotechnic properties. From this perspective, the electric field assisted assembly methods are preferable, since they leave no material footprint on the sample. However, their use requires particular precautions due the high sensitivity of nanothermites to electrostatic discharge.

2.6.1. *Chemical methods*

The chemical methods used for assembly consist of using various substances as "hyphens" between nanoparticles: polymers, functionalized organic molecules, polyelectrolytes or colloidal particles.

Shende *et al.* assembled nanorods of cupric oxide (CuO) with spherical aluminum nanoparticles, according to a three-stage method. In a first stage, oxide nanorods (0.5 g) are dispersed in 500 mL of a poly-4-vinylpyridine (P4VP, 1 g/L) solution. This macromolecule is a nearly universal binding agent, since the pair of electrons carried by the nitrogen atom can form a covalent bond with metals or hydrogen bonds with polar species. The suspension is submitted to prolonged ultrasonication (4 h), after which the oxide is separated from the solution by centrifugation. The polymer in excess is eliminated through a cycle of washings (four to five), consisting of re-dispersion of the oxide in isopropanol (200 mL), sonication of the suspension (2 h), then separation of the oxide by centrifugation. The sample is submitted to heat treatment (120°C, 1.5 h) in order to eliminate the solvent and fix the polymer to the surface of the oxide. The second stage consists of dispersing CuO nanorods coated with P4VP (0.4 g) in isopropanol (1.5 mL) with aluminum nanoparticles by applying sonication for several hours. The last stage consists of dispersing the slurry in isopropanol by means of a final ultrasound treatment. The nanothermite is separated by centrifugation, then dried (95°C, 10 min) [SHE 08].

Cheng *et al.* have used P4VP to assemble iron oxide nanotubes (Fe_2O_3) with aluminum nanoparticles. P4VP is deposited on the oxide nanoparticles by dispersing them by prolonged (4 h) ultrasonication in a solution of this polymer in isopropanol. The excess P4VP is eliminated by washing cycles. When the oxide nanotubes thus modified are assembled with aluminum nanoparticles, the resulting materials show that the two phases are intimately associated. By contrast, when powders are just randomly mixed in hexane, the resulting compositions exhibit a coexistence of the two phases in the form of independent aggregates of same nature particles [CHE 10].

Malchi *et al.* describe a method for assembling aluminum nanoparticles (~38 nm) and cupric oxide nanoparticles (~33 nm), allowing to form nanothermite microspheres (1–5 µm). The principle consists of grafting long hydrocarbon chains (in C11) on aluminum and on copper oxide, bearing at their ends groups that can bond by molecular recognition. In this way, fuel and oxidizer nanoparticles are organized in a sort of particular lattice, in which stoichiometry is respected at the scale of elementary particle size. An ω-aminated acid is grafted on the aluminum nanoparticles by its carboxylic end, the laminated part being salified in the form of chloride. The copper oxide is first functionalized with 11-mercaptoundecanoic acid, which bonds on its surface by a sulfide bond. The carboxyl bond is then salified by reacting with tetramethylammonium hydroxyde. The assembly is carried out by mixing suspensions of functionalized nanoparticles. In order to eliminate the ammonium salts formed in the assembling reaction and produce well-defined microspheres, an additional treatment is needed and consists of redissolving the precipitate in a mixture of dimethylsulfoxide and acetonitrile, then evaporating the acetonitrile at 80°C. The efficiency is then very weak and the recovered quantity is not sufficient for conducting combustion tests with the purified product [MAL 09]. The results of combustion tests carried out by Malchi *et al.* on the unpurified precipitate are disappointing. The combustion rate of the composite material, which stabilizes at 0.25 m/s, is three orders of magnitude lower than that of CuO/Al nanothermite prepared by simple physical mixture (285 ± 48 m/s). These results are a good illustration of a problem that is inherent to the method of assembly by grafting of organic ligands. The organic molecules used for organizing nanoparticles are generally introduced in a non-negligible proportion in the aluminothermic composition. Their pyrolysis by the aluminothermic reaction produces carbon. The carbon film is difficult to oxidize, it limits the contacts between oxide and aluminum and it is often involved in parasitic reactions, such as aluminum carburizing

[PIC 15]. The assembly-based fabrication of high-performance nanothermites implies the reduction of the carbon content of structuring molecules. The use of ligands that gasify when submitted to moderate heat treatment could also be a path worth exploring. Finally, the development of this fabrication technique at the industrial scale would require better efficiency and simpler preparation methods.

Yang *et al.* synthesized manganese oxide (α-MnO_2) nanowires by a hydrothermal method. The surface of these fibers is covered with hydroxyl groups, which confer it a negative charge. A cationic polyelectrolyte, polydiallyldimethylammonium chloride, is grafted by electrostatic interaction on the MnO_2 fibers so that the oxide surface acquires a positive charge. The modified oxide is mixed with aluminum (~80 nm) in an aqueous solution of ammonium dihydrogenophosphate (0.1 wt%), which inhibits the oxidation of aluminum by water. The aluminum nanoparticles assemble with oxide fibers, as a result of the electrostatic attraction between the surfaces, which are charged with opposite charges. Nanothermites thus produced are used for the fabrication of flexible membranes with centimeter sizes, by a method that is as simple as it is elegant, consisting of using the effect of pump aspiration during the filtration of the suspension in order to give the retentate a form. The membranes of MnO_2/Al nanothermite are finally rendered hydrophobic by the deposit of polydimethylsiloxane in gas phase at 235°C. Hydrophobicity is preserved in time; it protects the aluminum against the effects of atmospheric humidity or liquid water. The contact angle of a water droplet with the surface of a hydrophobic nanothermite (>150°) is practically independent of pH (1–14) [YAN 13a].

Thiruvengadathan *et al.* used graphene oxide (GO) for assembling bismuth oxide nanoparticles (90–210 nm) with aluminum nanoparticles (80 nm). GO results from nanosized graphite platelets being oxidized by permanganic anhydride dissolved in a sulfonitric mixture, then by a treatment with hydrogen peroxide. GO is purified by cycles of washing with water. It is finally extracted from the colloidal solution by simple heating (80°C, 30 min), which triggers the aggregation of nanoparticles in the form of a supernatant membrane. The liquid in which the various stages of synthesis by assembly take place is composed of dimethylformamide and isopropanol in equal volume proportions. GO is dispersed (0.5 wt%) in a mixture of solvents, by means of ultrasound treatment, in a sonication bath. The aluminum and bismuth oxide powders are separately dispersed through a similar treatment, but with a shorter duration (4 h). In a first step, the aluminum suspension is added to the GO suspension. The bismuth oxide suspension is added in a second step. After each mixing, ultrasound treatment is applied for 1 h. Various GO/Al/Bi_2O_3 nanocomposite materials have been prepared by varying the graphene mass content between 0 and 10 wt% and the equivalence ratio between 1.0 and 1.8. Irrespective of the composition, the mass of sample produced (400 mg) remains constant [THI 14].

Thiruvengadathan *et al.* noted that when the aluminum and bismuth oxide suspension is at rest, the oxide settles while the aluminum remains dispersed, but in the presence of GO all phases settle simultaneously. The precipitation time depends on the proportion of GO in the nanocomposite material (see Table 2.3). Rapid precipitation is an indicator of close interactions between nanoparticles and efficient assembly process.

GO (wt%)	≤2	3.5–5	>10
Precipitation time	24–36 h	2 min	∞

Table 2.3. *Precipitation time for nanocomposite materials made up of graphene oxide, bismuth oxide and aluminum as a function of the proportion of graphene oxide [THI 14]*

2.6.2. *Biological methods*

The biological methods of assembly consist of using complex chemical substances that are involved in certain mechanisms for molecular recognition in living organisms for arranging nanoparticles.

Slocik *et al.* use ferritin to synthesize bionanothermites by assembly. Ferritin is a molecule that controls the rate of Fe^{3+} cation in the body. This cage-shaped protein has an inner cavity with a diameter of 6 nm, which shelters a nanoparticle of iron oxyhydroxide, FeO(OH). The external surface of native ferritin is covered by carboxylate groups and has negative zeta potential (-32.9 ± 8.7 mV). Cationic ferritin is produced by the chemical modification of ferritin; its surface charge is positive ($+23.6 \pm 8.0$ mV). Apoferritin results when the iron oxyhydroxide core is eliminated from ferritin by reductive dissolution operated with mercaptoacetic acid diluted in a buffer solution and successive dialyses. The external surface of apoferritin is functionalized by a peptide with affinity for aluminum. To prepare hybrid nanothermites, Slocik *et al.* first coat the aluminum nanoparticles (80 nm), whose surface is negatively charged, with a monolayer of cationic ferritin. Then, making use of electrostatic attraction, they build a deposit with alternating layers of ferritin and cationic ferritin; this method allows to gradually increase the oxide content of the hybrid material in order to reach the stoichiometric proportion. The combustion of a hybrid material, which contains a double layer of ferritin, is clearly more vivid than the one of the homologue material based on apoferritin, which proves that iron oxyhydroxide encapsulated in ferritin definitely reacts with aluminum. On the other hand, a nanomaterial containing two layers of ferritin burns less vividly than a nanomaterial with 12 such layers. Slocik *et al.* also prepared an aluminothermic composition by assembling a monolayer of cationic apoferritin filled with ammonium perchlorate with aluminum nanoparticles. This hybrid material generates a very vivid combustion, despite its low oxygen content [SLO 13].

Séverac *et al.* assembled copper oxide nanoparticles (50 nm) and aluminum nanoparticles (80 and 120 nm) by a molecular recognition method, which consists of grafting complementary oligonucleotides at the surface of these particles. Grafting these DNA strands on the surface of nanoparticles involves in each case a different strategy. Oligonucleotides that are grafted at the surface of copper oxide are modified by the introduction of a thiol group, which permits them to chemically bond to this oxide. Grafting of oligonucleotides on the shell of aluminum nanoparticles is non-specific, but more complex. The first stage consists of the adsorption of a protein, neutravidin, on the alumina layer coating the aluminum nanoparticles. The second stage consists of grafting the oligonucleotides modified by the introduction of biotin on the aluminum nanoparticles coated with neutravidin. The hydrodynamic diameter of the particles in the suspensions of nanoparticles functionalized by complementary strands of DNA increases in time, while in the suspensions of nanoparticles functionalized by non-complementary oligonucleotides it remains constant. Nanothermites thus produced are composed of aggregates with micron-sized diameter (~2 μm) containing several hundred nanoparticles of aluminum and copper oxide [SÉV 12]. This method permits to assemble nanoparticles of various chemical natures and sizes, and to produce nanothermites with homogeneous compositions. However, this method remains quite difficult to implement and requires the use of a non-negligible proportion of carbonaceous molecules among the reagents. The separation of reactive phases by carbon substances raises the sensitivity thresholds, but also lowers pyrotechnic performances [SIE 10].

2.6.3. *Electrical methods*

Electrical methods consist of applying a potential difference in order to organize the constituents of nanothermites at various scales: reduction of ions on a nanostructured medium, transport and assembly of nanoparticles or still development of energetic matter at the macroscopic scale.

2.6.3.1. *Electrolysis and electrophoresis*

Electrolysis is a method for activating a chemical reaction by applying an electrical current. It can serve for depositing strongly reductive metals, starting from water-free solutions, on metal oxides nanostructured at the surface of an electrode. Electrophoresis consists of using an electrical current to cause the nanoparticles to migrate and assemble them in a compact manner.

Dong *et al.* grew nanosized filaments of $WO_{2,9}$ oxide at the surface of a tungsten wire ($D = 0.5$ mm) by flame synthesis. The oxide nanowires form a dense deposit, in which they are vertically aligned. Their diameter (20–50 nm) is very small compared to their length (>10 μm). These nanostructures are coated with aluminum by applying an electrochemical method in anhydrous medium. For this purpose, the tungsten wire supporting the oxide nanowires is used as a cathode; the anode is an aluminum wire. The electrolyte is composed of a solution of anhydrous aluminum trichloride in an ionic liquid, 1-ethyl-3-methylimidazolium chloride. Electrolytic deposition is carried out by applying for 15 min a direct current at a potential difference of −1.5 V. In the composite filaments thus produced, oxide and aluminum are coaxially aligned. The aluminum coating is monocrystalline and its thickness is around 16 nm. There is ideal contact between oxide and metal, since the two phases are only separated by one alumina layer [DON 13].

Sullivan *et al.* fabricated deposits of Al/CuO thermite by electrophoresis, starting from micron-sized (2 μm) aluminum particles and nanosized (<50 nm) copper oxide. This original method consists of applying an electrical field to generate the migration of particles in suspension toward an electrode where they are deposited. The thermite films that are produced with this technique are adherent, highly uniform and have a homogeneous chemical composition. Their apparent density (2.6 g/cm^3) reaches 51% of the maximum theoretical density of the composition. The suspension that serves for the deposition of nanothermites is composed of a mixture of ethanol/water in volume proportion of 3/1. The volume concentration of Al and CuO particles in the suspension has to be as low as possible in order to minimize flocculation phenomena and subsequent sedimentation. It has been optimized at 0.2%. The mass deposited by electrophoresis is reproducible. It increases logarithmically with the duration of the experiment, as the resistance of the deposited layer increases with its thickness. On the other hand, it is directly proportional with the electric field applied in view of the deposition. However, the electric field should not be too strong, so that particles can reach an arrangement as compact as possible [SUL 12a]. Sullivan *et al.* subsequently used thinner (80 nm) aluminum particles to produce deposits of Al/CuO nanothermite of micron-sized thickness (~10–200 μm) by this method [SUL 12b]. In order to study the combustion of thermites obtained by electrophoresis, Sullivan *et al.* formed deposits on specific electrodes consisting of platinum bands ($L = 20$ mm; $l = 0.25$ mm) formed on the surface of glass strips.

2.6.3.2. *Bipolar coagulation*

Bipolar coagulation consists of mixing nanoparticles carrying opposite electrical charges in view of their assembly by electrostatic attraction.

Kim *et al.* used this original method to prepare an Al/Fe_2O_3 nanothermite. For this purpose, aluminum nanoparticles (~50 nm) are dispersed in ethanol with a concentration of 9.5 g/L. This first suspension is atomized with a first atomizer, then dried by diffusion. Nanoparticles are then brought in a mixing reactor either directly or after having first crossed through a unipolar charger. Ferric oxide nanoparticles are synthesized by this sol-gel method in aerosol droplets composed of a solution of ferric chloride ($FeCl_3$), epoxide and water in ethanol. Then, oxide nanoparticles are either directly sent to the mixing reactor or charged by crossing through a unipolar charger. This ingenious system permits to produce the Al/Fe_2O_3 composition by Brownian coagulation or dipolar assembly [KIM 04].

2.6.3.3. *Electroatomization*

Electroatomization is a method that consists of spraying a liquid containing a solute or/and particles in suspension through a nozzle with typically submillimeter section by applying a voltage between the nozzle and a metallic plate on which particles are deposited.

Wang *et al.* set the distance between nozzle and target at 10 cm and applied a voltage of 19 kV to assemble with nitrocellulose (NC, 1–10 wt%) aluminum primary nanoparticles (~50 nm) in a three-dimensional structure. The size of the porous microspheres thus formed (2–16 μm) evolves as a linear function of the concentration of aluminum nanoparticles in suspension in the solution of NC [WAN 13a]. In a further study, Wang *et al.* used the same method to assemble aluminum nanoparticles (~50 nm) and cupric oxide nanoparticles (~50 nm) with NC, in the form of spherical microparticles [WAN 14d]. According to these scientists, the main advantages of the assembly by electroatomization are as follows: (1) a better contact between fuel and oxidizer with homogeneous distribution of the phases in the composite materials thus prepared, (2) the possibility to control the diameter of the microspheres thus produced and to obtain very narrow size distributions and (3) the possibility to incorporate polymer binders in small proportion or thermo-labile molecules. Finally, the absence of free nanoparticles in these powders of micron-sized granulometry allows them to be considered as classical micron-sized powdery substances for industrial pyrotechnic applications. Though this method may be of high interest for the assembly of structures according to an ascending approach (bottom-up), it has two weaknesses that can potentially compromise its use at a larger scale. First of all, the experiments described by Wang *et al.* indicate that the rate at which structured matter is produced is below 1 g/h. This problem may be partially solved by multiplying the number of unit systems. Then, the use of high electric potential in a medium that encloses not only flammable solvent vapors but also energetic matter that is particularly sensitive to this type of stress is practically an impossible challenge for the development of an electroatomization method for large-scale production of nanothermites.

Wang *et al.* used the electroatomization method to assemble aluminum nanoparticles (~50 nm) and metal iodates in the form of ultra reactive microspheres. Copper iodate (~65 nm) is synthesized by milling the cupric nitrate trihydrate with potassium iodate. Bismuth iodate (~90 nm) and iron iodate (~70 nm) are prepared by precipitation, by mixing the solutions of the nitrates of these metals in nitric acid with iodic acid diluted with water. The suspensions used in this electroatomization method are prepared by dispersing iodate nanoparticles in ethanol, then adding aluminum nanoparticles accompanied, in some cases, by collodion. In each of these stages, the medium is subjected to ultrasonication (60 min). The suspension produced is finally maintained under agitation for 24 h. Electroatomization takes place at a flow rate of 4.5 mL/h, through a needle of 0.43 mm, by applying a voltage of 19 kV between the needle and the substrate, which are separated by a distance of 10 cm. The microspheres obtained with NC (5 wt%) have diameters ranging from 3 to 5 μm for $Al/Bi(IO_3)_3$, from 2 to 4 μm for $Al/Cu(IO_3)_2$ and from 5 to 7 μm for $Al/Fe(IO_3)_3$ [WAN 15b].

2.6.3.4. *Electrospinning*

Electrospinning is a similar method to electroatomization. The solutions employed contain polymer in sufficient proportions so as to become organized into strands rather than subdivided into droplets. When the spun solution contains nanoparticles, the polymer serves as binding matrix that ensures the cohesion of the material. The filaments fabricated by this technique are organized at the macroscopic scale as felt structures.

Yan *et al.* developed an electrospinning method for producing fibers of pure NC or NC loaded with aluminum or CuO/Al thermite nanoparticles. The wires produced are wound by means of a rotating system in order to form by entanglement nonwoven fabrics. This technique permits to produce nanocomposites with up to 50% in mass of CuO/Al nanothermite in the form of wires with submicron cross-section (~300–1,000 nm) [YAN 12].

Electrospinning has also been used by Li *et al.* to elaborate NC fibers of submicron diameters. Nanoparticles of aluminum (~40 nm) and iron oxide α-Fe_2O_3(~30 nm) are incorporated in weak proportions in NC. Li *et al.* used a mixture of acetone and *N,N*-dimethylacetamide to dissolve NC and disperse the aluminum and oxide nanoparticles [LI 13].

2.7. Structuring at the surface of substrates

Structuring of nanothermites at the surface of substrates is an ascending synthesis method that consists of the deposition in various forms of fuel and oxidizer at the surface of a flat material, such as glass strips or silicon platelets. The nanosized structures are produced either by the chemical treatment of a metal layer previously formed on the substrate or by the deposition of a metal or oxide by methods such as cathode sputtering or evaporation under vacuum. The supported nanothermites have original morphologies: flat leaf-like or alveolar structures, coated wires or wires that interact with the substrate in which they are embedded. The treatment of the substrate is often required for improving the adherence of the energetic coating.

Menon *et al.* prepared nanothermites in the form of brushes composed of iron oxide (Fe_2O_3) filaments implanted in an aluminum frame. This original morphology was obtained by replicating a porous alumina matrix. The fabrication method involves 10 stages. Porous alumina is formed at the surface of an aluminum leaf by anodization in a solution of oxalic acid (3 wt%) by applying a voltage of 40 V. The pores (~50 nm) are filled with iron by an electrochemical method, and then sealed by an organic coating. The mass of aluminum that serves as a substrate for the porous alumina is then dissolved with a solution of mercuric chloride (3 wt%). After dissolving the organic coating in ethylic alcohol, the alumina matrix is treated with phosphochromic acid in order to reveal and oxidize the upper end of iron nanowires. An aluminum film ($E = 25$ nm) is deposited by thermal evaporation on iron oxide stalks, then the alumina matrix is fully dissolved in phosphochromic acid. This last treatment also results in the oxidation of that part of the iron filaments, which had been until then protected by alumina. The brush formed is finally annealed at 100°C in order to improve the contact between phases at the interface. Iron oxide wires have a diameter of 50 nm. Their surface density nears 10^{10} wires/cm^{-2}, which represents a coating rate of 20% of the aluminum surface. The method described by Menon *et al.* allows us to control the diameter (8–200 nm) and the length of iron oxide wires, as well as the thickness of the aluminum layer in which they are implanted [MEN 04].

Blobaum *et al.* produced a thermite by the deposition of alternating layers of copper oxide (CuO_x) and aluminum on a silicon substrate by cathode sputtering. The substrate is first covered by spin coating with a film of resin soluble in acetone to enable detachment of the laminated thermite by simple washing. Successive layers are deposited starting from targets of aluminum and cupric oxide (CuO) using, respectively, DC (150 W) and RF (200 W) sputtering in a chamber maintained under high vacuum (1.9×10^{-7} Torr). To avoid the initiation of an aluminothermic reaction, the sputtering guns are masked in order to confine plasma within a small volume above the target and the substrate is cooled with water. The final thermite deposit has a thickness of 14 μm. The Al/Cu_4O_3 bilayers have micrometer thickness (1 μm), except for the first and the last ones (0.5 μm) [BLO 03].

Petrantoni *et al.* fabricated CuO/Al nanothermites in the form of wire structures at the surface of silicon sheets. For this purpose, a bond coat of titanium (30 nm) is first deposited on the silicon by evaporation under high vacuum. Copper is next deposited, either by thermal evaporation or by electrolysis. In this particular case, the substrate should be made conductive by the deposition of a copper film (50 nm) produced by thermal evaporation. The superficial oxidation layer covering the copper surface is chemically stripped with a solution of diluted hydrochloric acid. Oxide nanowires are produced by oxidation under air (450°C, 5 h) of the copper deposit. Petrantoni *et al.* show that these filaments, composed of pure cupric oxide (CuO), are formed independently of the method having been used for copper deposition, but a critical thickness of this metal (300 nm) is required for observing the growth of oxide in this form. Nanothermites are finally produced by the deposition of aluminum by thermal evaporation on the oxide layers [PET 10].

Ohkura *et al.* used galvanoplasty for the deposition of a copper layer with thickness of 1 μm on a steel sheet. Cupric oxide (CuO) nanowires have been formed by annealing at 500°C, applied either for 5 h or for 24 h. A prolonged thermal treatment increases the length and surface density of the oxide nanowires, but does not significantly modify their diameter. Aluminum is then deposited in layers of various thicknesses (100–425 nm) on CuO nanowires by cathode sputtering [OHK 11b].

Kim *et al.* conducted a comparative study of the morphology of CuO/Al nanothermite deposits, produced either in powdery form or as a layer of composite fibers. The powder is deposited by spin coating of a suspension of aluminum nanoparticles (80 nm) and cupric oxide nanoparticles (30 nm) in isopropanol. Nanothermite fibers are produced by a method that is similar to the one described by Petrantoni *et al.* A glass strip, being used as substrate, is first coated with a silica layer (100 nm). A tantalum film (50 nm) and a copper layer (1 μm) are then successively deposited by cathode sputtering. The respective functions of these deposits are to facilitate the adhesion of nanothermite to the substrate and to serve as precursor for the copper oxide. The layer resulted from the superficial oxidation of copper is then eliminated by washing with a solution of hydrochloric acid, followed by water rinsing. Copper oxide nanowires are produced by oxidation under air of the metal layer (500°C, 4 h), and then coated with aluminum by cathode sputtering. The diameters of the nanothermite filaments and their oxide core are close to 200 nm and 50 nm, respectively [KIM 11].

Zhou *et al.* coated a silicon substrate with nanosized rodlets of magnesium by the glancing angle deposition technique. Metal is evaporated under a high vacuum (2×10^{-6} Torr), and then condenses on the substrate whose temperature is below 80°C. The rodlets formed have a diameter of about 200 nm and a length near 1.5 μm. They are then coated with a layer of copper oxide (CuO), deposited by reactive cathode sputtering, starting from a copper target. The oxide layer, whose thickness ranges between 100 and 200 nm, coats in a very regular manner the underlying magnesium, which it protects against oxidation [ZHO 13]. Using the same synthesis method, Zhou *et al.* coated nanosized rodlets of magnesium with a continuous layer of polytetrafluoroethylene [ZHO 14a].

Zhang *et al.* produced a NiO/Al nanothermite layer at the surface of a silicon wafer. The first stage of the preparation method consists of coating the silicon with a nickel film deposited by thermal evaporation of the metal under high vacuum (7×10^{-6} mbar), then condensation on the substrate (45°C). This compact coating, with a thickness of about 200 nm, is then oxidized by a heat treatment (450°C, 2 h) conducted in a mixture of nitrogen (80 vol%) and oxygen (20 vol%). Under these experimental conditions, nickel oxidation is complete; the oxide (NiO) deposit has alveolar, honeycomb morphology. The creation of surface porosity is accompanied by a thickening of the coating, whose average height goes from approximately 200 nm (Ni) to over 500 nm (NiO). Aluminum is finally deposited by thermal evaporation (5×10^{-6} mbar) on the oxide, which ensures excellent contact between the two phases [ZHA 09].

Xu *et al.* coated nanorods of cobalt oxide (Co_3O_4) with aluminum, formed at the surface of a silicon substrate. The oxide rodlets are synthesized by the immersion of a silicon wafer in a concentrated solution of cobalt nitrate, alkalinized with ammonia, and maintained at a temperature of 90°C for 13 h. The rodlets produced during this first stage contain not only the desired oxide but also cobalt hydroxide $Co(OH)_2$ that can be oxidized into Co_3O_4 through subsequent calcination (250°C, 4 h). The filaments produced have an average diameter of 500 nm, and a length between 5 and 15 µm. They are fixed on the underlying silicon by a thin cobalt interface (~200 nm). Aluminum is deposited by thermal evaporation under vacuum (4.5×10^{-6} mbar) in the form of layer with adjustable thickness (360–2320 µm), which fits perfectly the surface of coated oxide wires. The absence of metallic cobalt in the thermite proves that the oxide does not react with aluminum during the deposition stage [XU 12].

The synthesis of nanothermites at the surface of substrates allows for the integration of these energetic materials in microelectromechanical systems [ROS 07] and enables for the development of addressable pyrotechnic systems. Nevertheless, since the quantity of deposited matter is very low, the pyrotechnic effect produced is limited to a vivid combustion. This is not powerful enough to activate certain pyrotechnic effects such as the detonation of secondary explosives, which requires a far more significant quantity of matter (0.1–1.0 g) that is distributed throughout the volume with high porosity, rather than taking a flat compact form.

2.8. Conclusions and perspectives

The preparation of nanothermites by physical mixing of powders in a liquid or dry medium is currently the only method that provides a simple route for obtaining these energetic materials in significant quantities. Over the past two decades, many other more advanced methods of preparation, such as coating, sol-gel synthesis, impregnation of porous solids, assembly and surface nanostructuring, have been explored. Though they each have their own advantages, none of them seem to be applicable at the industrial scale.

The methods that will be used in the future for the synthesis of nanothermites will consist of assembling nanosized "building blocks" of metals, metalloids, oxides and metal salts in an ordered manner and under optimal safety conditions. These methods should enable the production of significant quantities of matter, from kilogram to ton, depending on the intended applications, under safe and environmentally-friendly operating conditions. Morphology control, throughout the scale from elementary particle to pyrotechnic system, is essential for ensuring the stability of materials and guaranteeing the invariability of their performances. Preparation methods should aim to render the composition of nanothermites homogeneous at the scale of nanoparticles, while preserving a high level of porosity between them, as such as possible. Presently, the assembly and aerosol synthesis seem to be the methods that are best adapted for mixing and structuring nanothermites.

The fabrication of macroscopic pyrotechnic systems integrating nanothermites has so far been little studied. It represents "the missing scale" between the raw energetic material and the application. As a result, there is large room for progress in this field of research.

3

The Experimental Study of Nanothermites

3.1. Introduction

The experimental study of nanothermites and their energetic compositions is complex. It includes the morphological analysis of materials, but above all the characterization of their pyrotechnic properties.

Nanothermite morphology is studied using experimental techniques that are used to characterize non-reactive nanomaterials. The adsorption of gas, particle size measurements, X-ray diffraction (XRD), electron or atomic force microscopy (AFM) techniques and spectroscopic methods may be used – provided the energetic character of nanothermites are taken into consideration, and the necessary protective measures are taken to guarantee the safety of the operator and the proper conduction of the experiment. The morphological study of nanothermites must consider the isolated "bricks", which are the nanoparticles of metal, oxide and metallic salts, and how they are combined in the mixture. Measured properties include the shape, surface area and the particle size distribution, their bulk and surface chemical composition, their mode of aggregation, the degree of homogeneity of the mixtures, the nature of interfaces and the pore volume.

The pyrotechnic properties of nanothermites are studied through very specific experimental methods, which are developed by research teams working in this field by combining high-tech metrology systems. The most commonly measured properties are the ignition delay time, the propagation velocity of the unconfined or confined reaction, the pressurization rate, the heat of the reaction and the light emission. The determination of these characteristics helps to explain the reaction mechanisms of different compositions and, in certain cases, to compare their performance.

The reactive properties of nanothermites depend on many intrinsic factors, such as the chemical nature of the composition, the size and shape of the particles, the degree of homogeneity in the mixture, the presence of inert or reactive additives and the degree of porosity. They also depend on the environment in which the nanothermite reacts, and in particular the device containing the composition as well as the nature and pressure of the gas in which the firing takes place. Finally, they depend on the power of the ignition, which must be capable of activating the strong reaction mode of the nanothermite.

The challenge of the characterization of nanothermites is in measuring as accurately as possible extremely short-lived energetic events, occurring on very small scales. Sophisticated analysis methods, such as the technique known as "T-Jump" [ZHO 09b], are used to clarify the basic mechanisms of reaction from small samples of nanothermites (1–10 mg). Tests at mesoscopic scale, such as firing in fitted tubes or in a manometric bomb, require larger quantities of materials (0.05–5 g) and provide information on the overall reactivity of the material. The effects of the explosion of macroscopic nanothermite loads (> 0.1 kg) have not yet been studied. It is likely that this kind of experiment has not been conducted due to regulatory constraints, which make it virtually impossible to study large quantities of these rather sensitive nanomaterials.

This chapter is concerned with, first and foremost, the characterization and properties of nanostructured fuels, which have been or will be used to prepare nanothermites. An important place has been reserved for aluminum, which is by far the most frequently used nanometric fuel. The special properties of the compositions obtained with various oxidizing substances, such as oxides or metallic oxidant salts, will be described subsequently. This chapter also deals with reactive and morphological characterization techniques used to study nanothermites. The conclusion focuses on methods that, in the near future, will increase the already outstanding performance of nanothermites.

3.2. Study and properties of main fuels

The maximum values of the enthalpy of combustion produced by several fuels of interest in the formulation of nanothermites are shown by Dreizin as a comparative diagram [DRE 09]. The numerical values extracted from this study are presented in Table 3.1.

In comparison, methane has a heat of combustion of 55.6 kJ/g. The heat of combustion of alkanes decreases with the length of their carbon chain and tends toward a limit close to 45 kJ/g, the density of solid alkanes being around 0.7 g/cm^3, and their volumetric heat of combustion approximately 30 kJ/cm^3. With the exception of boron, all fuels of interest for nanothermites have heats of combustion lower than alkanes per unit mass. Conversely, the volumetric combustion heats of these fuels are much larger than those of alkanes because of their higher densities. Diamond, which is the densest form of carbon, has a heat of oxidation of 33 kJ corresponding to116 kJ/cm^3.

	B	Al	Mg	Ti	Zr
Maximum enthalpy (kJ/g)	58	30	22	20	11
Maximum enthalpy (kJ/cm^3)	143	81	39	89	73

Table 3.1. *Maximum values of the enthalpy of combustion, taken from the comparative diagram published by Dreizin [DRE 09]*

Furthermore, the analysis of tables published by Fischer and Grubelich [FIS 96], which classify thermite and intermetallic compositions based on their theoretical thermochemical characteristics, demonstrates that the thermites whose combustion emits the most heat per unit of mass and volume are, respectively, the compositions Mg/B_2O_3 (8.92 kJ/g) and Be/PbO_2 (26.72 kJ/cm^3). These values are well below the heat released through oxidation reactions of fuels in pure forms. In other words, mixing a fuel and an oxidant always produces a heat of reaction smaller than that of the fuel alone.

Finally, the actual performance of a fuel depends not only on the heat released by its oxidation, but more importantly on the kinetics of this reaction. Thus, despite their attractive thermodynamic properties, nanostructured boron and diamond are in fact inferior to nanometric aluminum as fuels.

3.2.1. *Nanometric aluminum*

Nanometric aluminum is unquestionably the most important fuel in the formulation of high performing nanothermites. This material is composed of particles with a very particular morphology of a core–shell type. The core, composed of crystallized aluminum, is covered with an amorphous alumina shell, which provides fairly good protection for the metal against oxidation by air at room temperature. The thickness (E) of the alumina layer is often between 2 and 5 nm; it must be sufficient to prevent the oxidation of the underlying aluminum. An insufficiently passivated nanometric aluminum sample ($E < 2$ nm) is not stable when exposed to air. The heat generated by the slow oxidation of a powder composed of particles coated by an alumina layer of less than 1.5 nm thickness can accumulate through self-containment, and lead to a fire [PES 04]. The risk associated with the pyrophoric nature of nanometric aluminum is controlled by passivating it with an oxide layer of sufficient thickness, and by not storing insufficiently passivated powders in large quantities.

The alumina content of nanometric aluminum powders increases very significantly when the mean diameter of particles decreases (see Figure 3.1). As alumina is an inert substance from a pyrotechnic standpoint, the heat released by the oxidation reaction of aluminum relative to the mass of material decreases, and the quantity of nanometric aluminum necessary to achieve stoichiometry in aluminothermic compositions is greater.

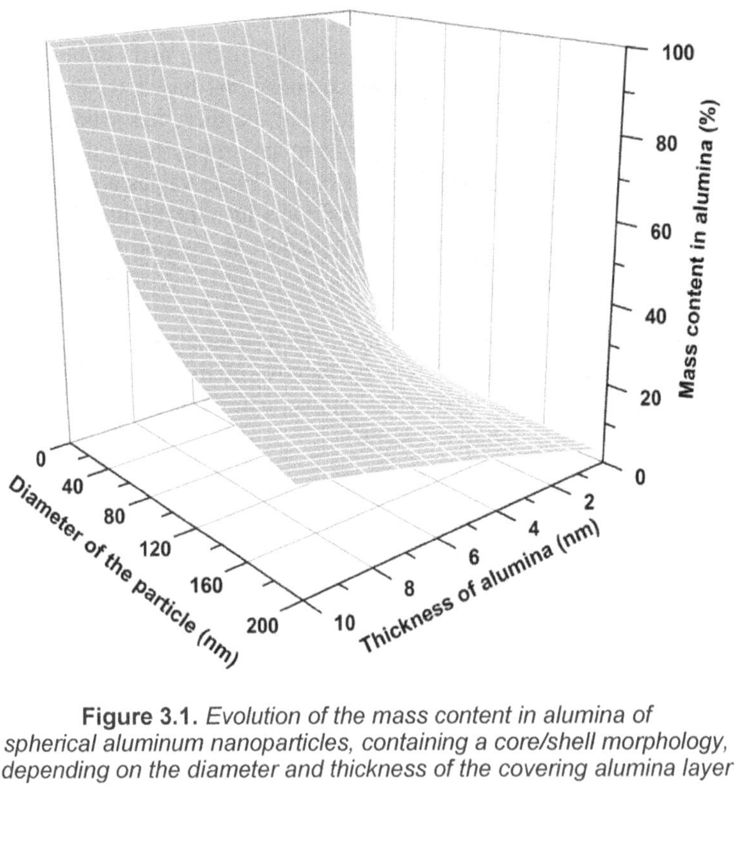

Figure 3.1. *Evolution of the mass content in alumina of spherical aluminum nanoparticles, containing a core/shell morphology, depending on the diameter and thickness of the covering alumina layer*

The controlled oxidation of aluminum nanoparticles is the only effective way to sustainably protect the metal core. To prevent "necks" being formed by cold welding between the aluminum nanometric spheres, they must undergo pre-oxidation in the moments following their creation and before they make contact with each other. The strong aggregation of passivated aluminum nanoparticles by a late oxidation treatment, for example, has been observed by Aumann *et al.* [AUM 95]. Pesiri *et al.* have overcome this technical difficulty, by directly introducing a limited quantity of oxygen in a newly formed aluminum aerosol. The thin layer of alumina produced is thickened through a process of post-passivation, which aims to increase its stability [PES O4]. The passivation of aluminum in the presence of gaseous species such as water vapor or nitrogen produces compounds of aluminum, hydride and nitride, which readily hydrolyze into hydrogen and ammonia upon contact with atmospheric moisture [KWO 03]. The organic coatings of the non-oxidized aluminum nanoparticles do not protect them against long-term oxidation. Indeed, observation through transmission electron microscopy (TEM) reveals that the aluminum particles coated in this way are covered by a layer of oxide that is formed over time between the aluminum core and the surrounding organic coating. Chemical instability, arising from the incompatibility of coating materials with aluminum, is equally to be observed. Thus, stearic and oleic acids react with aluminum to form aluminum carbide (Al_4C_3), a compound sensitive to moisture and hydrolyzing in methane and aluminum hydroxide. Nitrocellulose (NC) acts as a strong oxidizing agent with regards to aluminum: an Al/NC material loses 10% of its original mass of aluminum in a year [KWO 07].

Hammerstroem *et al.* [HAM 11] have tried to passivate incipient aluminum nanoparticles by polyether type structures in the form of loops, forming a bridge between different aluminum atoms from the surface. The grafting process is carried out in solution by epoxides with hydrocarbon chains of various lengths. The aluminum is formed in situ, in anhydrous toluene, by the decomposition of aluminum hydride (alane), catalyzed by titanium (IV) isopropoxide and in the presence of epoxides introduced in various proportions. Hammerstroem *et al.* showed that nanoparticles thus functionalized are more stable if they have been prepared with epoxides substituted by long alkyl chains. The size of coated aluminum crystallites from 1,2-epoxydodecane was determined by the Scherrer method in relation to the alane/epoxide molar ratio used to synthesize them. It is clear from this study that the mean diameter of the aluminum core (13–15 nm) is hardly dependent on this proportion, which nevertheless has a very clear influence on polydispersity. The size of the coated nanoparticles, measured by TEM, is between 20 and 30 nm. Consequently, the organic layer formed is relatively thick in view of the small size of the aluminum core. As such, the overall mass content of aluminum is rather modest (~35%), although it is true that most of the metal is in a non-oxidized state (~94%). Finally, thermal analysis shows that the epoxide coating of the aluminum begins to degrade before the temperature reaches 200°C [HAM 11].

Jouet *et al.* prepared aluminum nanoparticles in diethyl ether by decomposing complexed forms of aluminum hydride through the catalytic action of titanium isopropoxide (IV). The coating materials are the following perfluorinated monocarboxylic acids: C_8F_{17}-COOH, $C_{10}F_{21}$-COOH and $C_{13}F_{27}$-COOH, of which the reaction with aluminum is accompanied by a release of hydrogen. Grafting interrupts the growth of the aluminum nanoparticle by forming external defects. This method of passivation protects aluminum from oxidation, but the mass content of active metal is very low [JOU 05].

Kaplowitz *et al.* covered unoxidized aluminum nanoparticles with perfluoroheptanoic acid by a gas phase method. The layer formed (1–2 nm) is composed of alumina and perfluoroheptanoic acid. Analysis of the coating by X-ray photoelectron spectrometry (XPS) reveals that 56% of the surface aluminum is bonded to the oxygen of the perfluoroheptanoic acid and that the remaining 44% is combined with oxygen in the form of alumina. Infrared spectroscopy also shows that carboxylate groups form bridges between different atoms of aluminum from the surface. Although it is not as thick as the layer of alumina spontaneously forming at the surface of the aluminum nanoparticles exposed to air (~4 nm), the fluorinated coating effectively inhibits oxidation of the metal core. The mass content in non-oxidized aluminum is logically higher (80% vs. 63%). The fluorination of nanoparticle surfaces makes them more sensitive to ignition, in composition with the copper oxide, than conventional aluminum nanoparticles. Kaplowitz *et al.* suggest that the alumina reacts with fluorine to form aluminum trifluoride (AlF_3), which has the effect of exposing the metal to oxidation and of lowering its ignition temperature [KAP 13].

Wang *et al.* have coated aluminum nanoparticles, prepared by the method of electrical explosion of wires (EEW), with a polytetrafluoroethylene (PTFE) layer. This polymer deposits onto the metal cores in the form of islets, whose number increases with the deposition time, till it covers the whole aluminum surface. The PTFE layer protects the metal against oxidation and makes the surface hydrophobic. The coating makes the material less affected by moisture and considerably slows down metal dissolution in an alkaline solution. The Al@PTFE compositions with a nanosized core–shell morphology react more strongly than the corresponding physical mixing [WAN 16].

The effect of ultraviolet light on the oxidation of aluminum is an interesting subject, though the properties of this metal in a divided state are sparsely documented. Cabrera *et al.* produced transparent layers of aluminum through vacuum evaporation. The oxidation of the thin films thus obtained was then followed in transmission and reflection over time, in the presence and absence of ultraviolet light. The oxidation rate of the film exposed to ultraviolet light is higher. The authors demonstrate that this phenomenon is not due to the formation of ozone [CAB 47a]. It would be interesting to study whether the effect observed by Cabrera *et al.* on the thin aluminum deposits also occurs on the nanoparticles of the metal. Walter *et al.* measured the thickening of the oxide layer covering the nanoparticles of an aluminum powder ($D_{NP} \approx 39$ nm) stored in a non-airtight transparent bottle. The thickness of the oxide layer, initially equal to 1.8 nm, increases continuously to 2.9 nm at the end of the experiment (d = 4,000 h). Walter *et al.* also observed that the thickening is mainly produced in the first 2,000 h and follows a linear evolution as a function of the square root of time, according to Fick's law of diffusion [WAL 04].

3.2.1.1. *Characterization of nanometric aluminum*

Information on the metallic aluminum content of a nanometric aluminum powder is essential in order to formulate nanothermites. Puszynski reports having tested four methods to measure the content of reactive aluminum of nanometric aluminum powders: (1) thermogravimetric analysis (TGA), (2) a volumetric method, (3) calorimetry and (4) laser-induced breakdown spectroscopy (LIBS) [PUS 09].

Thermogravimentic analysis

This first method consists of using thermogravimetry (TG) to determine the gain in mass from the complete oxidation of a nanometric aluminum sample. The active metal content is 1.125 times the measured gain between the levels (2) and (6) observed on the thermogravimetric curve of aluminum oxidation (see Figure 3.2). This value (1.125) corresponds to the mass ratio between aluminum and oxygen in the alumina. The mass of alumina is derived by subtracting the calculated mass of aluminum from the mass of the sample measured in (2).

TGA is not only used for determining the proportion of Al/Al_2O_3 in the sample, but also provides crucial information on the aluminum characterized. Thus, the loss of mass measured in (1) corresponds to the desorption of water adsorbed on the surface of aluminum nanoparticles and to the dehydroxylation of the alumina layer. The dry material (2) is oxidized in two or three stages, depicted by the "steps" appearing in (3)–(5) on the thermogravimetric curve. The oxidation of aluminum in the solid phase, observed in (3), begins between 350 and 400°C. Expansion of the core of aluminum nanoparticles through gradual heating of the sample causes the cracking of the alumina shell, which thus becomes permeable to gas and no longer protects the enclosed metal against oxidation. Aluminum exposed to oxygen in the cracked areas reacts almost instantaneously, which results in the "healing" of the alumina layer. This explains the decrease in the rate of oxidation between domains (3) and (4) of the thermogravimetric curve. The melting of aluminum corresponds to the observed inflection on the curve at 650°C, at the beginning of the second step in (4). Oxidation of liquid aluminum is slower because the metal clogs the cracks which it flows through. Slowing of the oxidation as observed in (5) is attributed to this effect, which makes more difficult the oxidation of the internal aluminum of the particle. It should be noted that the third step of oxidation is not observed with finer particle sizes of aluminum (50, 100 nm). The splitting of the alumina layer from the aluminum melting temperature and the outpouring of molten metal were observed by Rai *et al.* by using TEM to follow the evolution of aluminum nanoparticles subjected to a gradual rise in temperature produced by a hot stage [RAI 04]. More recently, Egan *et al.* used dynamic TEM to examine the effects of rapid temperature rises (10^6–10^{11} K/s) on aluminum nanoparticles (80 nm). They came to the conclusion that, under rapid heating, adjacent nanoparticles are transformed – by a coalescence mechanism – into particles of a much larger size. The study of aggregates of 50–100 nanoparticles shows that a temperature rise of 15 ns is sufficient to destroy the initial nanostructuring, and that the substance is completely sintered after only 50 ns. The temperature threshold (~1,300 K) above which this transformation occurs is situated between the melting temperature of aluminum and that of alumina. According to Egan *et al.* the coalescence of aluminum nanoparticles is probably facilitated by the fracturing or softening of the alumina layer. The process of aluminum thickening is rapid enough to modify the kinetics of the metal combustion reaction. The influence of coalescence on the reaction of a nanostructured aluminothermic composition depends on the distribution of aluminum nanoparticles within it; if nanoparticles are grouped together in large size aggregates, the destruction of the divided state is prejudicial to the kinetics of the reaction. Conversely, if nanosized aluminum is intimately mixed with the oxidizing phase, coalescence can facilitate the contact between metal and oxide and improve performance [EGA 14]. Observation with a scanning electron microscope (SEM) of a sample of Al/WO_3 nanothermite subjected to rapid heating (~1 ms), from ambient temperature to 1,473 K, shows that the aspect of the mixture at microscopic scale is little modified. Sullivan *et al.* noted that changes in morphology occurred essentially in areas where aluminum and oxide were very close to one another. They have therefore concluded that the exothermic reaction between these phases was responsible for the fusion and sintering observed between adjacent particles, and that there was no indication that a violent reaction mechanism, with alumina spallation and aluminum ejection, takes place under these experimental conditions [SUL 10b].

Nanometric aluminum is characterized by the total amount of oxygen that it can bind, which is proportional to its active metal content. It is also defined by the mass ratio of the oxygen fixed in liquid and solid phases, which constitutes its "fingerprint".

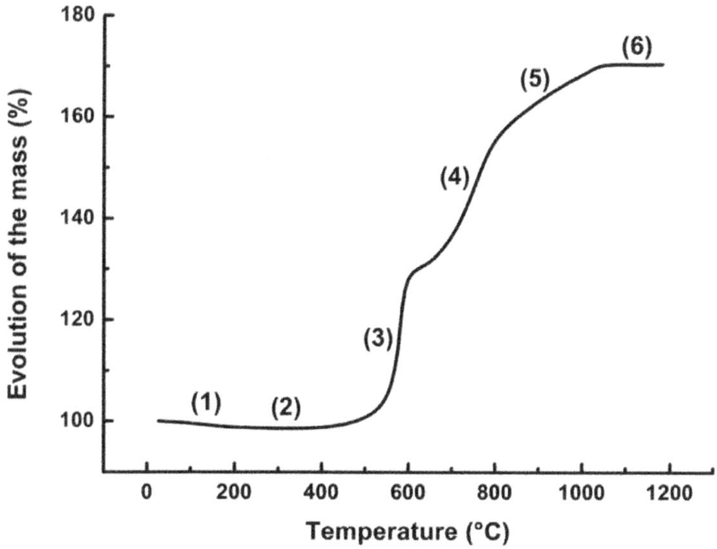

Figure 3.2. *Typical curve obtained by thermogravimetric analysis (TGA) of the gradual oxidation of a nanometric aluminum powder, type Alex, formed of particles with an average size of 160 nm*

TGA of nanometric aluminum powders requires well-defined conditions for experiments. First, it must be carried out in a gaseous environment completely free of nitrogen, because aluminum reacts with this gas forming a nitride (AlN). Mench *et al.* have also shown that Alex ($d \approx 0.184$ µm) becomes fully nitrided and at a much lower temperature (580–680°C) than an aluminum powder of conventional size ($d \approx 20$ µm), for which the nitriding occurs between 970 and 1,020°C [MEN 98]. Second, the oxygen must be diluted in argon to a volume concentration of a few percent to avoid an uncontrollable reaction. The use of argon is justified not only because this gas is chemically inert, but also because it moderates the reaction of oxygen with aluminum [MEI 03].

The progress of the reaction is also controlled by varying the mass of the sample being characterized as well as the heating rate. The apparent volume of the alumina produced by progressive oxidation is comparable to that which initially occupied the nanometric aluminum powder. However, the abrupt oxidation is characterized by the formation of an alumina drop that contains a significant amount of non-oxidized aluminum.

Volumetric-method

The second method of dosage of active aluminum is to measure the volume of hydrogen released by attacking the aluminum powder with a potassium hydroxide solution. This method has the advantage of being easily implemented with relatively few resources, available in all chemical laboratories.

Calorimetry

The heat released by the oxidation of the nanometric aluminum powder can be measured through a calorimetric method. The measured value is then compared with the heat produced by the oxidation of pure aluminum, which enables a calculation of the proportion of aluminum and alumina in the sample. The validity of this measuring technique is based on the assumption that the heat released by the aluminum oxidation is the same at the macroscopic level and at the nanoscale.

But the use of calorimetry is not limited to the dosage of aluminum. Thus, Sun *et al.* used differential scanning calorimetry (DSC) to study several nanometric aluminum powders of average particle size ranging between 25 and 92 nm, and coated with a thick layer of alumina whose thickness was between 3 and 5 nm [SUN 07]. These authors demonstrated a net lowering of the melting point of aluminum, with the decreasing radius of the particles. This effect follows the Gibbs–Thomson law, but the value of the solid–liquid surface tension calculated is low in comparison with the experimental and theoretical values reported in the literature. This is attributed by the authors to the effect of aluminum confinement by the oxide layer, which exerts a compression on the metallic core that melts at a temperature above that which would be expected for an aluminum particle of this size. The results published by Mei *et al.* allow us to estimate the increase in the melting temperature of the aluminum core according to the pressure exerted on it by the oxide shell, at around 55 K/GPa [MEI 05]. In other words, the lowering of the melting point is less pronounced than it should be in the absence of oxide. On the other hand, the melting enthalpy decreases significantly with the size of aluminum nanoparticles: it represents, respectively, 88% and 20% of the melting heat of massive aluminum for aluminum powders whose average diameter is 92 nm and 25 nm. In essence, nanometric aluminum melts much more easily than solid aluminum. Sun and Simon proposed that this effect is linked to an increase in energy within the aluminum, emanating from defects or irregularities in the crystal structure [SUN 07].

Laser-induced breakdown spectroscopy (LIBS)

LIBS involves vaporizing very small amounts of material with an Nd:YAG laser pulse focused on the sample being tested. During its expansion, the microplasma produced cools and the elements that it contains are characterized by their atomic emission lines. Despite its qualities, the use of this technique for analyzing the metal in nanometric aluminum powders is rather anecdotal.

The average thickness of the alumina layer can be measured locally by TEM on a population of nanoparticles representative for the sample. A second method involves measuring the specific surface area (A, m^2/g) and the true density of aluminum powder (ρ, g/cm^3) by adsorption of nitrogen and helium pycnometry, respectively. Under the assumption that the sample is composed of spherical particles of low polydispersity and that the alumina shell is of constant thickness, it is possible to calculate the average thickness of alumina ($E_{ox.}$, nm), the diameter of the aluminum core (R_{Al}, nm) and thus derive the diameter of the particle (D_{NP}, nm). The rationale for establishing the necessary equations for calculating these values has been described by Pesiri *et al.* [PES 04]. These authors did however make an error in the formulas they offer, which is corrected in the equations given below.

The first step is to calculate the constant (β) according to equation [3.1], using the mass gain (Δm) measured by TGA (see Figure 3.2) and the densities of aluminum ($\rho_{Al} = 2.7\ g/cm^3$) and of alumina ($\rho_{ox.} = 3.2\ g/cm^3$). The density of alumina varies considerably depending on whether it is hydrated (AlOOH: $3.01\ g/cm^3$), amorphous ($3.40\ g/cm^3$) or crystallized (α-Al_2O_3: $3.96\ g/cm^3$). The value used by Pesiri was measured by helium pycnometry; it is between that of hydrated alumina and amorphous alumina. The values (M_O) and (M_{Al}) are the atomic masses of molecular oxygen and aluminum:

$$\beta = \frac{\rho_{Al}}{\rho_{ox.}}\left(\frac{3M_O}{2M_{Al}} \times \frac{100}{\Delta m} - 1\right) \approx \frac{75}{\Delta m} - 1 \approx \frac{75}{0.89 \times \%m_{Al}} - 1 \qquad [3.1]$$

The thickness of the alumina layer and the diameter of the aluminum core are then calculated using equations [3.2] and [3.3], using the specific surface area (A) in a spherical model:

$$E_{ox.} = \frac{3(1+\beta)^{\frac{2}{3}}}{A(\rho_{Al} + \beta\rho_{ox.})}\left[(1+\beta)^{\frac{1}{3}} - 1\right] \qquad [3.2]$$

$$D_{Al} = \frac{6(1+\beta)^{\frac{2}{3}}}{A(\rho_{Al} + \beta\rho_{ox.})} \qquad [3.3]$$

The total diameter of the aluminum nanoparticle (D_{NP}) is finally calculated with equation [3.4]:

$$D_{NP} = \frac{6(1+\beta)}{A(\rho_{Al} + \beta\rho_{ox.})} = D_{Al} + 2E_{ox.} \qquad [3.4]$$

These formulas cannot be used for characterizing aluminum powders in which the nanoparticles are highly aggregated and/or covered by an irregular morphology passivation layer (see Figure 3.3).

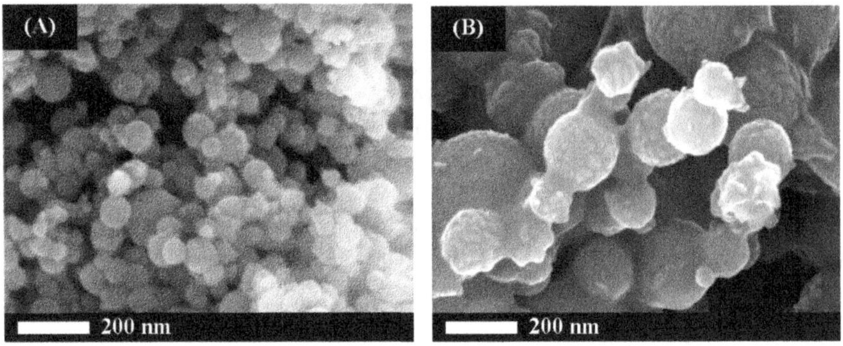

Figure 3.3. *Electron micrographs of nanometric aluminum samples whose characteristics a) can; and b) cannot be calculated by Pesiri's method*

3.2.1.2. *Reaction mechanisms of nanometric aluminum*

The reaction mechanism of aluminum nanoparticles is a very controversial subject, the various aspects of which have been studied extensively and been the subject of numerous publications over the past 20 years. The main weakness of these studies, both theoretical and practical, is in describing the behavior of an isolated aluminum nanoparticle, without taking its environment into account. In a review article, Sundaram *et al.* suggests that three processes control the combustion of aluminum nanoparticles: (1) the diffusion of matter in the gaseous mixture, a mechanism dismissed by the authors; (2) the diffusion of matter through the alumina layer; (3) the kinetics of chemical reactions. The authors also mention "alternative" mechanisms such as the explosion of the aluminum nanoparticle proposed by Levitas [LEV 06]. In reality, the combustion of aluminum in nanothermites has three components: diffusion, explosion and kinetics. The particular conditions for the experiment in which the combustion of metal occurs define the mechanism.

3.2.1.2.1. The mechanism of oxidation by the diffusion of chemical species

The tarnishing of metals exposed to air is a fairly common occurence, which originates from the formation of a very thin oxide layer on the surface. The theorizing of this effect goes back to Wagner's work during the decade preceding World War II [WAG 33]. Mott distinguishes two categories of metal: (1) those which are covered with an oxide layer whose thickness increases proportionally to the square root of time, such as copper or iron, and (2) those which develop a protective oxide layer whose thickness remains constant, such as aluminum, chromium or zinc. Oxidation is always carried out by the diffusion of metal dissociated into cation and electron in the oxide layer and not by the diffusion of oxygen [MOT 40]. In the case of aluminum, the aluminum atoms are soluble in alumina. The migration of metal atoms proceeds from the core of the particle toward the oxide/gas interface where the reaction with oxygen occurs. This process requires sufficient energy to transfer a metal electron into the oxide conduction band and to bring the metal ion thus formed into an interstitial position in the oxide. However, aluminum ions traverse the alumina films produced by anodization without difficulty, of which the thickness is considerable ($E > 1{,}000$ nm). Electrons, however, are not capable of crossing the alumina, which is an excellent electrical insulator. Mott attributes the initial formation of the alumina layer to the fact that electrons can cross it by the tunnel effect, until its thickness reaches 4 nm [MOT 39]. Mott also believes that exchanges occur between the metal ions, which migrate by occupying the interstitial sites and the metal ions of the oxide layer [MOT 40]. In later works, Mott proposes a new theory to explain the thickening of the oxide layer covering the aluminum when the temperature rises. This model is based on two assumptions: (1) the electron work function is low enough for the thermionic emission to facilitate the migration of electrons from the metal to the conduction band of alumina; (2) the solubility of metal ions in the oxide is too low at the temperature considered for the metal to migrate by diffusion. Under these conditions, the oxygen adsorbed on the external surface of the alumina is reduced to oxide anions by the electrons emitted by the metal. The electric field that is established between the ionized metal and the surface anions causes the migration of aluminum ions and is the driving force of oxidation [MOT 47]. By applying this model, Cabrera *et al.* deduced that we can expect a very rapid increase in the rate of oxidation of aluminum from 200°C. By studying and experimenting on aluminum layers deposited by vacuum evaporation, these authors show that the speed of oxidation of aluminum increases from 200°C and becomes very high at 400°C [CAB 47].

As we have seen, the first studies on the oxidation of aluminum were undertaken on sheets or thin films of this metal. Contemporary research is greatly inspired by the models devised by the pioneers of research in this area, but they are rather more interested in the oxidation of aluminum in the form of nanoparticles.

Henz *et al.* conducted simulations of the mechanochemical behavior of aluminum nanoparticles using a molecular dynamics method [HEN 10]. Modeled particles have an aluminum core with a diameter of 5.6 or 8 nm, which is covered by a shell of a thickness of 1 or 2 nm in the first case, and of 2 nm in the second case. Different hypotheses about the nature of the alumina layer have been studied, depending on whether it is (1) crystallized (α-Al_2O_3), (2) amorphous with an oxygen deficit compared to the stoichiometry and (3) amorphous and porous with a stoichiometric composition.

The size of the modeled nanoparticles seems rather to have been defined in terms of the calculating capabilities that Henz *et al.* had available to them, than by taking into account the particle size of the aluminum nanoparticle powders used in the experiments to formulate the nanothermites (40–120 nm). The modeled particles have too small a thickness of the alumina layer (1 nm), or barely sufficient (2 nm) to be stable with regard to the oxidation. Despite this, their alumina content is significant because of their small size and because the proportion of active aluminum they contain is very low (15–35%). These "theoretical nanoparticles", nevertheless, allowed Henz *et al.* to highlight several interesting effects to explain the oxidation of aluminum nanoparticles of ultimate size.

The nanoparticles thus defined are subjected to a very rapid theoretical temperature rise (10^{11}–10^{13} K/s). Vacuum heating of a nanoparticle between 300 and 1,000 K shows that around 900 K the aluminum core expands rapidly due to the melting of the aluminum. Migration of the aluminum cations through the oxide shell to its surface is therefore observed. The alumina shell is not fragmented; its conservation can be explained by a better elasticity than that of solid alumina, or by an insufficient expansion of the aluminum. The diffusion of aluminum undoubtedly helps to relax the pressure exerted by the core onto the shell.

Simulations by Henz *et al.* show that a gradient of negative charge is formed in the oxide shell, which causes an influx of aluminum cations to the core–shell interface. The mass flow is mainly produced by the action of this electric field, which is all the more intense when the temperature that the aluminum nanoparticle is subjected to is low and when the covering oxide layer is thick and well organized.

Finally, Henz *et al.* placed an aluminum nanoparticle in oxygen in order to explain how the hollow alumina spheres obtained by Rai [RAI 06] and Nakamura

[NAK 07] are formed by oxidizing the aluminum nanoparticles. The simulations undertaken using the Henz *et al.* model show that the diffusion of aluminum through the oxide shell under the action of the electric field is much higher than that of the oxide anions to the core of the particle. The oxidation reaction actually occurs in the volume of the shell near to the oxide/atmosphere interface. The diffusion of aluminum toward the periphery of the nanoparticle leaves the space originally occupied by the metal vacant, producing a hollow particle. If, conversely, the oxygen was diffused toward the core of the particle and the oxidation reaction took place at the core/shell interface, the pressure produced would be such that it would burst the particle, a phenomenon that Rai and Nakamura did not observe.

By studying the oxidation of aluminum nanoparticles at different temperatures between 500 and 1,100°C, Rai *et al.* observed through TEM that hollow alumina nanoparticles were formed above 800°C. These results were confirmed by density measurements, which showed that the density of oxidized nanoparticles increased gradually with the temperature (500–800°C), until it reached that of alpha alumina. When the oxidation is performed at a higher temperature (800–1,100°C), the density of the nanoparticles decreases through the formation of a central cavity. Rai *et al.* deduced from these observations a two-step mechanism, in which the oxidation is first produced by the diffusion of oxygen into solid aluminum, and then by a simultaneous diffusion of oxygen and aluminum as the metal melts [RAI 06]. Modeling these experimental results led to Rai *et al.* proposing a power law linking the oxidation time ($D_{ox.}$) of an aluminum nanoparticle to its radius (R_{Al}), taking into account the fact that the core exerts a positive pressure on the shell, and that conversely, the shell is subjected to a negative pressure:

$$D_{ox.} \propto \left(R_{Al}\right)^{1.6\pm0.1} \qquad [3.5]$$

Wang *et al.* have used molecular dynamics simulations to study the atomistic mechanisms involved in the oxidation reactions of an aluminum nanoparticle placed in oxygen, whose core is heated by a laser pulse [WAN 09]. The diameter of the core (48 nm) and the thickness of the oxide shell (4 nm) of the studied nanoparticle are realistic, but the defined temperature conditions are significantly less so. Wang *et al.* assume that the oxide shell remains at room temperature, while the metal core is heated to 3,000, 6,000 and 9,000 K. These working hypotheses are essential to maintain the structure of the nanoparticle, which would otherwise be destroyed by the vaporization of its constituents at the temperature studied. The heating of the core to 3,000 K does very little damage to the oxide shell through which the oxygen atoms are shallowly diffused into the core of the aluminum nanoparticle. The application of a higher temperature (6,000 K) first produces a strong expansion of the nanoparticle and the formation of a central cavity. The implosion that follows is accompanied by a considerable penetration of oxygen into the core of the nanoparticle, and the formation of a solid body. The oxide layer is damaged and its composition is close to $AlO_{1,1}$. At 9,000 K, the swelling of the nanoparticle so strongly distends the oxide shell that the shell becomes porous. The opening of the pores allows the ballistic transport of atoms and a very rapid oxidation of aluminum.

Park et al. studied the oxidation of aluminum nanoparticles according to their size by single particle mass spectrometry (SPMS) [PAR 05a]. This very advanced method of analysis was initially developed by Reents *et al.* [REE 00]. It consists of vaporizing a single particle using an Nd:YAG laser. The affected particle is atomized and all its contained elements are ionized in cationic form, which enables them to be determined by mass spectrometry. Determining the number of atoms and their nature helps in evaluating the size of the particle analyzed, under the assumption of a spherical geometry. Reents *et al.* used this technique to characterize fine particles of sodium chloride (70 nm), ammonium sulfate (70 nm) and silica (40 – 2,000 nm) [REE 01]. Mahadevan *et al.* have modified this technique in order to study the kinetics of the decomposition of metal nitrates (Al, Ca, Ag, Sr). This involves the aerosol passing through a tubular furnace maintained at a steady temperature before being characterized [MAH 02].

Park *et al.* used SPMS to follow the kinetics of oxidation of aluminum nanoparticles. These are produced *in situ* by arc discharge or by laser ablation; they are also introduced by spraying/drying of a suspension in methanol. The aluminum nanoparticles are then passed through a tubular oven, whose temperature is fixed between 25 and 1,110°C, where they oxidize in the air before being ionized and then characterized by mass spectrometry. The authors identified a conversion factor equal to the quotient of the volume oxidized by the total volume of the particle, and then studied its evolution according to the particle sizes and the oxidation temperature. It appears from this study that the smallest aluminum nanoparticles oxidize more completely, and more quickly, than larger nanoparticles. Park *et al.* then indicate that the kinetics of oxidation of the aluminum nanoparticles is controlled by diffusion. The oxidation occurs in two stages: the aluminum particle is covered very quickly by a fine alumina shell. From that moment, the reaction is limited by the diffusion of species through the oxide layer. The SPMS technique shows that the oxidation of isolated aluminum nanoparticles begins when the metal melts, while it is observed at a much lower temperature in TGA. This lowering of the oxidation temperature is attributed to the effects of transfers of heat and mass that occur within a bulk sample [PAR 05a].

The low thermal conductivity of the nanoaluminum powder (see Figure 3.4) promotes confinement of the heat within the material and thereby lowers the temperature of reaction. This is why incorrectly passivated aluminum powders may ignite when exposed to an oxidizing atmosphere. Solid aluminum has a high thermal conductivity ($\lambda_{Al} = 237\ W{\cdot}m^{-1}{\cdot}K^{-1}$) making it one of the best conductors of heat known. However, in a finely divided state, its thermal conductivity ($0.046\ W{\cdot}m^{-1}{\cdot}K^{-1}$) is about the same order of magnitude as that of air. Accordingly, the nanometric aluminum powder should therefore be considered – on a macroscopic scale – as an insulating material.

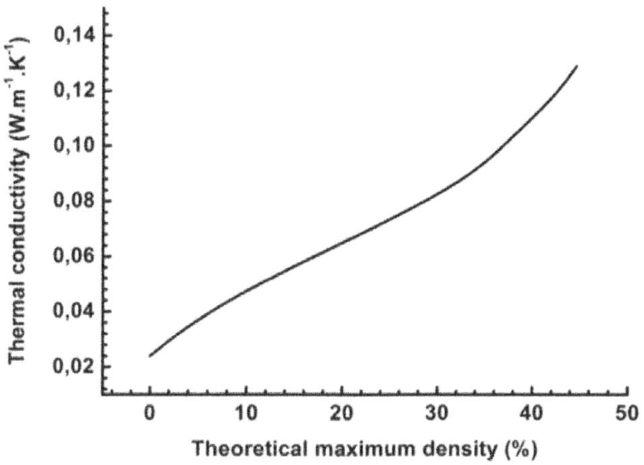

Figure 3.4. *Evolution of the thermal conductivity measured through experimentation of a nanometric aluminum powder (≈ 100 nm) containing 74 wt% aluminum; the conductivity for a percentage of TMD equal to zero is that of air*

3.2.1.2.2. The mechanism of oxidation by explosion

The existence of dynamic effects produced by the oxide layer covering the surface of aluminum nanoparticles has been demonstrated by Wang *et al.* [WAN 05]. In order to do this, the authors used aluminum nanoparticles with a diameter close to 60 nm, coated with a thick oxide layer of between 2.5 and 6 nm. These particles are incorporated in different mass concentrations in a nitrocellulose film of micrometric thickness (2–3 µm) deposited on a glass substrate. The composite films are subjected to laser pulses of variable duration (0.1, 10 and 25 ns). The authors calculate the average distance (d_{avg}) separating the aluminum nanoparticles dispersed in the film. If the distance at which the effects produced by the reaction of the aluminum nanoparticles are manifested (d_{rxn}) is greater than the space separating them ($d_{avg} > d_{rxn}$), the combustion spreads and consumes a continuous circular area. In the opposite case ($d_{avg} < d_{rxn}$), the reaction remains localized. The surface density of energy required to move from a discontinuous combustion to a continuous combustion is the initiation threshold of the reaction. This quantity is used to calculate the absorbed energy (E_v) per unit volume in the aluminum. Wang *et al.* then show that the mean distance of reaction propagation (d_{rxn}) increases linearly with E_v when the laser pulse is short (0.1 ns) and sublinearly when the latter is longer (10 and 25 ns). In other words, the reaction of aluminum nanoparticles is much stronger when it is induced by a stronger stress. Wang *et al.* also note that for a comparable absorbed energy, the aluminum particles coated with a thicker oxide layer have a greater mean distance of reaction propagation. Accordingly, the oxide layer plays an active role in the explosion of aluminum nanoparticles when they are subjected to a temperature rise fast enough to cause a bursting of the core–shell structure. Wang *et al.* also noted through time-resolved microscopy that the reaction of a film containing 1% mass of aluminum nanoparticles, initiated using a laser pulse of 165 µJ for 25 ns on the sample, produces a hemispherical shock front moving at approximately 1.4 km/s. The gaseous species carry along solid fragments moving with an average speed of about 0.7 km/s.

The mechanism of oxidation by dispersion of molten aluminum has been proposed by Levitas *et al.* to explain the very particular properties of the compositions containing nanometric aluminum [LEV 06]. Levitas *et al.* note that the rise in temperature and pressure in the reaction front of nanothermites is much too fast (≈10 µs) for the aluminum and oxygen to diffuse through the alumina layer. The authors also note that the flame front propagation velocity and the ignition delay time of nanothermites are independent of the size of aluminum particles when their radius is less than 40–50 nm. These observations achieved through experimentation lead them to develop a mechanochemical model [LEV 07] to describe the specific reactions of aluminum nanoparticles subjected to ultra-fast heating. The ensuing mechanism, called the "melt dispersion mechanism" (MDM), is probably the most appropriate to describe the specific reactivity of nanostructured aluminothermic compositions. Under the effect of a rapid rise in temperature, the aluminum core melts. Levitas estimates that the rate of heating required to induce MDM is between 10^6 and 10^8 K/s. This is attained by an aluminothermic composition, when the combustion front propagates at over 10 m/s [LEV 09]. The expansion accompanying the melting of the aluminum produces a very high pressure (0.1–4 GPa) on the oxide layer, which is entirely ejected by spallation. At this moment, the pressure in the aluminum drop remains unchanged, but the pressure at the surface of the liquid aluminum falls abruptly to 10 MPa. The difference in pressure thus produced creates a shock wave that propagates to the center of the particle. The aluminum drop is dispersed into drops of smaller size, driven at a speed of 100–250 m/s, whose oxidation is not limited by diffusion.

Campbell *et al.* have demonstrated, through molecular dynamic simulations, that the oxidation of an aluminum nanoparticle (20 nm) in oxygen at a temperature of

300 K is very rapid. The initial rate of thickening of the alumina layer is approximately 58 m/s. The oxide shell reaches its final thickness of 3.3 nm in 0.1 ns [CAM 99]. The aluminum droplets spread by the MDM have a much smaller diameter than the aluminum core from which they originate. On the other hand, because the MDM occurs by definition at the melting temperature of aluminum, the oxidation depth of the ejected droplets is greater than that calculated by Campbell *et al.* Under these conditions, the "lifetime" of droplets spread by MDM would be from 0.01 to 0.1 ns and the distance they would travel would not exceed a several dozen nanometers before the aluminum is fully oxidized. On the other hand, Bazyn *et al.* have studied the combustion of nanometric aluminum (80 nm) caused by the reflection of a shock. These authors determine the duration of the combustion of aluminum by measuring the change in the intensity of light that produces this phenomenon. They show that the combustion time of nanometric aluminum in an equimolar mixture of O_2/N_2 decreases as the gas pressure and temperature both increase. However, at high temperature ($T > 1{,}600$ K), the pressure has a far less significant effect on the combustion length, the value of which – calculated by extrapolation to 2,560 K – is of a few dozen microseconds [BAZ 06]. Assuming that MDM occurs under these experimental conditions, this means that the distance covered by the projected aluminum before being completely oxidized is approximately a few millimeters. The distance travelled before the nanometric aluminum is completely oxidized therefore depends largely on the experimental conditions: nevertheless, it seems possible to estimate it between 10^{-7} and 10^{-3} m.

In a nanothermite, nanometric aluminum is mainly surrounded by air filling the interparticle porosity, and more locally by oxide nanoparticles. A basic calculation, carried out under normal temperature and pressure conditions, shows that the oxidation and nitriding of a cubic centimeter of aluminum requires 1.2 L of air. At the reaction temperature of nanothermites, the volume of air necessary to achieve this result is roughly 10 times larger. The air trapped in the pores of the nanothermites, in a state of loose powders, therefore plays a role – albeit very minor – in the overall aluminothermic reaction. It should be noted here that aluminum nanothermites reach their optimum performance when they contain an excess of metal of 20–60% with respect to stoichiometry. This gap has been observed by Prentice *et al.* with nanothermites composed of aluminum nanoparticles (80 nm) mixed with tungsten oxide aerogels [PRE 06]. Oxidation by air is not the only cause: water also plays an important role. In nanothermites, water can come from hydroxyl groups contained in undercalcined metal oxides, from the crystalline structure of certain oxidizer salts, or simply from the adsorption of atmospheric humidity on the surface of nanoparticles. It is worth noting that water is an oxidant of choice for nanometric aluminum: according to Risha *et al.* wet powders and n-Al/H_2O pastes have a mass combustion velocity that exceeds those of several high-energy propellants [RIS 07]. However, the combustion velocities of these mixtures (1–10 cm/s) are much lower than those of nanothermites prepared with aluminum of comparable grain size [PUS 11]. Risha *et al.* observed that the combustion velocity of n-Al/H_2O mixtures increases with pressure, and that it changes in inverse proportion to the diameter of the aluminum nanoparticles. In other words, the reaction depends on both the diffusion of the species and on the kinetics of the chemical reactions. The study of the n-Al/H_2O composition was furthered by Tappan *et al.* who showed that the combustion velocity of nanometric aluminum (80 nm) in a pressurized chamber (6.9 MPa) is lower in heavy water (D_2O, 1.43 cm/s) than in water (H_2O, 1.92 cm/s). These authors also studied the evolution of the combustion velocity of such mixtures, with aluminum nanoparticles of different particle sizes. Finally, Tappan *et al.* concluded from their experiments that the step limiting the kinetics of the reaction is the proton abstraction from the water molecule, and that aluminum reacts at the gas phase [TAP 14].

The results obtained by Risha and Tappan provide experimental evidence that MDM does not occur in compositions of nanometric aluminum and water. However, these compositions differ from nanothermites in two very important properties, which help explain why MDM is formed and maintained in deflagrating aluminothermic compositions. By capillary diffusion in the aluminum powder, water partially or completely fills the interparticle porosity that it contains; the surface forces established tend to bring the aluminum nanoparticles together. The combination of these effects helps to explain why the n-Al/H_2O compositions have a higher bulk density than that of nanothermites, in which the oxide nanoparticles contribute to their porosity, instead of filling – as water does – the porosity presented by aluminum. Finally, in the n-Al/H_2O, compositions the aluminum nanoparticles are closer together than in the nanothermites and are joined by a continuous film of water. Upon ignition, the water absorbs the heat supplied to the composition, facilitates its diffusion in the sample, then its dissipation through the walls of the quartz tube used for confinement. The heating rate is lowered by this heat sink effect, and becomes insufficient to activate the MDM. Thereafter, the propagation of the aluminothermic reaction is too slow (1–10 cm/s) for the heating rate to reach the necessary threshold for the MDM to take place. The propagation occurs step by step by a conduction mechanism. The filling of all or part of the porosity of the n-Al/H_2O compositions by water limits or prevents the convection of gas in the uncombusted composition. In this particular case, the transition from a slow combustion to an explosive regime seems therefore quite difficult.

The secret to the very high velocities of combustion front propagation in nanostructured aluminothermic compositions lies in their porosity, which promotes the convection of gases, and strengthens the efficiency of MDM. To understand this we can draw an analogy between the combustion of a nanothermite and a falling line of dominoes (see Figure 3.5). The most fragmented configuration of the line of dominoes represents a nanothermite in loose powder form, and the most compact configuration is the compressed nanothermite. The fall time (t) of the two systems described is the same, but the distance covered is clearly much higher in the first case than in the second ($D_1 > D_2$). The falling velocity is therefore greater when the dominoes are distributed over a greater distance ($V_1 > V_2$). The total energy produced by the fall of an arrangement of dominoes therefore depends on the number of dominoes in it, and not on their arrangement. However, the energy related to the distance travelled is higher when the dominoes are arranged compactly. This is the same for nanothermites, for which the volumetric heat reaction is much higher when they are strongly compressed. Prentice *et al.* have observed that the combustion front propagation velocity in compositions based on tungsten trioxide and nanometric aluminum decreases by several orders of magnitude when the loose powders are compacted between 20 and 50% of the theoretical maximum density (TMD) [PRE 06].

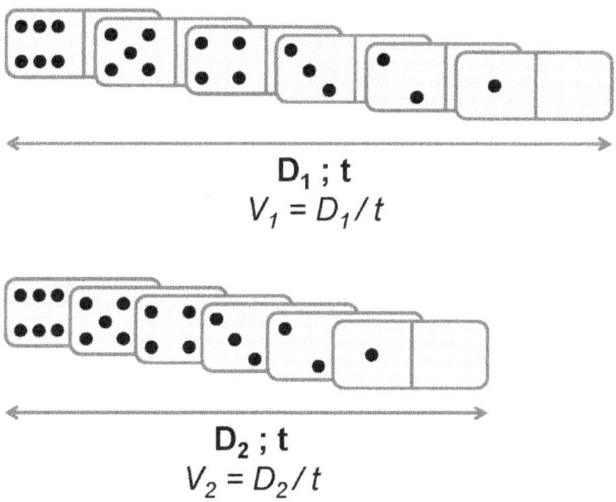

Figure 3.5. *Analogy of the propagation of combustion in nanothermites with the falling of dominoes*

The stages of initiation and reaction by MDM of a nanostructured aluminothermic composition are illustrated below (see Figure 3.6). The explosion of the aluminum nanoparticle is the first step of the aluminothermic reaction. The molten metal droplets are projected in all directions and are oxidized upon contact with the gaseous mixture in which the nanothermite is located. The flight time of the aluminum droplets – from the nanoparticle source to the surrounding oxide particles – defines the second step of the reaction. The impact of aluminum on the metal oxide is accompanied by heat transfer from the metal to the oxide, leading to the activation of the aluminothermic reaction. The hot gases produced by the reaction then diffuse into the porosity of the composition and enable the MDM over a longer distance. The transfer of the heat produced by the reaction toward the uncombusted composition ensures the self-propagation of the combustion. The three conventional modes of heat transfer are conduction, radiation and convection of gas. Using calculations based on sometimes very simplified hypotheses, Egan *et al.* demonstrated that none of these three mechanisms can explain the significant transfer of heat required for the propagation of the reaction in a nanostructured CuO/Al composition. These authors then demonstrate that the projection and cooling of the condensed phases ($Cu + Al_2O_3$) produced by the reaction play a fundamental role in the propagation of heat [EGA 15]. This interesting mechanism is similar to MDM, of which it is in some ways the generalization. The very particular reaction of nanostructured aluminothermic compositions could well arise from the propagation of branched chain nanoexplosions. This action would explain the regularity of the combustion, which is in a sort of pyrotechnic no man's land between deflagration and traditional detonation.

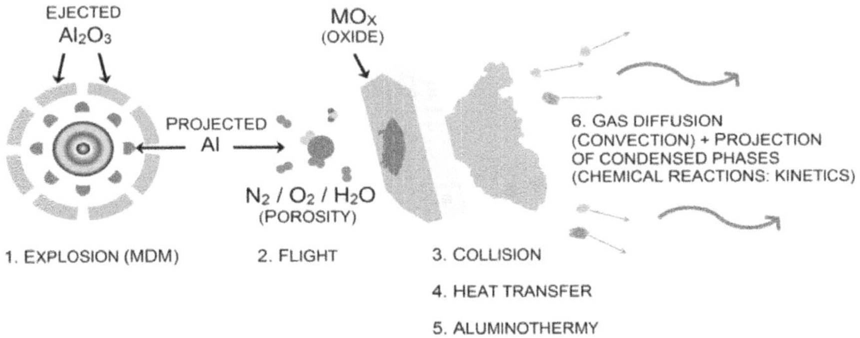

Figure 3.6. *Steps governing the reaction kinetics of nanostructured aluminothermic compositions*

Levitas introduced a factor (*f*) corresponding to the fraction of aluminum having melted upon the bursting of the aluminum nanoparticle. The highest reaction rates of nanothermites are produced with aluminum nanoparticles in which all of the metal has melted before the explosion ($f = 1$). Finally, Levitas defines a dimensionless quantity (*M*), equal to the ratio of the radius of the aluminum core (R_{Al}) to the thickness of the alumina shell ($E_{ox.}$). The entirety of the aluminum melts before the explosion of the nanoparticle, if the condition expressed in equation [3.6] is verified:

$$M = \frac{R_{Al}}{E_{ox.}} \leq 19 \qquad [3.6]$$

The flame propagation velocity of MoO_3/Al nanothermites then reaches its maximum value, regardless of the values of M, R_{Al} and E_{ox}. Conversely, if $M > 19$, only a part of the aluminum melts before the breaking of the oxide layer and is dispersed during the bursting of the particle. The flame propagation velocity is proportional to the fraction of projected molten metal. The oxidation of the remaining solid aluminum is slower and has no influence over the velocity [LEV 08a]. In more recent work, Levitas *et al.* apply thermal treatment to aluminum nanoparticles in order to induce in the core–shell structure opposite constraints to those produced by MDM. The nanoparticles are prestressed by an isothermal treatment (105 or 170°C), applied for 10 min, followed by a rapid cooling. In composition with MoO_3 nanoscale rods, the modified aluminum nanoparticles produce combustions whose propagation velocities are higher. A similar effect is observed with micron-sized aluminum particles (3–4.5 µm) [LEV 11]. Although this method of strengthening the MDM is attractive, it seems to have limited practical interest. Prestressed nanoparticles are in effect metastable materials, whose properties may vary with the temperature changes to which the energetic substance containing the thus treated aluminum is subjected. The mechanical properties of the alumina layer play a crucial role in the MDM; they do not depend only on its state of stress, but also on its chemical nature and potential defects.

Levitas demonstrated that MDM also occurs with micron-sized (1–3 µm) aluminum particles. The velocities of the aluminum clusters projected by their explosion are, however, significantly lower than those of the droplets produced by the explosion of nanoparticles. This effect is explained by the fact that the spallation of the alumina layer occurs at a much lower internal pressure in the case of micrometric particles. Levitas then demonstrates that the low ejection velocity of aluminum clusters has less influence on the performance of Teflon compositions than on compositions based on molybdenum trioxide (MoO_3). Levitas attributes this different behavior to the reaction kinetics of the ejected aluminum clusters: in the first case, gaseous fluorine under high pressure reacts very rapidly with aluminum. In the second case, the flame propagation velocity is limited by the kinetics of the reaction of aluminum with the oxygen provided by MoO_3 [LEV 08b].

Mixtures of nanometric (80 nm) and micrometric (4 or 20 µm) aluminum particles, in different proportions, were used as fuels in compositions based on molybdenum trioxide [MOO 07]. Moore *et al.* note that the addition of a nanometric aluminum fraction facilitates the ignition of the thermites, whose reductive phase is, moreover, essentially composed of micrometric aluminum. The combustion velocity of Al/MoO_3 compositions, in the form of loose powders or pellets, increases almost linearly with the proportion of nanostructured metal contained in the aluminum powder. Moore *et al.* show that micrometric aluminum particles do not participate in the rapid reaction responsible for the pressure peak produced by the combustion of Al/MoO_3 compositions. The slower kinetics of the reaction of micrometric aluminum particles is also reflected in the extended incandescence of the aerosols produced by the combustion of Al/MoO_3 compositions which contain them. However, the addition of microstructured aluminum in a proportion of less than 30% of the metal mass barely alters the performance. This process not only reduces the mass of inert alumina in the composition but also saves the nanometric aluminum, which is more expensive.

The flash ignition experiments on nanometric aluminum carried out by Ohkura *et al.* confirm the relevance of MDM. Spherical aluminum nanoparticles that have an average diameter of 60–96 nm ($E_{ox.}$ = 2 nm) ignite in the air when subjected to camera flash. Under the same experimental conditions, micrometric aluminum particles (20 µm) do not react. The analysis of these experiments through a physical model led Ohkura *et al.* to identify two important parameters: (1) the diameter of the aluminum particles and (2) the bulk density of the powder. The most readily flammable aluminum particles have a diameter of 75 nm. The heating produced by the flash on isolated particles of this size ($\Delta T < 1$ K) is insufficient to cause a reaction. However, if these same particles are assembled in a loose powder with a bulk density greater than 1%, the heat produced by the flash ($\approx 10^6$ K/s) raises the temperature above 1,100 K, a value sufficient to initiate the combustion of aluminum nanoparticles. Transmission electron microscope observation of nanoparticles that have been exposed to flash in air shows unquestionably that the aluminum core exerts a pressure on the alumina layer, which causes its breakdown and the dispersion of molten aluminum [OHK 11a].

Although MDM theory is supported by numerous experimental observations and helps to explain the very specific reactivity of nanostructured aluminothermic compositions, it is not unanimously accepted by the scientific community. In order to invalidate MDM and demonstrate the mechanism of oxidation by diffusion, Chowdhury *et al.* studied the ignitions of Al/CuO nanostructured compositions. For this, the oxide shell ($E_{ox.}$ = 2 nm) covering the nanoparticles (50 nm) of an original aluminum powder is first thickened by an oxidation treatment at 500°C. The modified aluminum nanoparticles ($E_{ox.}$ = 3 and 4 nm) are then used to prepare the Al/CuO nanothermites. The ignition device is a platinum wire, heated by the Joule effect with a diameter of 76 µm and a length of 12 mm. The composition suspended in hexane is then deposited on the central zone of the wire (3–4 mm) using a micropipette. The ignition delay time corresponds to the time elapsing between the end of the electrical discharge and the appearance of the luminous phenomenon produced by the reaction of the nanothermite. The experiments undertaken with this setup lead Chowdhury *et al.* to conclude that: first, the ignition temperature is not dependent on the heating rate; second, the reaction occurs around 1,250 K, in other words well above the melting temperature of aluminum (993 K); third, the ignition delay time increases with the thickness of the alumina layer, even though the heating has stopped before the reaction is initiated. These observations, which contradict the theory of MDM, lead Chowdhury *et al.* to reject this mechanism and conclude in favor of a process of oxidation by diffusion [CHO 10]. It seems, however, that the heating rates used by these scientists are too slow (<10^6 K/s) to induce MDM. In addition, the masses of nanothermites tested are probably too low for MDM to activate. The research conducted by Chowdhury *et al.* does however have the merit of highlighting a particularly interesting point. Presumably oxidation by diffusion occurs during the ignition of nanothermites but MDM takes over as soon as the temperature increase produced by the combustion of the composition reaches the necessary threshold for its activation (see Figure 3.7). The ignition of nanostructured aluminothermic compositions must be sufficiently powerful to enable MDM, without first passing through a phase of slow combustion and then acceleration. One technique for achieving both quickly and over shorter distances the necessary "cruising speed" for the combustion front of nanothermites consists of using nanostructured compositions of powdered copper oxide and aluminum as ignition materials for the main aluminothermic reaction.

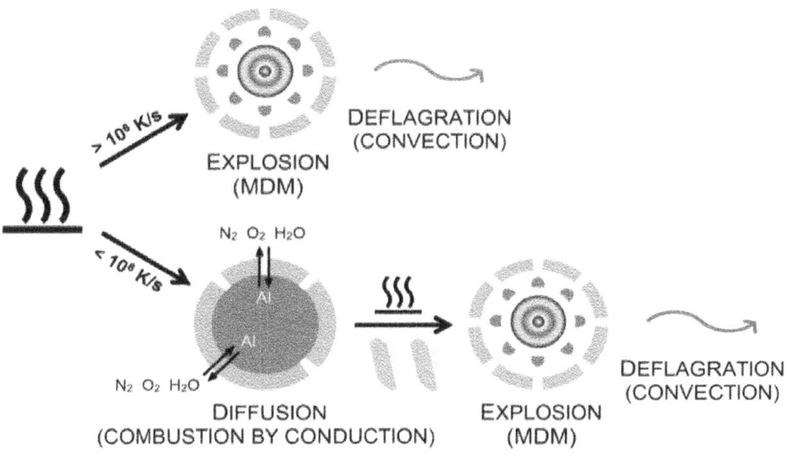

Figure 3.7. *Diagram of the direct or indirect activation of MDM in nanostructured aluminothermic compositions*

As we have seen, the explosion of aluminum nanoparticles is one of the reasons for the high reactivity of nanostructured aluminothermic compositions. Levitas suggests that the molten metal explosion and projection mechanism that he proposes for aluminum could also be applied to titanium and magnesium [LEV 07]. As it stands, there are few experimental studies on nanothermites made from other nanostructured fuels than aluminum. However, Fischer and Grubelich have identified a series of fuels that are of interest for nanothermites, considering thermodynamic parameters such as temperature and heat of the reaction [FIS 96]. Among these substances, there are a certain number of metals (Al, Be, Hf, La, Li, Mg, Nd, Ta, Th, Ti, Y, Zr) as well as two metalloids (B, Si). To this series, we should also add calcium (Ca), having been used in the past by Goldschmidt to manufacture some of the first thermite compositions; ytterbium (Yb) whose pyrotechnic behavior is similar – according to Koch *et al.* – to that of magnesium [KOC 12] and zinc (Zn). Their ability to react by explosion is evaluated in a semi-quantitative manner in terms of their specific properties and those of the overlying oxide (see Table 3.2): melting temperature (*Mp*) and boiling temperature (*Bp*); coefficients of linear expansion of solids ($\alpha_{Sol.}$), variation of metal density, accompanying the transition from the solid state ($d_{Sol.}$) to the liquid state ($d_{Liq.}$). The corresponding variation in volume ($\Delta V_{Sol.\rightarrow Liq.}$) is calculated according to equation [3.7]:

$$\Delta V_{Sol.\rightarrow Liq.} = 100 \times \left(\frac{d_{Sol.}}{d_{Liq.}} - 1 \right) \qquad [3.7]$$

For a metal or metalloid nanoparticle to explode, the matter which composes its core must remain strongly confined by oxide until it melts. This, for example, is not the case for boron or tantalum, whose fusion occurs at a much higher temperature than that of their respective oxides (see Table 3.2). The explosion of nanoparticles from these elements is therefore impossible. This situation is obviously not the most common because the metal almost always melts at a lower temperature than the melting temperature of the covering oxide layer. The Tamman's temperature, calculated in Kelvin according to equation [3.8], provides valuable information to assess whether the explosion may or may not occur:

$$T_{Tamman}(K) = \frac{Mp(°C) + 273}{2} \qquad [3.8]$$

This quantity represents, from a physical point of view, the temperature from which the molecules that make up a solid become mobile. The oxide, therefore, maintains its mechanical properties all the better when the gap between the metal melting temperature and the Tamman's temperature is significant. The ratio (Γ_K) of the Tamman's temperature of the oxide to the melting temperature of the metal, expressed in kelvin, provides an interesting measure of this gap, calculated according to equation [3.9]:

$$\Gamma_K = \frac{T_{Tamman}\ \text{(Oxide)}}{Mp\ \text{(Metal)}} \qquad [3.9]$$

When $\Gamma_K > 1$, the mechanical properties of the oxide layer are not degraded by heating, and the oxide effectively confines the metal until it melts. This situation is the most favorable to the explosion of the metallic nanoparticle. Aluminum, calcium, lithium, magnesium, ytterbium and zinc are examples of nanoparticles that can react by explosion. Conversely, when $\Gamma_K < 1$, the resistance of the oxide layer is altered by the heat. These circumstances are unfavorable to the sudden bursting of the metallic nanoparticle. So although we cannot completely exclude the possibility that nanoparticles of titanium, zirconium, or hafnium may explode, there is no doubt that the effect is much less strong than for metals for which the (Γ_K) ratio is significantly higher than 1.

	Mp (°C)	**Bp (°C)**	**T_{Tamman} (K)**	**$\alpha_{Sol.}$ (ppm.K^{-1})**	**$d_{Sol.}$ - $d_{Liq.}$(g/cm^3) $\Delta V_{Sol.\rightarrow Liq.}$ (%)**	**Explosion (MDM) Criteria Γ_K**
Al	660.32	2519	193.7	23.1-37.4 (20-627°C)	2.70-2.375 (+ 13.7)	Yes
α-Al_2O_3	2053	≈3000	890	5.4-11.0 (20-1627°C)	–	(1.25)
B	2075	4000	901	5[(a)] (20-200°C)	2.34-2.08 (+ 12.5)	No
B_2O_3	450	–	88.5	–	–	(0.15)
Be	1287	2471	507	11.3-23.7 (20-1227°C)	1.85-1.690 (+ 9.5)	Uncertain
BeO	2577	–	1152	6.3-15.1 (20-2027°C)	–	(0.91)
Ca	842	1484	284.5	22.3-24.5 (20-352°C)	1.54-1.378 (+ 11.8)	Yes
CaO	2898	–	1312.5	11.2-19.3 (20-2127°C)	–	(1.42)
Hf	2233	4603	980	5.9-8.4 (20-1027°C)	13.3-12 (+ 10.8)	Uncertain
HfO_2	2774	–	1250.5	3.8-11.2 (20-1700°C)	–	(0.61)
La	918	3464	322.5	5.2-11.3 (20-727°C)	6.15-5.94 (+ 3.5)	Uncertain
La_2O_3	2304	3620	1015.5	10.8-16.3 (20-927°C)	–	(1.08)
Li	180.5	1342	-46.3	46-56 (20-177°C)	0.534-0.512 (+ 4.3)	Yes
Li_2O	1570	–	648.5	33.6±0.8[(b)] (25-1000°C)	–	(2.03)

Mg	650	1090	188.5	24.8-37.6 (20-627°C)	1.74-1.584 (+ 9.8)	Yes
MgO	2852	3600	1289.5	10.5-16.7 (20-1427°C)	–	(1.69)
Nd	1021	3074	374	6.9-11.0 (20-727°C)	7.01-6.89 (+1.7)	No (0.97)
Nd_2O_3	2233	3760	980	11.0-15.3 (20-1127°C)		
Si	1414	3265	570.5	2.57-4.33[(c)] (20-727°C)	2.329-2.57 (- 9.4)	No
SiO_2 (vitreous)	1713	2950	720	0.49-0.37 (20-727°C)	–	(0.59)
Ta	3007	5458	1367	6.3-24,4 (20-2927°C)	16.4-15 (+ 9.3)	No
Ta_2O_5	1784	–	755.5	–	–	(0.31)
Th	1750	4788	738.5	11.0-14.7 (20-1027°C)	11.7-10.35[(d)] (+ 13)	Uncertain
ThO_2	3390	4400	1558.5	7.7-12.1 (20-1727°C)	–	(0.91)
Ti	1670	3287	698.5	8.6-13.5 (20-1327°C)	4.506-4.11 (+ 9.6)	Uncertain
TiO_2	1843	–	785	7.5-11.0 (20-1127°C)	–	(0.54)
Y	1522	3345	624.5	11.3-12.8 (20-927°C)	4.47-4.24 (+ 5.4)	No
Y_2O_3	2438	–	1082.5	7.3-11.1 (20-1727°C)	–	(0.76)
Yb	819	1196	546	25.1-34.2 (20-727°C)	6.90-6.21 (+11.1)	Yes
Yb_2O_3	2355	4070	1314	6.1-9.7 (20-1277°C)	–	(1.20)
Zn	419.53	907	73.3	30.2-34.0 (20-417°C)	7.14-6.57 (+ 8.7)	Yes
ZnO	1974	–	850.5	4.3-8.0 (20-1227°C)	–	(1.62)
Zr	1854	4409	790.5	5.7-11.3 (20-1527°C)	6.52-5.8 (+ 12.4)	Uncertain
ZrO_2	2709	–	1218	8.8-11.6 (20-1127°C)	–	(0.70)

Table 3.2. *Evaluation of the ability to react by MDM of certain fuels of interest in the preparation of nanothermites; sources: temperature of phase change Mp, Bp and densities $d_{Liq.}$, $d_{Sol.}$ [LID 05]; linear expansion coefficients ($\alpha_{Sol.}$) for the metals [TOU 75] and the oxides [TOU 77]; point values from [a][KIR 08], [b][KUR 82], [c][WAT 04], [d][KIR 63]*

According to Rosenband, the differences between the coefficients of linear expansion of metal and solid ($\alpha_{Sol.}$) and the variation of volume accompanying the transitions or the metal phase transformations are responsible for the breakdown of the oxide film [ROS 04]. Certain metals such as aluminum or zinc undergo a much greater expansion than their oxides, which can induce a cracking of the oxide layer occurring before the melting of the metal, if the heating of the nanoparticle is too slow. For other metals such as titanium, zirconium or hafnium, whose expansion coefficients are very close to those of their oxides, the heating produces more moderate mechanical stress in the oxide shell. The latter probably maintains its integrity up to the melting of the metal, despite the variation of ($\alpha_{Sol.}$) accompanying the allotropic transitions affecting titanium (883°C) and zirconium (864°C). With the notable exception of silicon, which contracts when melting, all fuels undergo a marked expansion when they liquefy. The low gains in volume observed for lanthanum and neodymium are not favorable for a reaction mechanism by explosion of these metals.

3.2.1.3. *Particular forms of aluminum*

Eapen *et al.* studied the combustion of aerosolized aluminum particles in air in a reactor with a constant volume. Using this system, they were able to compare the combustion of spherical particles of micrometric size (2–5 µm) with those of aluminum flakes with a nanometric thickness (20–200 nm). These experiments showed that despite their lower surface area, through burning in air the spherical aluminum particles produced a higher rate of pressurization than the flakes, and that the fraction of oxidized metal was greater in the first case than in the second. The slower and less complete combustion of aluminum flakes is attributed by Eapen *et al.* to their aggregation in the air vortex, which serves to disperse them [EAP 04]. The behavior of the aluminum flakes in a nanothermite is probably quite different, since they are then separated from each other by the layers of oxide nanoparticles that cover their surface.

Wang *et al.* prepared porous aluminum microspheres (2–16 µm) by assembling nanoparticles of the metal (~50 nm), using nitrocellulose (1–10 wt%) as a binder, using the electrospraying process described in Chapter 2. Increasing the nitrocellulose content in the pseudo-micrometric aluminum has the dual effect of shortening the ignition delay and extending its combustion duration. It is worth noting that aluminum powder, composed of massive particles of comparable size (3.0–4.5 µm) to nanometric aluminum "cookies", does not react when tested under the same experimental conditions. The research undertaken by Wang *et al.* is a good example of initially structuring the subject on the long road from nanoparticle to "customized" pyrotechnic object [WAN 13a].

Young *et al.* studied the hydrogen release under vacuum and the combustion under different atmospheres (air, CO_2, Ar/O_2) of aluminum hydride (α-AlH_3), subjected to rapid heating (7×10^4–6×10^5 K/s). They showed that the temperatures of dehydrogenation and combustion are higher when the hydride is quickly heated. Dehydrogenation is produced between 650 and 1,200 K and combustion – which is hardly dependent on the composition of the oxidizing atmosphere – between 900 and 1,500 K [YOU 10]. Thereafter, Young *et al.* have used this aluminum hydride (α-AlH_3) of micrometric particle size (< 25 µm) as an additive to the nanometric aluminum (~50 nm) in nanothermites based on copper oxide (CuO, <50 nm), iron oxide (Fe_2O_3, <50 nm) and bismuth oxide (90–120 nm). The effect of adding hydride on the rate of pressurization in a pressure vessel has been studied based on the molar content of this substance in the total aluminum. The performance of the CuO/Al nanothermite is enhanced when the aluminum contains less than 30 mol% of hydride. For the Bi_2O_3/Al composition, the performance is hardly dependent on the addition of hydride, as long as the proportion of this compound in the aluminum remains lower than 50 mol%. Finally, increasing the hydride content of the aluminum alters the reactivity of the Fe_2O_3/Al nanothermite. In the three cases, high levels of hydride (40–60 mol%) cause a fall in the rate of pressurization. Young *et al.* attribute this inhibition to the lowering of the adiabatic temperature of the nanothermite reaction, induced by the decomposition of hydride. The reaction environment is no longer hot enough for the reaction to take place homogeneously in the gas phase. The transition to a heterogeneous combustion mode, involving condensed phases, explains the collapse of the reactivity. In conclusion, the incorporation of hydride, in small proportions, in the nanometric aluminum used as fuel in the nanothermites could, under certain conditions, improve their reactivity [YOU 14]. For the hydrogen released by the thermolysis of hydride to effectively increase the rate of pressurization, the aluminum hydride should be added in the form of nanoparticles to nanothermites with the highest possible reaction heats. Finally, it should be noted that hydrogen is the ideal substance for convection-based transport of energy and, therefore, for combustion propagation. Due to its small size, hydrogen can easily infiltrate into any available pore spaces. This gas also has higher thermal conductivity and heat capacity than those of other gases supplying the combustion or produced from it.

Noor *et al.* studied the morphology and oxidation in air of aluminum particles alloyed with copper (~211 nm), produced by electrical explosion of wires (EEW). The properties of these particles were compared to those of aluminum particles (~150 nm), synthesized by the same method. The alloy nanoparticles contain about 56 wt% aluminum and 38 wt% copper, mainly as $AlCu_2$, but also with traces of Cu_9Al_4. The oxide layer covering the alloy (~3–4 nm) is thinner than that covering the pure aluminum nanoparticles (~4–5 nm). The study of the oxidation by thermal analysis shows that the alloy melts and is oxidized at a lower temperature than the pure metal [NOO 15].

3.2.2. *Other fuels of interest*

In a review article, Yetter *et al.* examined various aspects of the combustion of metallic particles usable as fuels, and of their integration in nanometric form in different energetic materials [YET 09].

For a substance to be usable as a fuel in nanothermites, it must first exhibit a very strong affinity for oxygen; in other words, it must have marked reducing properties. The most oxophilic elements are mainly located towards the left side of the periodic table. Many oxidizable elements cannot be used as fuels due to the impossibility of passivating them by an oxide layer that is stable over time. Finally, the toxicity or the radioactivity of some elements, such as beryllium or thorium respectively, inevitably prohibits their use. Fuels that exist in the nanometric state often have spherical morphology and, more rarely, platelet morphology.

3.2.2.1. *Nanometric boron*

Boron is the non-metallic solid element with the smallest molecular weight in the periodic table. Its very high oxidation enthalpy (~58 kJ/g) and its excellent physical and chemical stability make it, in theory, an excellent choice of reducing agent for nanothermites. Due to its low molecular weight, the mass proportion of boron required to reduce the metal oxide is quite low. As such, the theoretical maximum density (TMD) of borothermic compositions in which the metallic oxide accounts for most of the mass is relatively high. Despite these intrinsic qualities, boron is less suitable as a fuel than aluminum in pyrotechnic compositions, because of its refractory nature. The melting temperature (2,075°C) and the evaporation point (4,000°C) of boron are very high, and the quantities of heat required to operate these phase changes are significant (50.2 and 480 kJ/g).

Sullivan *et al.* have shown, through tests carried out in a pressure cell, that the addition of nanometric boron (62 nm) to a CuO/Al nanothermite increased the rate of pressurization produced by combustion, and shortened the ignition delay time. However, this performance improvement is only observed when the molar boron content of the fuel (Al + B) remains below 50%. The addition of boron in larger proportions alters the performance. The beneficial effect of boron on the reactivity is not observed with a coarser powder (700 nm). In order to explain these experimental results, Sullivan *et al.* have developed a heat transfer model in a boron particle exposed to a high temperature (>2,350 K). The temperature produced by the aluminothermic composition must be high enough to not only evaporate the oxide (B_2O_3) but also to melt the boron. In addition, these phenomena must occur in a shorter time scale than the aluminothermic reaction, so that the molten boron can react with the cupric oxide. The temperature required for boron with a diameter of 62 nm or 700 nm to participate in the reaction is, respectively, equal to 2,370 K and 2,800 K [SUL 09]. According to this model, nanometric boron could be used to enhance the performances of compositions with high reaction heats, such as sulfate-based nanothermites [COM 15a], by increasing the proportion of gas in the products of combustion.

Martirosyan *et al.* noted that the addition of nanometric boron (~50 nm) in low mass ratios (2.53 wt%) to a MoO_3/Al nanothermite dramatically increased the maximum pressure produced by combustion, from 0.48 to 1.3 MPa [MAR 09c].

Comet *et al.* compared the ejection velocity and reactive powers of CuO/B and Bi_2O_3/B nanothermites. In order to do this, they are arranged in loose powder form in small cups (10–15 mm^3), in which they are ignited by an electrostatic discharge. Observation of the combustion by a high-speed video camera helps to determine the reaction time, which corresponds to the time taken for the bulk of the mass to react and get ejected. The ejection velocity is then calculated by dividing the reacted mass by the reaction time. The energy released by the reaction of the compositions is determined by multiplying their mass by their specific heat of combustion. The reactive power is obtained by dividing the energy by the reaction time. The study of the evolution of the ejection speed and reactive power in terms of the boron content of the compositions showed that the Bi_2O_3/B nanothermite had the best performance. Comet *et al.* also used this method to show that copper and bismuth oxides reacted much more strongly with aluminum than with boron [COM 14b].

3.2.2.2. *Nanometric zinc*

Zinc melts and evaporates at a particularly low temperature for a transition metal (see Table 3.2). Although it is theoretically less reducing than aluminum, zinc is more easily oxidized in solution because the covering oxide layer (ZnO) is less protective than alumina. Zinc is therefore very sensitive to its chemical environment. The combination of zinc with substances whose dissolution produces species with acidic, alkaline, oxidizing or complexing natures produces unstable compositions. For instance, a mixture of zinc with ammonium nitrate and sodium or ammonium chloride ignites in contact with water or even atmospheric moisture. The precarious stability of compositions containing zinc explains why this metal has not been used as a nanostructured fuel in nanothermites.

Mixed with sulfur, or with oxidizing salts such as potassium permanganate or chlorate, micron-sized zinc powder provides combustible compositions. The combustion of zinc is characterized by the abundant emission of incandescent fumes, which are initially yellow before fading to white. This behavior is due to the oxidation of the metal in vapor form and the thermochromism of zinc oxide.

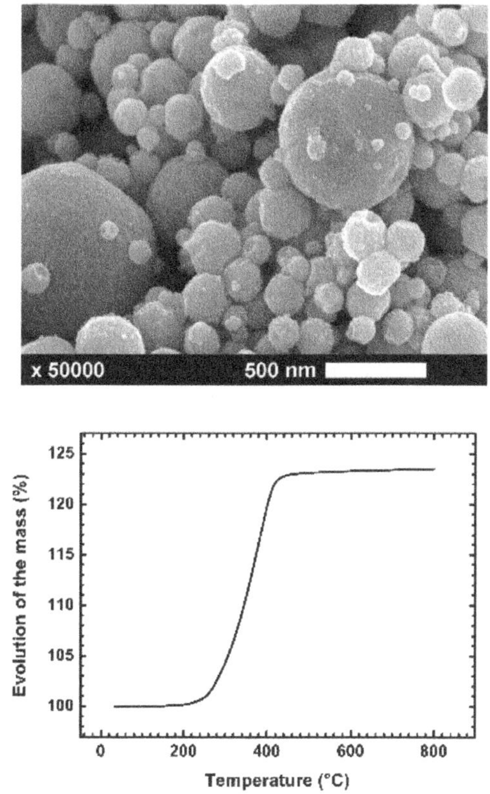

Figure 3.8. *Electron microscope observation of submicron-sized particles of zinc; typical curve obtained by thermogravimetric analysis (TGA) of the progressive oxidation of zinc powder*

Submicron-sized zinc powders, such as those with the morphological and thermal characteristics shown in Figure 3.8, are commercially available. These particles generally have a spherical shape, but their surface is faceted. Their size distribution is quite wide, ranging from 50 nm to 1 μm. These particles are not strongly aggregated; they fit together in clusters with cohesion ensured by surface forces. The clusters of particles dissociate easily in liquid by applying ultrasonic treatment. The metal surface is covered with a granular layer formed by the recrystallization of the oxide [MOT 40]. TGA shows that the studied zinc powder binds a quantity of oxygen representing 23.5% of its original mass, which corresponds to a high mass content of metal (96 wt%). The zinc binds a much lower mass of oxygen than aluminum and nanometric boron. In other words, the mass of fuel required to reduce the oxidizer is relatively high, representing one-third to two-thirds of the mass of the composition.

3.2.2.3. *Nanometric magnesium*

Magnesium is a highly reducing metal of low density (1.74 g/cm^3) whose melting point (650°C) is close to that of aluminum (660°C), but which has a very low boiling point (1,090°C). This latter feature is very favorable for increasing the reactivity of magnesiothermic compositions, because the vaporization of metal would increase the initial pressurization and promote combustion with gaseous oxygen emitted by the decomposition of the oxidant. The main weakness of magnesium is its high oxidability in powdered form. Zhou *et al.* have shown, for example, that the surface of magnesium nanorods oxidizes very significantly through simple contact with the atmosphere, and that after 1 month of conservation in a drying cabinet, the proportion of Mg/O becomes equal to the stoichiometry of the oxide [ZHO 13]. Magnesium reacts with all species containing oxygen, including carbon dioxide. In other words, finely divided magnesium powders are not stable and can be ignited by the self-heating produced by the spontaneous oxidation.

Ricceri *et al.* have used Mg/B_2O_3 and Mg/WO_3 magnesiothermic compositions to produce, through a mechanosynthesis process, submicron-sized particles of boron (80–400 nm) [RIC 03a] and tungsten polycrystalline aggregates (70–100 nm) [RIC 03b]. In these compositions, magnesium is included in the form of micron-sized powder (125–350 µm). The analysis of the ground composition at various points in time shows that in both cases, after a short induction period, the reaction is produced almost instantaneously.

Zhou *et al.* stabilized magnesium nanorods deposited on a silicon substrate by coating them with cupric oxide. Analysis by DSC showed that the reaction of the Mg/CuO composition provides a substantial amount of heat (~3.4 kJ/g) and that it occurs without the magnesium melting. The reaction produces magnesium oxide (MgO), copper and cuprite (Cu_2O) [ZHO 13]. Coating of magnesium rods by PTFE makes them insensitive to water. The angle of contact of a drop of water deposited onto a matrix of Mg@PTFE rods in some cases reaches 162°, a characteristic value of superhydrophobic deposits. Thermal analysis showed that the Mg@PTFE compositions produced a very high reaction heat (6.35–8.65 kJ/g) and that their onset temperature was particularly low (~270°C). The flame propagation velocity of an Mg@PTFE composition has been measured in a capillary 10 cm in length, filled only with a short composition column (1.8 cm). The average flame propagation velocity is between 450 and 600 m/s in the composition, and reaches 1,500 m/s thereafter in the unfilled part of the tube [ZHO 14a].

3.2.2.4. *Nanometric titanium*

Titanium is a highly reactive metal which, when in contact with air, is covered with a protective oxide layer, which makes the metal less sensitive to its chemical environment once passivated [MIL 12]. According to Billy, the chemical properties of titanium are often comparable to those of silicon. The metal is violently attacked at high temperature by permanganate, copper oxide and lead dioxide, and more moderately by potassium chlorate, potassium nitrate and sodium carbonate [BIL 32]. These characteristics suggest interesting pyrotechnic properties for nanostructured titanothermic compositions.

Titanium exists in the form of a commercially available powder composed of spherical particles with diameters between 30 and 300 nm (see Figure 3.9). The study of the slow oxidation of these particles in a mixture of argon (98% vol) and oxygen (2% vol) shows that oxidation starts at a very low temperature (100–200°C), and is not completed until above 1,000°C. The mass of oxygen fixed by the titanium nanoparticles (48.5 wt%) is lower than it should be (66.8 wt%), which indicates that the sample only contains 72.6 wt% of metallic titanium. The final product of oxidation is of a very light orange color, and when characterized by XRD, only contains titanium oxide (TiO_2), in the form of rutile.

3.2.2.5. *Red phosphorus*

The use of red phosphorus to reduce metallic oxides and form combustible powders was patented by Hale *et al.* [HAL 49]. Although these authors did not specify the size of the oxides used, the combustion velocities they mention typically correspond to compositions formed from micrometric particles. Busky *et al.* patented a method of forming metal hydroxides (0.1-2 wt%) by precipitation from metal salt solutions on the surface of red phosphorus particles. This method makes red phosphorus less sensitive to oxidation by water vapor and oxygen, and consequently reduces the amount of phosphine (PH_3) and phosphoric acid formed during storage [BUS 10]. The use of metallic oxide nanoparticles to formulate thermites based on red phosphorus was proposed by Comet *et al.* [COM 10a]. Red phosphorus is an ideal fuel, which reacts easily with oxygen even when found in stable salts like calcium sulfate. According to Davies, the energy released by the phosphorus oxidation reaction is between 18.2 kJ/g and 24.3 kJ/g, depending on the amount of oxygen present, resulting in the formation of P_4O_6 or P_4O_{10}, respectively [DAV 99]. The mixture of strongly oxidizing salts with phosphorus systematically leads to explosive compositions.

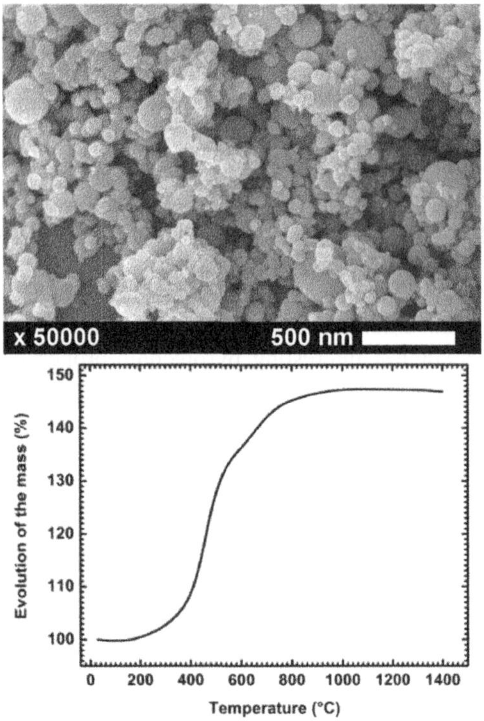

Figure 3.9. *Observation by SEM of a titanium powder containing submicron-sized and nanometric particles; a typical curve obtained by TGA of the progressive oxidation of titanium powder*

The weaknesses of red phosphorus are its high flammability ($T > 260°C$) and the production of phosphine (PH_3), which is very toxic when in contact with moisture. The use of phosphorus in smoke compositions [KOC 08] for military use led to the development of stabilized red phosphorus powder, in order to limit the emission of phosphine.

The conversion reaction of white phosphorus into red phosphorus is exothermic (0.56 kJ/g): it is carried out in a cell at a temperature between 220 and 280°C. Red phosphorus is obtained in solid form, which requires grinding to reduce it to micrometric powder. This process explains the angular morphology of commercial red phosphorus particles (see Figure 3.10).

Red phosphorus does not exist in the form of nanoparticles, because it cannot be dissolved in a liquid nor developed by conventional methods used to structure materials. A preliminary study helped to classify several compositions formed by mixing metallic oxides and red phosphorus, according to their apparent reactivity:

$$NiO / P < Fe_2O_3 / P < Bi_2O_3 / P < CuO / P < PbO_2 / P \text{ (explosion)}$$

Figure 3.10. *Observation by electron microscope, a) of the morphology of red phosphorus particles; and b) of thermites formed by mixing these particles (15.7 wt%) with a nanometric CuO powder [COM 10b]*

The reaction mechanism of n-CuO/P thermites has been studied by calcining a mass of 0.7 g of material in a tubular furnace. The samples were subjected to slow heating (2 K/min), applied from room temperature to 500°C. The experiment is carried out in an inert atmosphere (Ar, 300 mL/min); the argon flow is maintained during cooling in order to avoid oxidation of the reaction products after combustion.

Calcining a thermite containing 15.7 wt% of red phosphorus is not accompanied by any mass variation in the sample, which shows that the phosphorus does not sublime and that, furthermore, the reaction produces no phosphorus anhydride. XRD analysis of the calcined product shows that it is composed of a mixture of copper and copper salts: phosphide (Cu_3P), metaphosphate ($Cu(PO_3)_2$) and pyrophosphate ($Cu_2(P_2O_7)$). These results enable us to write a balanced reaction assessment of 13.0 wt% phosphorus that seems in fairly good agreement with the proportions from the experiment of the original composition (15.7 wt%):

$$13\ CuO + 5\ P \rightarrow 7\ Cu + Cu_3P + Cu(PO_3)_2 + Cu_2(P_2O_7)$$

The calcination of a CuO/P thermite with a higher phosphorous content (30 wt%) is accompanied by a loss of mass of 82 wt%, corresponding to the ejection of reaction products onto the walls of the quartz tube (see Figure 3.11) and the deposit of a pale red film with a shade of yellow accumulating in the coldest area, at the end of the tube, located downstream of the alumina crucible. The phase splashed onto the tube wall above the crucible strongly adheres to it; it is composed of metallic copper and copper combined as a phosphide (Cu_3P) and cyclotetraphosphate ($Cu_2P_4O_{12}$). The sublimate deposit, whose hydrolysis makes an acidic solution, essentially consists of phosphorous anhydride. Its specific coloring is attributed to the presence of excess phosphorus in the calcined composition.

Figure 3.11. *Image of the products projected during the calcination of a CuO/P thermite containing 30 wt% of red phosphorus*

Combustion residues have a spherical morphology, regardless of the observation scale. Copper is first formed as nanosized droplets, assembled into micron-sized particles that arrange themselves into macroscopic spheres when the n-CuO/P composition is compressed [COM 10b].

In light of the results obtained, the n-CuO/P reaction mechanism can be described as follows:

1) The heating of the mixture triggers the sublimation of phosphorus, whose vapors cross the layer of copper oxide nanoparticles (see Figure 3.10). The subsequent reduction of CuO generates nanosized copper droplets and vapors of phosphoric anhydride (P_2O_5).

2) Oxygenated copper salts are then formed by the reaction between the phosphoric anhydride (acid) and the copper oxide (base) according to the following reactions:

$$2CuO + P_2O_5 \rightarrow Cu_2(P_2O_7)$$

$$CuO + P_2O_5 \rightarrow Cu(PO_3)_2$$

$$2Cu(PO_3)_2 \rightarrow Cu_2(P_4O_{12})$$

The selectivity of the reaction is apparently determined by its kinetics: copper pyrophosphate results from slow decomposition, while the cyclotetraphosphate is formed by vivid combustion.

3) Copper phosphide, which is unstable at high temperature, results from the reaction of phosphorus vapors with metallic copper droplets as they cool:

$$3Cu + P \rightarrow Cu_3P$$

Given that its melting temperature ($T_F \approx 900°C$) is below that of copper, phosphide probably forms in liquid state on the surface of copper droplets that solidify first. Since the density of Cu_3P is lower than that of copper, phosphide remains at the surface of copper droplets. The coating of copper nanoparticles with phosphor bronze explains the fact that they aggregate without coalescing and preserve their spherical form. In contrast to most phosphides (e.g. Ca, Al) whose hydrolysis releases phosphine, the copper phosphide exhibits good stability in water, since it is used for the treatment of copper surfaces exposed to corrosion in marine environments. This is an important characteristic for the development of pyrotechnic systems whose operation does not generate toxic substances.

The vivacity of combusting n-CuO/P thermites is reduced by adding nanostructured iron oxide (n-Fe_2O_3). By varying the composition of n-CuO/n-Fe_2O_3/P ternary nanothermites, it is possible to modulate the combustion propagation velocity, and also the threshold of sensitivity to various forms of stress [COM 11b].

The high flammability of red phosphorus-based thermites can be explained by the vaporization of phosphorus, which then readily reacts with surrounding oxygenated species. It may be that the presence of air from the surrounding atmosphere or contained within the internal porosity of the material plays a significant role in the ignition of these materials. By comparison, the break-up of the core–shell structure of aluminum nanoparticles requires much higher levels of activation temperature and initiation power.

3.2.2.6. *Other fuels*

In a review, Zhou *et al.* point out that porous silicon has been used as nanosized fuel in compositions based on perchlorates of alkaline or alkaline-earth metals [ZHO 14b]. On-chip tests have, for example, indicated that the combustion of porous silicon (840 m^2/g) impregnated with sodium perchlorate had a propagation velocity of 3,050 m/s [BEC 11]. In a review where they quote the research of Laucht *et al.* [LAU 04], Rossi *et al.* report particularly high values of the heat of reaction, measured in bomb calorimeter, for $Na(ClO_4)/Si$ (9.2 kJ/g) and $Ca(ClO_4)_2/Si$ (8.6 kJ/g) compositions [ROS 07]. These energetic compositions cannot be considered nanothermites in the strict sense of the term, since the metallic elements they contain are only present in very feeble proportions, and do not play an active role in their pyrotechnic reaction. Therefore, they will not be explored in further depth in this book.

Koch *et al.* have proposed the use of rare earth particles whose reaction with oxygen-based oxidizers or halocarbons generates bright flames. The metals mentioned by Koch *et al.*, i.e. europium, samarium, thulium and ytterbium, meet the Glassman's criterion, meaning that their boiling temperature is below the melting temperature of their oxide. From this point of view, these metals are comparable to magnesium [KOC 11]. Moreover, these metal fuels have quite high densities and good oxidizability, which makes them an interesting choice for the formulation of nanothermites. However, their use for this purpose will be conditional upon the development of techniques that allow for their efficient stabilization in a divided state by protecting them against oxidation. Furthermore, due to the rarity of these elements, their use will most certainly be limited to very specific applications.

3.3. Oxidizers of interest for nanothermites

The oxidizing substances that are part of the composition of nanothermites are often metal oxides, and less often metalloid oxides. Current research is increasingly focused on oxygenated metal salts, which present the advantage of being more reactive than the oxides, but also the drawback of being less stable. The oxidizing phases are used in the form of submicron-sized (<1 μm) or nanosized (<100 nm) particles. The basic principle of all these reactions is the transfer of oxygen from the oxidizer to the fuel. While the exchange of other elements, such as sulfur or fluorine, is theoretically possible, metal sulfides and fluorides have apparently not yet been used in the form of nanostructured oxidizers in nanothermites.

3.3.1. *Metal or metalloid oxides*

Inorganic chemistry provides the pyrotechnician with a wide variety of metal oxides. Which among these would enable the preparation of nanothermites? To answer this question, let us consider the periodic table of elements. Highly reducing metals, such as alkaline or alkaline earth elements, are very easily combined with oxygen. Consequently, their oxides are very stable and cannot be used to prepare thermites. The same applies to all oxides containing a metal that could serve as fuel in thermites (Al_2O_3, TiO_2, ZrO_2, HfO_2, Ta_2O_5, Nb_2O_5, ZnO). Finally, the oxides of choice for the formulation of nanothermites are those of metals whose electronegativity on the Pauling scale is between 1.5 and 2.5. Fischer *et al.* have determined by calculation a ranking of energetic compositions based on metal oxides, depending on their thermochemical characteristics [FIS 96].

The metal oxides that are currently the most frequently used to prepare nanothermites are bismuth oxide (Bi_2O_3), copper oxide (CuO), iron oxide (Fe_2O_3), molybdenum oxide (MoO_3) and tungsten oxide (WO_3). The general properties of several aluminothermic compositions prepared from these oxides shall be reported in the following.

Weismiller *et al.* studied the influence of the oxide and aluminum granulometry on the reactivity of two compositions, Al/CuO and Al/MoO_3. This research indicates that the compositions in which the two phases are nanostructured have higher combustion rates. In the combined mixtures that are formed by associating micron-sized and nanosized particles, the nanostructuration of oxide has a more significant effect than that of aluminum. According to Weismiller *et al.* this result can be explained by the fact that nanosized aluminum contains a significant quantity of alumina, which leads to a dilution of the composition and a deterioration of its performances. On the other hand, a reduction in the size of oxide particles facilitates the generation of gas by oxide sublimation (in the case of MoO_3) or decomposition (in the case of CuO). The diffusional distances of oxidizing gaseous species are also shortened. These effects contribute to stronger pressurization and enhance the propagation of the reaction by the convection of gaseous species [WEI 11a]. Puszynski has compiled the combustion rates for nanostructured aluminothermic compositions based on oxides of bismuth (Bi_2O_3), copper (CuO), molybdenum (MoO_3) and tungsten (WO_3), measured by various authors in unconfined system, depending on the size of aluminum nanoparticles [PUS 11]. This comparison reveals that the most rapidly reacting compositions are obtained with aluminum nanoparticles whose size ranges between 50 and 80 nm.

The pressure cell study conducted by Puszynski *et al.* in relation with the performances of various aluminothermic compositions that were prepared starting from the same type of nanosized aluminum (~40 nm) and from several metal oxides with comparable granulometries allows for a classification of three commonly studied nanothermites depending on the dynamic pressure produced by their reaction: Al/CuO < Al/MoO_3 < Al/Bi_2O_3. The Al/Bi_2O_3 nanothermite also exhibits the shortest ignition delay time [PUS 06]. More recently, Glavier *et al.* studied the evolution of maximum pressure and pressurization rate generated in a pressure cell by the reaction of these compositions, compressed at 10, 30 and 50% of their TMD. The highest maximum pressure and rate of pressurization are obtained with Al/CuO composition and Al/Bi_2O_3 mixture, respectively, compacted at 50% and 30% of their own TMD [GLA 15].

Sanders *et al.* have shown that aluminothermic compositions based on metal oxides (Bi_2O_3, CuO, MoO_3, WO_3) burn more rapidly in a confined environment than in an open system. Moreover, they have observed that tube combustion of loose powders of Al/Bi_2O_3 and Al/MoO_3 nanothermites is faster than that of pellets. On the other hand, the pressure produced by combustion increases when these nanothermites are compacted [SAN 07]. Trebs *et al.* have determined by calculation that the above-described aluminothermic compositions are highly sensitive to humidity, since a water mass content of several percentage points is enough to induce strong modification of the pressure signal generated by their reaction [TRE 10].

Jian *et al.* have noted that the ignition of nanothermites is sometimes concomitant with the release of oxygen by the oxide, while in other cases the reaction is activated before the emission of oxygen (Al/Bi_2O_3, Al/SnO_2), or on the contrary, after the latter (Al/Co_3O_4). In other nanothermites (Al/MoO_3, Al/Sb_2O_3, Al/WO_3), the reaction is initiated without release of oxygen. These results have led Jian *et al.* to the conclusion that nanothermites that ignite before the release of oxygen or in its absence get initiated due to the direct interfacial contact between fuel and oxidizer [JIA 13c].

The following sections offer a description of the properties of nanothermites depending on the nature of the oxide from which they have been prepared.

3.3.1.1. *Silver oxide (Ag_2O)*

Silver oxide exhibits poor thermal stability, since its breakdown into silver and oxygen starts at only 160°C. This thermolysis reaction is autocatalytic, as it speeds up in the presence of silver. Silver oxide attacks carbon monoxide, which releases heat, and strongly absorbs carbon dioxide [OLM 33]. In other terms, the use of silver oxide as an oxidizer in nanothermites is a subject of study that presents interest from a purely chemical point of view. But it seems doubtful that this oxide is ever going to be used in the development of nanothermites that integrate pyrotechnic systems.

Russell *et al.* have studied the reaction of a thermite composed of micron-sized particles of silver oxide (20 μm) and aluminum (40 μm), confined in a tube. The corresponding flame propagation rate is high (531 ± 32 m/s) [RUS 11], despite the coarse granulometry of constituents.

Sullivan *et al.* have measured the reactivity in a pressure cell of nanothermites based on nanostructured silver oxide (Ag_2O). These scientists proceed from the observation that the reaction of a binary composition (Ag_2O/Al) generates a rate of pressurization that is 4,500 lower than that provided, under the same conditions, by a CuO/Al composition. This result is all the more surprising as the thermodynamic properties of these two compositions are comparable [FIS 96]. In situ time of flight mass spectrometry shows that the reaction of the two nanothermites yields a significant quantity of oxygen, and that in the case of the Ag_2O/Al composition this phenomenon takes place at a lower temperature. That has led Sullivan *et al.* to conclude that gas escapes before the reaction develops. On the other hand, the combustion of the CuO/Al composition generates a flame whose adiabatic temperature is higher (2,843 K) than that of the Ag_2O/Al (2,436 K) composition. The fact that this temperature is close to the melting point of the alumina (2,327 K) coating the aluminum nanoparticles may explain its slower and incomplete melting. Solid alumina acts as a barrier that slows down the diffusion of species feeding the reaction. In order to improve the performances of the Ag_2O/Al composition, Sullivan *et al.* have thus prepared ternary nanothermites, by associating silver oxide either to cupric oxide (CuO) or to silver iodate ($AgIO_3$). The rate of pressurization produced by the Al/$AgIO_3$/Ag_2O composition diminishes regularly with the mass proportion of silver oxide contained in the oxidizing phase (0–45 wt%), then plummets beyond it (>45 wt%). The rate of pressurization produced by the Al/CuO/Ag_2O composition remains practically unchanged when the Ag_2O content is between 0 and 70 wt%, then diminishes continuously for higher content values (70–95 wt%) [SUL 13]. This study points out that the mechanism of the aluminothermic reaction of silver oxide is similar to that of copper oxide. The addition of copper oxide with a mass content above 25 wt% in the Ag_2O/CuO fraction enables us to assess the performance of silver oxide.

3.3.1.2. *Bismuth oxide (Bi_2O_3)*

Bismuth oxide gives the most vigorously reacting aluminothermic compositions. This oxide also reacts, though less vividly, with nanosized boron and zinc, as well as with red phosphorus. Bismuth oxide is photosensitive: its lengthy exposure to light induces a change in its color, from lemon yellow to gray. The aluminothermic composition produces a lot of gas, as its adiabatic temperature of reaction (3,253 K) [FIS 96] is higher than the melting temperature of bismuth (1,837 K) and close to that of the alumina (~3,273 K) [LID 05] produced by the reaction.

Wang *et al.* have studied the influence of various parameters on the reaction of Bi_2O_3/Al nanothermites through tests conducted in an instrumented tube and in a pressure cell. They have also shown that the combustion propagation velocity and the maximum pressure generated by the reaction diminish strongly with the increase in size of aluminum particles used for the formulation of the nanothermite (see Table 3.3).

D_{Al} (µm)	P_{max} (MPa)	V_P (m/s)
0.1	12.6	2,500
3	8.9	1,900
20	1.6	1,000
70	0.7	90

Table 3.3. *Maximal pressure (p_{max}) and combustion propagation velocity (V_P) in nanothermites composed of crystallized bismuth oxide (40–50 nm) and aluminum of various granulometries, depending on the diameter (D_{Al}) of aluminum particles [WAN 11]*

The combustion propagation velocity in the composition prepared with nanosized aluminum depends on the duration of the attrition treatment. It reaches a maximum value (~2,500 m/s) for a 7 h duration. An insufficiently short treatment does not mix the phases well, while extended milling causes them to aggregate. In both cases, combustion velocities are strongly lowered. The nature of the atmosphere (air, N_2, vacuum) in which the Bi_2O_3/Al composition reacts has little impact on the maximal pressure produced by combustion, which indicates that the observed pressurization is essentially caused by the vaporization of bismuth. Moreover, the experimentally measured temperature of reaction reaches 2,700°C for the composition containing nanosized aluminum and 1,700°C for the thermites containing micron-sized aluminum. These values are enough to trigger the vaporization of bismuth (1,564°C). Wang *et al.* have also observed that the rate of pressurization increases with a finer granulometry of aluminum. Finally, the maximal pressure produced by the reaction is indirectly correlated with the thickness of the alumina shell coating the aluminum nanoparticles [WAN 11].

Thiruvengadathan *et al.* have prepared Bi_2O_3/Al nanothermites using a method of assembly of nanoparticles with graphene oxide (GO) and have proposed a mechanism that explains this synthesis. In the liquid used for the assembly, composed of a mixture of isopropanol and dimethylformamide, the surface charges of aluminum and bismuth oxide are positive, while the charge of GO is negative. In suspension, long-range attraction manifests between the particles of aluminum and GO, as an effect of their opposite charges. Aluminum nanoparticles then become attached by covalent C-O-Al and O=C-O-Al bonds to the surface of GO platelets. Thiruvengadathan *et al.* have shown that the surface of GO is functionalized by various oxygenated groups: hydroxyls, carboxyls, epoxides, and ketones, but only the first two types of groups react with the hydroxyl functions carried by the alumina shell covering the aluminum nanoparticles. Bismuth oxide, whose surface is free of hydroxyl groups, associates in a non-covalent manner to GO/Al composite particles. The bricks of the GO/Al/Bi_2O_3 composition are finally assembled into macrostructures, either randomly or layer by layer, in an orderly manner. The mode of assembly depends on the flatness of particles, which facilitates the organization of matter. Consequently, the most orderly structures are obtained within the higher size composite particles population [THI 14]. In subsequent research, Thiruvengadathan *et al.* have compared the performances of an Al/Bi_2O_3 nanothermite prepared by physical mixing to those of GO/Al/Bi_2O_3 nanocomposite materials produced by assembly, depending on their GO content (see Table 3.4). The enhancement of performances results from a better homogeneity of the mixture in the assembled compositions [THI 15].

GO (wt%)	P_{max} (MPa)	dP/dt (MPa/µs)	Vp (km/s)	I_S (s)
0	86	5	1.15	44
1	93	6	1.12	46
2	140	13	1.24	48
3.5	176	17	1.54	62
5	214	11	1.26	71

Table 3.4. *Maximal pressure (P_{max}), rate of pressurization (dP/dt), flame front propagation velocity (V_P) and specific impulse (I_S) produced by the reaction of GO/Bi_2O_3/Al nanothermites depending on their graphene oxide content. The corresponding data are taken from the diagrams published in [THI 15]*

Pichot *et al.* have shown that coating the bismuth oxide particles with a nanodiamond layer diminishes the combustion propagation velocity and reduces the heat generated by the aluminothermic reaction (see Table 3.5). The deterioration in performance is explained by the absorption of heat by the diamond. The temperature of reaction is then too low to allow the bismuth formed to gasify. The shift from a reaction controlled by gas convection to a reaction that propagates by conduction explains the fall in performances. Moreover, the carburizing of aluminum in contact with nanodiamond absorbs heat and creates a refractory shield (Al_4C_3) between metal and oxide [PIC 15].

nD (wt%)	0	0.5	1.2	1.8	4.5
FPV (m/s)	492	360	254	140	14
Q_R (J/g)	2,322	2,064	1,916	–	1,815

Table 3.5. *Evolution of the flame propagation velocity (FPV) and heat of reaction (Q_R) of the Bi_2O_3@nD/Al compositions, depending on the mass proportion of nanodiamond (nD) in the coated oxide [PIC 15]*

3.3.1.3. *Cobalt oxide (Co_3O_4)*

Xu *et al.* prepared Co_3O_4/Al nanothermites in the form of filaments with core–shell morphology, at the surface of a silicon substrate. Thermal analyses conducted on samples gathered by surface scratching have shown that the aluminothermic reaction takes place at a relatively low temperature (~530°C) and produces a significant heat release. The differential thermal analysis (DTA) curves exhibit two exothermic signals, between which lies the endothermic signal corresponding to aluminum melting (660°C), whose contributions to the overall heat of the aluminothermic reaction vary with the proportion of metal in the thermite. The reaction of Co_3O_4/Al compositions characterized by thermal analysis yields only cobalt in crystallized form [XU 12]. This result shows that cobalt oxide releases the oxygen it contains more readily than most oxides, whose slow reaction with aluminum often produces mixed compounds.

3.3.1.4. *Chromium oxide (Cr_2O_3)*

Chromium oxide is a refractory ceramic, and is highly stable both physically and chemically. The reduction of this oxide by aluminum is nevertheless possible, but it is one of the rare aluminothermic reactions that do not produce gas [FIS 96].

Comet *et al.* studied the combustion of Cr_2O_3/Al nanothermite pellets, prepared starting from chromium oxide of various morphologies. The first variety of Cr_2O_3 is a porous matrix, with high specific surface area (~44 m^2/g), composed of merged nanoparticles. The second variety of Cr_2O_3 is prepared by detonation: it consists of independent nanoparticles, with lower specific surface area (~20 m^2/g). Under identical pressing conditions, the composition containing porous chromium oxide densifies less than the one containing oxide prepared by detonation. The higher porosity of the first nanothermite facilitates its ignition, but renders its combustion highly irregular. The second, less porous nanothermite, in which aluminum and oxide are better mixed, is less sensitive to ignition and is consumed in a more regular manner (1.45 cm/s) [COM 11c].

Gibot *et al.* studied the influence of the granulometry of chromium oxide on the combustion of pelleted Cr_2O_3/Al compositions. The two sorts of oxide used are a powder formed of nanoparticles (113 m^2/g), and an oxide with coarser granulometry (3.1 m^2/g). Compared to the composition that contains submicron-sized oxide, the composition formulated with the finest oxide exhibits an ignition delay time that is three times shorter and a combustion velocity that is 10 times higher. The reaction rate (340 ± 10 mm/s) remains, nevertheless, low compared to that of other nanothermites (e.g. WO_3/Al, CuO/Al) tested under similar conditions [GIB 11a].

The partial filling of a porous nanostructured matrix of chromium oxide with a secondary explosive such as hexogen (RDX) results in RDX@Cr_2O_3 nanocomposite materials [COM 08b]. By mixing the explosive-loaded oxide with aluminum nanoparticles (~50 nm), gas generating nanothermites are obtained. The combustion of these hybrid materials is driven by a mechanism of synergy between the explosive combustion and the aluminothermic reaction [SPI 10].

3.3.1.5. *Copper oxide (CuO)*

Cupric oxide is one the most reactive oxides in the preparation of nanothermites, but unfortunately it is also one of the most toxic. The melting temperature of this compound appears to be quite poorly defined, since we can find very different values in literature, ranging between 1,064°C [ISA 33] and 1,446°C [LID 05]. The TGA of a fine powder of cupric oxide under a flow of argon shows that it loses around 10% of its initial weight between 750°C and 885°C. The mass loss of the sample accurately corresponds to the one produced by the oxygen release accompanying the transformation of cupric oxide (CuO) into cuprous oxide (Cu_2O). This release of oxygen is one of the causes of the strong reactivity of nanothermites containing cupric oxide.

By using an on-chip characterization method, Apperson *et al.* observed that the energy needed to induce the ignition of nanothermites produced starting from CuO nanorods and aluminum nanoparticles diminished with the mean size of aluminum nanoparticles, from 32 mJ (Al: 120 nm) to 8.8 mJ (Al: 80 nm). The addition of a polymer binder to the CuO/Al (80 nm) composition, in mass proportions of 5% and 10%, induces a change in the ignition energy from 8.8 mJ to 48 mJ and 229 mJ, respectively. The low binder content composition preserves the performances of the pure nanothermite, with a combustion velocity of 1,500 m/s. On the other hand, the combustion rate of the most binder-rich composition is reduced to only 300 m/s [APP 06]. These important results prove that it is possible to raise the threshold of nanothermite sensitivity to a given stress without deteriorating its performances, simply by varying its composition.

Using a specific experimental device, Apperson *et al.* have shown that the reaction of a CuO/Al nanothermite prepared by an assembly method produced faster combustion and pressure waves in comparison with a nanothermite prepared by simple physical mixing. Flame propagation velocities are of the order of kilometers (2,200 ± 300 m/s vs. 1,500 ± 250 m/s). The shock wave propagates more slowly, with sub-kilometric celerities (831 ± 44.4 m/s vs. 766 ± 8.1 m/s). Apperson *et al.* have also shown that the flame propagation velocity in a CuO/Al nanothermite prepared by physical mixing diminishes when it is compacted, while on the contrary, the shock wave celerity and the maximal pressure value increase strongly [APP 07].

Shende *et al.* have studied the influence of various parameters on the combustion wave propagation velocity in CuO/Al nanothermites. The highest velocities are attained with compositions rich in aluminum, whose equivalence ratio, as defined by Granier *et al.*, is equal to 1.6 [GRA 04]. On the other hand, for a given copper oxide, a nanosized aluminum of intermediate granulometry (80 nm) enables to reach the highest reaction rate (1,650 m/s), compared to a finer aluminum (18 nm; 260 m/s), or conversely, to a coarser one (120 nm; 720 m/s). Shende *et al.* also note that the reaction of Al/CuO nanothermites is all the more rapid as the oxide particles have an elongated morphology, in the following ascending ranking: spheres, rods, wires. Finally, the performances of the compositions produced by the assembly of oxide and aluminum are far better than those of the corresponding physical mixtures, in which the nanoparticles of these two bodies are randomly distributed [SHE 08].

By providing the copper oxide with a particular form, Jian *et al.* have prepared CuO/Al nanothermites whose performances surpass those of chemically similar compositions, produced by simple mixing of nanoparticles. For this purpose, these scientists have synthesized spherical and hollow cupric oxide nanoparticles. The outer diameters of the particles composing this oxide with a very particular morphology range between 25 and 400 nm, with an average value of around 85 nm. The central cavity is bounded by an oxide shell formed by the assembly of primary nanoparticles with a diameter of around 10 nm. The CuO/Al composition formulated with hollow CuO nanoparticles generates a rate of pressurization that is three times higher (0.747 MPa/μs) than that of the homologous composition, elaborated starting from full spherical nanoparticles (~50 nm) (0.250 MPa/μs). The analysis of the quantity of oxygen released in time by rapid heating (5×10^5 K/s) of various tested copper oxides shows that gas is always released at around 1,000 K, irrespective of the oxide morphology. On the other hand, the hollow spheres of CuO lose the oxygen they contain far more rapidly than other forms of this oxide. Jian *et al.* consider that the higher "fugacity" of oxygen is due to the small size of elementary crystallites of CuO [JIA 13b]. In reality, besides the fact that it enables the stabilization of CuO nanoparticles of ultimate size, this method also enables the formation of an intraparticle porosity in the oxide (here, ~70%).

Wang *et al.* have used the electrospray method, with nitrocellulose (NC) as a binder, to assemble cupric oxide nanoparticles (~50 nm) and aluminum nanoparticles (~50 nm), in order to form porous micron-sized particles (3–6 μm). The minimal mass content of NC required to produce spherical aggregates is around 5%. Beyond this threshold, there is an increase in the diameter of these microspheres with their NC content. The maximal pressure and the rate of pressurization generated by the combustion of composite microparticles reach a maximum for a mass content of NC of 5%. Then, performances fall rapidly and reach values that are even lower than those for the pure CuO/Al nanothermite when the mass content of NC exceeds 7%. Wang *et al.* have also compared the reaction of microspheres to that of the material with identical composition, elaborated by natural evaporation of the suspension used in the electrospray method. The performance of the material thus produced is not only lower than those of the pure nanothermite, but also diminishes with the increase in mass content of NC. Finally, Wang *et al.* show that the combustion of microspheres yields particles that are far smaller (0.2–1 μm) than those resulting from the reaction of simple composite material (1–10 μm). Based on these results, they conclude that the gas locally produced within the microspheres by the decomposition of NC limits the sintering of nanothermite components, which thus preserve a high surface area during the reaction. When microspheres contain NC in excess, the high pressure of the combustion gas triggers the early dispersion of aluminum and copper oxide nanoparticles, which are then too distant to interact.

Yan *et al.* used the electrospinning method to fabricate combustible nanocomposite mats, made of sub-millimeter diameter fibers (~300–1,000 nm). In this case NC is used as binding agent, at a high mass content (35–100%). Strands of nanocomposite material are produced starting from a solution of pure NC, or from a suspension resulting from the dispersion of nanoparticles in this solution. Fabrics of pure NC, aluminum-loaded NC (NC/Al) and nanothermite-loaded NC (NC/Al-CuO) are thus produced. The nanocomposite materials preserve a fibrous morphology as long as their mass content of nanoparticles does not exceed 50%. Combustion tests have been conducted in the air on strips of mats (3 cm × 1 cm). The average combustion velocity of the fabric made of pure NC is close to 12.4 cm/s. NC/Al nanocomposite fibers are consumed at a lower rate than NC (~4–9 cm/s), since aluminum slows down combustion, probably by a heat sink effect. A similar effect is observed with nanothermite mats, when aluminum and oxide particles are highly diluted in the NC matrix (NC/Al-CuO: 86/14 wt%). The metal and the oxide do not interact, their behavior from a pyrotechnical perspective being that of inert particles. On the other hand, as soon as fuel (Al) and oxidizer (CuO) particles are sufficiently close (NC/Al-CuO: 75/25 wt%), the aluminothermic reaction can take place. The combustion rate increases then very rapidly with the increase in the content of nanothermite, reaching 106 cm/s for a mat with 50 wt% nanothermite [YAN 12]. However, it should be noted that this combustion is very slow compared to those of CuO/Al nanosized compositions in powder form [PUS 11].

Blobaum *et al.* have measured the combustion propagation velocity in a dense foil of CuO_x/Al nanothermite, fabricated by cathode sputtering and composed of elemental layers of submicron thickness. Activated by a flame, the reaction propagates without extinction, despite the reduced thickness of the foil (14 µm), at a rate of 1 m/s [BLO 03].

Taton *et al.* have shown the essential influence of the substrate nature on the ignition delay time and on the combustion of a multilayer CuO/Al nanothermite. For this purpose, the nanothermite is deposited on a resistive titanium filament (E = 300 nm) featuring gold contacts, produced by photolithography at the surface of various substrates: polyethylene terephthalate (PET) membrane, glass sheet and silicon sheet. The polymeric membrane is built by an original photolithographic method. It is composed of a PET film, with a 75 µm thickness, which is glued to the surface of a thinner (E = 25 µm) resin layer, set on resin pillars which are in turn supported by a silicon substrate. This microsystem permits thermal isolation of the nanothermite from silicon. The nanothermite deposit (E = 4.5 µm) is composed of 15 bilayers of aluminum (100 ± 5 nm) and cupric oxide (200 ± 5 nm), deposited by cathode sputtering; it covers an area of 1.6 × 1.5 mm. Taton *et al.* have shown that the current intensity required for triggering the ignition of the nanothermite is 250 mA and 500 mA, respectively, depending on whether it is deposited on the PET membrane or on the glass wafer. For both devices, the ignition delay time has been measured as a function of the applied current intensity. A significant difference between the two values has been observed for low-intensity currents (0.5 A), but this is very rapidly reduced when current intensity increases (1 – 4 A). The light intensity of the sparks generated by the reaction of the nanothermite deposited on the polymeric membrane is much stronger than that obtained for the nanothermite backed by the glass sheet. It is impossible to ignite the nanothermite on the silicon sheet, even when applying a current of 4 A [TAT 13]. These results illustrate the essential role played by substrate heat drainage in reaction initiation and development in thin nanothermite layers. There appears to be an inverse correlation between the performances of the systems studied by Taton *et al.* and the thermal conductivity of the substrates used: silicon (149 $W \cdot m^{-1} \cdot K^{-1}$), Pyrex 7740 (1.1 $W \cdot m^{-1} \cdot K^{-1}$) and PET (0.15–0.24 $W \cdot m^{-1} \cdot K^{-1}$). The deposit of nanothermites on backings with even better heat isolation properties, such as silica aerogels (~0.03 $W \cdot m^{-1} \cdot K^{-1}$), would probably allow enhancement of the performance of this type of microsystem.

Malchi *et al.* have studied the influence of the dilution of a CuO/Al nanothermite with nanosized alumina, added in mass proportions varying from 0 to 20 wt%. The combustion tests carried out in a pressure cell show that the increase in alumina content in the Al/CuO/Al_2O_3 ternary composition increases the time for reaction induction, considerably decreases the rate of pressurization and reduces maximal pressure. When measured in open channel, the average combustion velocity falls from 380 m/s for pure nanothermite to 300 m/s and only 50 m/s for an addition of 1 wt% and 5 wt% alumina, respectively. The change in combustion mode is observed when the alumina content in the composition reaches or exceeds 4 wt%. The experiments carried out in tubes, in which convective effects are amplified, enable us to differentiate between three reaction propagation regimes, depending on the alumina content of the Al/CuO/Al_2O_3 composition. The addition of a small quantity of alumina (0–5 wt%) causes a drop in velocity, which nevertheless remains constant. In compositions that are slightly richer in alumina (10–15 wt%), the reaction propagates in a continuously accelerated manner. Finally, the reaction of the composition that is richest in alumina (20 wt%) propagates very slowly, before it transforms into rotating combustion. Malchi *et al.* show by calculation that the dilution of the nanothermite by alumina abolishes the production of gaseous species. By absorbing a part of the heat released in the reaction, alumina lowers the flame temperature to a level which is insufficient for the gasification of the reaction products. The transfer of energy by convection becomes impossible for lack of gas, and energetic performances are sharply lowered [MAL 08].

Weismiller *et al.* have shown that the pressurization of gas in which a CuO/Al nanothermite reacts leads to a drop in flame propagation velocity, from around 1 km/s (~0 MPa) to only 1 m/s (15 MPa). Propagation by convection, characterized by high and constant velocity, is only observed at low pressures. However, high pressures impede the convection of gases and the heat released in the reaction propagates by conduction. The flame then travels in the composition at constant but very low velocity. Between these two regimes is a transition zone, in which propagation is accelerated or oscillating (see Table 3.6). In helium, the pressure domain in which nanothermite combustion is unstable is narrower than in the other two gases. The specific effects observed in helium are attributed to its high thermal conductivity [WEI 09].

Modes of propagation	Argon pressure (MPa)	Helium pressure (MPa)	Nitrogen pressure (MPa)
Constant velocity (convection)	0–2.0	0–2.0	0–3.2
Acceleration	2.0–4.3	Not observed	3.2–4.8
Oscillation	4.3–9.2	2.0–4.2	4.8–8.2
Constant velocity (conduction)	>9.2	>4.2	>8.2

Table 3.6. *Domains of various modes of flame propagation in a CuO/Al nanothermite, depending on the nature and pressure of the gas in which it reacts; the values are taken from Figures 8–10 published in [WEI 09]*

Sullivan *et al.* used electrophoresis to deposit CuO/Al thermites at the surface of platinum microelectrodes. The compositions used for this purpose are prepared by mixing nanostructured copper oxide (<50 nm) and aluminum with either micrometric (3.2 μm) [SUL 12a] or nanometric (80 nm) [SUL 12b] granulometry. In the first case, the flame propagation velocity is low (~1.7 m/s) and independent of the deposit thickness (~10–104 μm) [SUL 12a]. In the second case, velocity varies significantly with the thickness of the nanothermite layer. The flame propagates at constant velocity (~5.1 m/s) in the thinner deposits (~13–44 μm), which increases with the thickness of the nanothermite layer (~44–105 μm), until finally reaching a velocity (~34.9 m/s) that does not depend on thickness (~105–180 μm) [SUL 12b]. These experimental results, which were obtained with one of the most reactive aluminothermic systems, show that when in the state of too thin layers, nanothermites cannot reach their optimal performances. On the other hand, Sullivan *et al.* have noted that massive deposits were cracking in a direction perpendicular to the microelectrode on which they formed [SUL 12b]. Consequently, the strongest pyrotechnic effect that can be produced by the reaction of a nanothermite deposited as a stable thin layer is at best a vivid combustion.

3.3.1.6. *Iron oxide (Fe_2O_3)*

Iron oxide is one of the substances used by J. W. Goldschmidt to formulate the first thermites containing aluminum powder. Since then, this aluminothermic reaction has been universally used for rail welding. It is worth noting at this point that iron oxide was also a component of the first nanothermites, which were accidentally formulated as result of the friction between aluminum parts and rusty iron surfaces, and were believed to generate the sparks that caused mining or industrial accidents [WAN 93]. Though iron oxide-based nanothermites are not considered among the best-performing, their advantage consists of being less toxic and less expensive than most of the other nanothermites.

Menon *et al.* have shown that the combustion of Fe_2O_3/Al nanothermites, formed of oxide wires (50 nm) implanted in an aluminum layer, is activated at 410°C. The temperature of the flame produced by the reaction (~4,000°C) does not depend on the ignition temperature [MEN 04].

Kim *et al.* have shown that the method of mixing nanoparticles of aluminum and iron oxide had a significant influence on the reaction of the nanothermite. Kim *et al.* contrast Brownian coagulation, consisting of randomly mixing aerosolized nanoparticles, to dipolar assembly, in which the oxide and aluminum particles, having opposite charges, are associated due to electrostatic attraction. The second mode of preparation produces nanothermites in which the stoichiometry is better respected at the scale of particles. Activated by a flame, combustion is consequently significantly more vivid. The same is applicable to the heat of reaction measured by DSC, which reaches 1.8 kJ/g instead of 0.7 kJ/g [KIM 04]. The homogeneity of the mixture does not increase the quantity of chemical energy stored by the nanothermite, but it does enable a more complete and rapid release of this energy.

Prentice *et al.* have shown that the addition of silica (SiO_2) to aluminothermic compositions based on iron oxide (Fe_2O_3) considerably reduces their combustion velocity. This effect is attributed to the drainage of reaction heat by the silica and also to its thermal resistance. Silica incorporated in the form of Fe_2O_3/SiO_2 aerogel slows down the reaction to a greater extent than when it is added in the form of independent particles [PRE 05].

Cheng *et al.* have shown that an Fe_2O_3/Al nanothermite fabricated by assembling iron oxide nanotubes with aluminum nanoparticles was more reactive than a composition fabricated by the simple random mixing of these two phases. The enhancement of performance observed by Cheng *et al.* is attributed to improved contact between the phases, which facilitates diffusion between reagents in the solid state [CHE 10]. It is also possible that the observed effect results from the formation of gaseous species produced by the pyrolysis of the organic phase used for the assembly of nanoparticles.

By incorporating a gas-generating non-energetic polymer, (acrylamidomethyl)-cellulose acetate butyrate (AAMCAB) in the interparticle porosity of a nanostructured iron oxide (Fe_2O_3), Bezmelnitsyn *et al.* managed to modify the characteristics of the combustion of a Fe_2O_3/Al nanothermite. The AAMCAB@Fe_2O_3/Al compositions have been prepared in such a way that the impregnated oxide to aluminum mass ratio is constant. Consequently, the increase in the content of AAMCAB induces an increase in the equivalence ratio in the sense of Granier *et al.* [GRA 04], from 1.4 for the pure Fe_2O_3/Al nanothermite to 1.75 for the AAMCAB@Fe_2O_3/Al composition in which the oxide contains 20 wt% polymer. Bezmelnitsyn *et al.* have shown that the velocity of combustion propagation in the tube decreases with the equivalence ratio. It sustains high values as long as the proportion of AAMCAB in the oxide is below 10%, but it collapses above this. The maximal pressure produced by the reaction in the pressure cell increases with the proportion of polymer contained in the nanothermite, which is explained by the increase in the quantity of gas generated by pyrolysis. On the other hand, the rate of pressurization strongly diminishes with an increase in the content of AAMCAB. The slowdown of the aluminothermic reaction is attributed to an aluminum excess, the absorption of energy by polymer thermolysis and to the less effective interfacial contacts between oxide and aluminum [BEZ 10].

Kaplowitz *et al.* have attempted to replace the alumina layer covering aluminum nanoparticles with an iron oxide (Fe_3O_4) coating, hoping that a substance that becomes reactive only at high temperature would protect the aluminum core against oxidation. In reality, however, the alumina layer that develops on the aluminum nanoparticles covered with iron oxide is thicker than the one which forms naturally at the surface of pure aluminum exposed to air. Kaplowitz *et al.* postulate that at high temperature, aluminum reduces iron oxide, which is consequently re-oxidized by fixing the oxygen from the air. The heat generated by the reaction facilitates the penetration of oxygen in the aluminum. The alumina layer grows thicker through this mechanism, which continues until oxygen diffusion becomes impossible. Kaplowitz *et al.* used aluminum nanoparticles coated with iron oxide (Al@Fe_3O_4) as fuel in a CuO/Al@Fe_3O_4 nanothermite and observed a significant drop in the ignition temperature compared to the CuO/Al@Al_2O_3 classical mixture (973 vs. 1,076 K). This result is attributed to the probable formation of an intermetallic composition, such as Fe_3Al [KAP 14].

3.3.1.7. *Iodine oxide (I_2O_5)*

Iodine oxide is a very powerful oxidizer which can react with many combustible materials, such as organic substances, carbon, sulfur and ammonia. It is worth noting that I_2O_5 can even oxidize certain toxic products formed during combustion reactions: carbon monoxide (CO), sulfur dioxide (SO_2), hydrogen sulfide (H_2S) and hydrogen chloride (HCl) [PAS 31]. Iodine oxide can serve as an oxidizer in compositions based on metal fuels such as aluminum, lanthanum, neodymium, yttrium or tantalum. The combustion of these mixtures generates extremely high temperatures and yields only gaseous species [FIS 96]. Nevertheless, the exceptional pyrotechnic performances of the iodine oxide should not obscure a significant drawback: its hygroscopicity. Air humidity transforms the anhydride (I_2O_5) into iodic acid (HIO_3). The resulting physical and chemical instability does not allow the stabilization of this oxide in its pure state in the form of nanosized particles. Pascal *et al.* also report that iodine oxide is sensitive to light, which breaks it down and releases iodine [PAS 31]. This photosensitivity is likely stronger when the oxide is finely divided.

The amount of heat released by the reaction of iodine pentoxide (I_2O_5) with aluminum (6.22 kJ/g) is higher than that produced by the combustion of thermites based on metal oxides [FIS 96]. Martirosyan *et al.* were the first to study aluminothermic compositions based on nanostructured iodine pentoxide and to show that flame propagation velocity reaches 2 km/s in these mixtures. In a pressure cell, the product of maximum pressure and chamber volume (~3.8 kPa·m^3) is comparable to that of a Bi_2O_3/Al nanothermite, tested under the same conditions (~3.9 kPa·m^3). The performances of I_2O_5-based nanothermites are attributed to the formation of iodine in gaseous state. The temperature of ignition of an I_2O_5/Al nanothermite, measured by DSC, ranges between 605°C and 620°C [MAR 09b]. In DSC, the value of the ignition temperature depends on the sample heating rate. Since the heating rates used by Martirosyan *et al.* (41–66°C/min) are high, they induce a significant shift of the ignition temperature toward higher values. This probably explains why the ignition temperature of the aluminothermic reaction measured by Martirosyan *et al.* is higher than that reported by Farley *et al.* [FAR 10].

Because of its hygroscopicity, iodine pentoxide cannot be used in its pure state in aluminothermic compositions that may be exposed to humid air. In contact with water vapor, this compound hydrolyzes and yields iodic acid (HIO_3), which triggers the agglomeration of I_2O_5 nanoparticles into larger particles, and the chemical attack of aluminum, which is not stable in an acidic environment. Feng *et al.* developed a method that enables the effective protection of nanostructured iodine pentoxide against air humidity by coating its surface with iron oxide. The I_2O_5@Fe_2O_3 nanocomposite materials have been used as oxidizers in compositions with nanosized aluminum. Measured in a pressure cell, the performances vary significantly with the Fe/I molar ratio of the I_2O_5@Fe_2O_3 material, and reach an optimum level when this ratio ranges between 1.8 and 4.3. Within this range of composition, the reaction mechanism is similar to that of a CuO/Al nanothermite, with rapid decomposition of the oxide (I_2O_5) and release of oxygen, followed by the much slower combustion of the aluminum. On the other hand, too low a Fe/I molar ratio (0.25) or, conversely, too high a ratio (6.2), facilitates a slower reaction, comparable to that of a Fe_2O_3/Al nanothermite. In this case, the decomposition of iodine pentoxide and the oxidation of aluminum take place simultaneously, over a longer period of time [FEN 13].

The measurements conducted in instrumented tubes by Russell *et al.* show that the velocity of flame propagation in an Al/I_2O_5 composition with micrometric granulometry is high (1,305 ± 32 m/s) [RUS 11].

3.3.1.8. *Manganese oxide (MnO_2)*

Manganese oxide has the advantage that it can be synthesized very easily in the state of ultimate-sized nanoparticles [COM 08c] by various solution-based methods. Its main characteristic is that it melts while decomposing, at low temperature (~535°C). Oxygen is released at the temperature at which nanosized aluminum starts to oxidize in solid phase (see Figure 3.2).

Siegert *et al.* have prepared nanothermites with manganese oxide, aluminum and hollow carbon nanofibers, either by directly mixing these three phases or by mixing the aluminum with nanofibers that were previously filled with manganese oxide. The cold compaction of these powders results in pellets with excellent cohesion, in which carbon fibers play a reinforcement role for the porous matrix of nanoparticles. The study of the combustion of these pellets has shown that the addition of carbon nanofibers delays the ignition and considerably lowers the combustion velocity, which only reaches several millimeters per second [SIE 10].

Yang *et al.* have modified manganese oxide (MnO_2) nanowires by growing at their surface nanorods of tin oxide (SnO_2). Ternary nanothermites that are fabricated by mixing these heterostructures with aluminum nanoparticles are organized into membranes by simple filtration [YAN 13b]. The entanglement of these nanosized wires, which carry a more or less developed secondary structure (see Figure 3.12), likely confers an excellent cohesion to the membrane, and enables the retention of aluminum nanoparticles, which are trapped in the particular meshes of MnO_2/SnO_2 heterostructures.

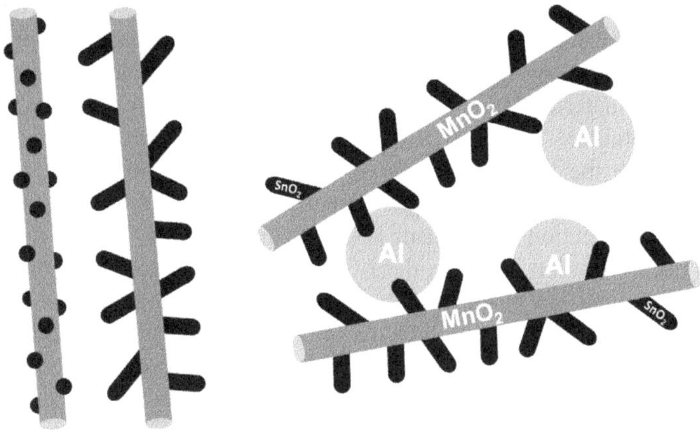

Figure 3.12. *Diagram of two forms of hierarchical MnO_2/SnO_2 heterostructures and a $MnO_2/SnO_2/Al$ ternary nanothermite, according to the results published by Yang et al. [YAN 13b]*

3.3.1.9. *Molybdenum oxide (MoO_3)*

Molybdenum oxide is one of the oxides that were most frequently used to formulate nanothermites. The MoO_3 particles have platelet morphology. Molybdenum oxide has a distinctive characteristic, which resides in the fact that it starts subliming before melting [PAN 06], which explains the good reactivity of nanothermites that contain it.

According to Moore *et al.* the bright yellow color of nanostructured molybdenum oxide shifts to greenish blue after lengthy (4 days) exposure to fluorescent light. This treatment changes the aspect of the oxide, but does not impair the morphology of MoO_3 nanoparticles. Thermal treatment at 400°C of the nanostructured oxide induces the sintering of nanoparticles into larger platelets. Molybdenum oxide has been exposed to various environmental conditions: (1) fluorescent light, (2) UV lamp and (3) relative humidity (99%). The modified oxide is then used to prepare MoO_3/Al thermites, whose combustion velocity is determined by open tray burn tests. These experiments have shown that the exposure of molybdenum oxide to humidity was more detrimental to the performances of the MoO_3/Al composition than the exposure to fluorescent or ultraviolet lighting. Moore *et al.* have shown that the calcination of oxide was an effective means to limit the detrimental effect of humidity [MOO 06].

The reaction in tube of a MoO_3/Al nanothermite has been studied by Asay *et al.* in various experimental configurations in order to determine the combustion propagation mechanism. In a first type of test, the composition that fills the tube is divided into two domains, separated by a sapphire window of low thickness (0.4 mm), transparent to infrared radiation. The combustion front is stopped by this obstacle up to the moment when the pressure reaches the value needed for breaking the sapphire membrane. This simple experiment shows that energy is not propagated by radiation. The measurement of pressure at the end of the tube, coupled with the observation of the propagation by high-speed video camera (62,500 frames per second), has allowed to estimate the thickness of the combustion front to around 10 mm, and the time of pressure rise to approximately 10 μs. Combustion tests have also been conducted under atmospheric pressure (~71 kPa) or under vacuum (3.3 Pa) by using in each case, tubes filled with loose (~0.27 g/cm^3) or compacted (0.4 g/cm^3) powder. These experiments show that neither pressure nor density modify to a measurable extent the combustion propagation velocity. On the contrary, the presence of a free space at the end of the tube induces an increase in the velocity, which occurs both under vacuum and under pressure. Based on these observations, Asay *et al.* conclude that the combustion of the MoO_3/Al nanothermite very likely occurs through a convection mechanism [ASA 04].

Granier *et al.* studied the combustion of Al/MoO_3 thermite pellets, depending on the equivalence ratio of compositions and on the granulometry of various aluminum powders (17.4–20 µm) used for preparing these mixtures. The molybdenum oxide used by Granier *et al.* is composed of platelet-like particles with micrometric length and nanometric thickness (~20 nm). These Al/MoO_3 compositions based on nanosized aluminum (17.4–202 nm) exhibit ignition delay times that are much shorter than those of compositions prepared with aluminum of coarser granulometry (3–20 µm). Nanosized aluminum thus increases the sensitivity to ignition of these mixtures. In the case of compositions based on nanosized aluminum, the shortest ignition delay times and the highest combustion velocities are obtained with 20% excess aluminum compared to the stoichiometry of the aluminothermic reaction [GRA 04]. The Al/MoO_3 compositions have significantly lower reaction rates under compacted form (~1–30 m/s) [GRA 04] than under unconfined, free powder form [PUS 11].

Dutro *et al.* studied the influence of aluminum and oxide proportions on the combustion of MoO_3/Al nanothermites. They have shown that the mass content of aluminum in these mixtures should range between 10 and 85 wt% for the aluminothermic reaction to be self-sustaining. The highest velocities (~1,000 m/s) are obtained for mixtures whose composition is close to stoichiometry. The combustion propagates at constant velocity (100–1,000 m/s) in compositions containing 10 to 65 wt% of aluminum. In the mixtures containing a large excess of aluminum (75–85 wt%), the reaction propagates also at constant, though lower (~ 0.1–1 m/s), velocity. In the domain of intermediary composition (Al: 70 wt%), the combustion is not stable and propagates in an oscillating and accelerated manner [DUT 09].

Bockmon *et al.* studied the influence of nanosized aluminum granulometry (44, 80, 121 nm) on the characteristics of tube combustion of MoO_3/Al nanothermites. They have shown that compositions made of aluminum with an average particle size of 80 nm had combustion velocities above those of the mixtures produced with less finer aluminum (120 nm). The use of even finer aluminum particles (44 nm) does not result in additional gains in performance. Bockmon *et al.* have also shown that the equivalence ratio for which the combustion of a MoO_3/Al (Al : 80 nm) mixture reaches its peak value is equal to 1.2. Finally, Bockmon *et al.* have observed that the combustion velocity of MoO_3/Al nanothermites is independent of the diameter (1–6 mm) of the tube in which they are characterized [BOC 05].

Son *et al.* showed that the combustion of MoO_3/Al nanothermites propagates easily through channels of submillimeter cross-section. In a cruciform geometry, combustion can even be transmitted along directions that are perpendicular to the initial propagation path. Son *et al.* think that the projection of molten molybdenum and its subsequent cooling play a significant role in the combustion propagation mechanism [SON 07].

Dikici *et al.* prepared two sorts of MoO_3/Al thermites, by mixing nanostructured molybdenum oxide in the form of nanorods (44 nm) with aluminum powders of micrometric granulometry (μ-Al: 3–4.5 μm) in one case, and nanometric granulometry (n-Al: 80 nm) in another case. By preheating these thermites between 80 and 170°C, the flame propagation velocity increases significantly, reaching values that are by 60% (n-Al) and 70% (μ-Al) higher than those measured at ambient temperature. This effect can be explained by the rise in the heat of reaction. However, Dikici *et al.* have also shown that a thermite first submitted to preheating and then to slow cooling (0.06°C/s) down to ambient temperature reacts as if it had not been subjected to heat treatment. On the other hand, a thermite subjected to more rapid cooling (0.13°C/s) yields during its reaction a flame propagation velocity that approaches that of the preheated thermite [DIK 10]. The reactive properties of nanothermites appear to depend on their thermal history. This form of "memory" is the result of the thermal modification of the stress state of the aluminum particles' core–shell structure [LEV 11].

Gesner *et al.* studied the influence of the alumina layer thickness on the flame front propagation velocity in a MoO_3/Al nanothermite. For this purpose, aluminum nanoparticles are oxidized through an isothermal treatment (480°C) conducted under pure oxygen atmosphere. The thickness of the alumina layer is defined by varying the oxidation duration (8–150 min). The original nanoparticles (95 nm) are covered by a native alumina layer with a thickness of 2.7 nm, which is gradually increased to 8 nm. The flame front propagation in MoO_3/Al compositions prepared with aluminum nanoparticles thus modified is much slower (315–392 m/s) than in the nanothermite produced with original aluminum (682 m/s). The deterioration of performances is attributed to the damage of the alumina shell, which can be due to (1) stresses built during cooling, (2) variations of volume induced by the reaction or (3) still phase change in alumina, which goes from amorphous to crystalline state (γ-Al_2O_3). Gesner *et al.* lastly note that velocity depends to a little extent on the alumina proportion, when the quantity *(M)*, which is the ratio of aluminum core radius to the alumina shell thickness, reaches values ranging between 6.1 and 13.4 (see equation [3.6]). The quasi-stability of velocity in this domain is in good agreement with the predictions formulated by applying the aluminum explosion/dispersion Levitas' mechanism [LEV 06]. On the other hand, excessive proportion of alumina (30 ± ~3%) causes a sharp drop in the flame propagation velocity by draining the heat and reducing the proportion of reactive matter in the composition [GES 12].

Micron-sized particles (~10 μm) of an Al/MoO_3 nanocomposite material prepared by reactive milling have been dispersed in a shock tube. Ignition is generated by the shock travelling through the particle aerosol. Tested under these conditions, the ignition of Al/MoO_3 nanocomposite particles takes place, depending on their composition, between 900 and 1,050 K. In a similar manner, Bazyn *et al.* have determined the ignition temperature of aluminum particles of 10 μm (2,000 K), 2 μm (~1,400 K) and 80 nm (1,150 K). The association of aluminum with molybdenum oxide facilitates its ignition. The metal then burns in contact with the air [BAZ 09].

3.3.1.10. *Nickel oxide (NiO)*

Very few studies have been dedicated to aluminothermic compositions based on nanostructured nickel oxide. The reason is that the low quantity of gas generated by the reaction leads to predict that this nanothermite has quite modest performances. Moreover, the nickel compounds' toxicity to the respiratory tracts and their probable (Ni) or acknowledged (NiO) carcinogenicity [BON 09] do not allows us to contemplate applications in which these substances would be aerolized.

By means of a thermochemical computation code, Dean *et al.* have shown that the adiabatic flame temperature of NiO/Al and CuO/Al nanothermites is similar, but what clearly differentiates these two systems is the proportion of gas they generate, which is significant for CuO (5.4 mmol/g), and negligible for NiO (0.108 mmol/g). Dean *et al.* therefore, postulate that, if combustion propagates in nanothermites only by hot gases convection, then the performances of the NiO/Al composition should be much lower than those of the CuO/Al composition. Conducting tests in instrumented tubes, the scientists have then measured the flame propagation velocity and the pressure produced by these two nanothermites, depending on their composition, defined according to Granier's equivalence ratio [GRA 04]. The values obtained (see Table 3.7) led Dean *et al.* to conclude that the performances of the NiO/Al composition are in reality higher than those that would be obtained if the reaction propagated by a classical mechanism of hot gases convection. Rapid propagation of the combustion wave in NiO/Al nanothermites would presumably stem from the interstitial gas (here, argon) being drawn in by the metallic aerosol projected by the explosion of aluminum nanoparticles [LEV 06]. This mechanism of advection-based combustion would presumably explain the commendable performances of nanothermites that generate little or no gas [DEA 10].

Nanothermite	FPV (m/s)	P_{max} (MPa)	T_P (GPa/s)
Al (50 nm)/NiO (~11 nm)	205.5 ± 71.2	1.83 ± 0.89	6.04
Al (50 nm)/CuO (30–50 nm)	582.9 ± 87.6	3.75 ± 0.85	19.3

Table 3.7. *Average values of flame propagation velocity (FPV), intensity of the pressurization peak (P_{max}) and the rate of pressurization (T_P) experimentally determined for NiO/Al and CuO/Al compositions [DEA 10]*

Wen *et al.* have synthesized nickel oxide nanowires and have used them to prepare NiO/Al nanothermites. These compositions have an original morphology, characterized by aggregates of aluminum nanoparticles resting on entanglements of nickel oxide nanowires. Wen *et al.* have studied the composition of products yielded by the reaction of an aluminum rich NiO/Al nanothermite. For this purpose, nanothermite discs obtained by compression are placed on a hotplate maintained at various temperatures, under argon atmosphere. The XRD analysis of combustion residues reveals the presence of nickel, aluminum, alumina and also an intermetallic compound (AlNi) formed due to excess aluminum. Molecular dynamic simulations, conducted at 1,000 K and 3,500 K, have confirmed that the aluminothermic reaction between NiO and Al can effectively yield an AlNi alloy by the diffusion of nickel and aluminum atoms [WEN 13].

Zhang *et al.* have coated a silicon substrate with a NiO/Al nanothermite layer produced by condensing aluminum on a nickel oxide deposit with honeycomb morphology. The reaction start temperature (~400°C) is measured under argon flow by DTA on a sample of nanothermite gathered by scratching. Among the three exothermic signals produced by the reaction, two are situated before the endothermic peak of aluminum melting (~610°C), which means that the reaction takes place by a mechanism of diffusion among solids [ZHA 09].

3.3.1.11. *Tin oxide (SnO_2)*

Tin oxide-based nanothermites have been little studied. According to the thermochemical data published by Fischer *et al.* the mixture of this oxide with yttrium would presumably have even more interesting properties than the aluminothermic composition, in particular in terms of gas generation and heat of reaction [FIS 96].

Ferguson *et al.* have produced Al/SnO_2 nanothermites by an atomic layer deposition method. Even though the compositions produced by this technique are highly underoxygenated, they have interesting pyrotechnic properties. The reaction of a sample (~20 mg), ignited by the electrical discharge produced by a Tesla coil, has been observed by conventional camera (30 frames per second). The ignition delay is below 0.07 s and the duration of reaction is close to 0.1 s. The combustion generates a highly intense light flash that is not accompanied by any sound effect. Pure nanosized aluminum, characterized under the same experimental conditions, ignites less rapidly (<0.17 s) and burns far more slowly (~6 s) [FER 05].

By resorting to a variant of the method described by Ferguson *et al.*, Qin *et al.* have succeeded in preparing Al@SnO_2 nanothermites that are richer in tin oxide. Being ignited by a laser beam, these nanothermites react far more vividly than the aluminum powder tested under the same conditions. The ignition delay time slightly increases with the content of SnO_2, it passes from 33 ms for pure aluminum to 38 ms for a mixture whose composition is close to stoichiometry. The reaction of the richest composition in tin oxide (~70 wt%) is very vivid, and is accompanied by an explosion. The flash emitted by combustion is five times shorter (19 ms) than the one generated by the reaction of a composition prepared by simple physical mixing (106 ms) [QIN 13].

Yang *et al.* prepared $MnO_2/SnO_2/Al$ ternary nanothermites, by using a very particular combination of the two oxides, in which manganese dioxide (MnO_2) nanowires are used as a substrate to enable the growth of tin oxide (SnO_2) "branches". The mass proportion of SnO_2 in these heterostructures varies between 0 and 66.6%. The addition of tin oxide in this form modifies the thermal properties of the MnO_2/Al nanothermite measured by DSC. According to Yang *et al.* the tin oxide barrier between manganese oxide and aluminum is a means to control the properties of these energetic compositions [YAN 13b].

3.3.1.12. *Titanium oxide (TiO_2)*

Valliappan *et al.* measured the velocity of combustion in the air of a loose powder of TiO_2/Al nanothermite by using an open-tray burn test equipment [VAL 03]. The slowness of this propagation (0.1 m/s) indicates that the reactivity of this nanothermite is very low. This type of composition can, nevertheless, be of interest for particular applications, such as pyrotechnic delay compositions.

3.3.1.13. *Tungsten oxide (WO_3)*

Tungsten oxide has reasonably good thermal stability, as it melts at 1,472°C [LID 05]. Because of this, nanothermites based on tungsten oxide react less vividly than those that are prepared with molybdenum oxide. The combustion of these compositions under confined conditions yields dense macroscopic fragments, composed of tungsten and alumina.

Perry *et al.* showed that WO_3/Al and WO_2/Al nanothermites had optimal energetic performances for compositions approaching theoretical stoichiometry, defined by exclusive formation of tungsten and alumina. At an optimum energetic level, the WO_3/Al composition generates maximum rate of pressurization (500 GPa/s) and pressure (1.45 MPa). In comparison, the performances of WO_2/Al composition are far more modest (6 MPa/s; 38 kPa). The flame propagation velocity in WO_3/Al nanothermites ranges between 170 and 250 m/s and also peaks at stoichiometry [PER 04].

In subsequent published research, Perry *et al.* have shown that the performances of a WO_3/Al composition are considerably augmented by using tungsten trioxide in its hydrated form ($WO_3 \cdot H_2O$). Thus, at energetic optimum, the amount of heat released by the aluminothermic reaction of the hydrated oxide is much higher

(1.8 MJ/kg) than that generated by the anhydrous oxide (1.1 MJ/kg). Likewise, the $WO_3 \cdot H_2O$/Al composition develops a level of reactive power (215 GW/m^2) that is higher than that of the WO_3/Al composition (130 GW/m^2) [PER 07].

Dong *et al.* have produced $WO_{2,9}$@Al nanothermites by coating tungsten oxide nanowires with aluminum. These deposits generate a vivid combustion when activated by the electrical heating of the tungsten wire on which they have formed. The original method used by Dong *et al.* enables an excellent contact between phases, which are separated only by a unique alumina layer [DON 13]. The smoothness of this interface confers to this material an apparent stability, which can be disrupted by the slightest stress bringing the metal in contact with the oxide. Such a composition may therefore exhibit a very low threshold of sensitivity to friction.

3.3.1.14. *Zinc oxide (ZnO)*

Qin *et al.* prepared Al@ZnO nanothermites, by coating aluminum nanoparticles with atomic layers of zinc oxide (ZnO). The combustion of a composite material with 80% mass content of ZnO is possible due to the core–shell morphology of the material, but it is much slower than that of an Al@SnO_2 nanothermite [QIN 13]. According to Zhou *et al.* zinc oxide releases oxygen with more difficulty than the iron or copper oxides [ZHO 10]. The modest reactivity of aluminothermic compositions based on ZnO may be related to the very strong reducing character of zinc. Since the oxidation of this metal is highly exothermic, it yields a quite stable oxide, which is difficult to be reduced by the aluminum.

3.3.2. *Oxidizing salts*

The first thermites were prepared by Goldschmidt starting with metal oxides, sulfides and chlorides [GOL 97]. Pyrotechnicians resorted to using oxygenated salts as oxidizers in aluminothermic compositions in order to augment the performances of the latter. Such compositions are even used on a wide scale in the field of solid propulsion. Thus, the famous Ariane V rocket, flagship of European airspace research, is propelled by a composition which essentially consists of ammonium perchlorate (NH_4ClO_4) and aluminum. It is now over a century since the properties of the mixtures of aluminum and strong oxidizers, in the form of powders with micrometric granulometry, started being studied. The “ammonals” are an example of explosive composition in which aluminum facilitates the detonation of ammonium nitrate, which is an explosive little sensitive to priming when it is in pure state [FUZ 00].

The pyrotechnic compositions that associate, on the one hand, aluminum nanoparticles and, on the other hand, submicron-sized or nanosized particles of oxidizing salts have, until now, been studied very little. Nevertheless, in comparison with oxides, the oxidizing salts have several advantages: first, they usually contain a more significant mass proportion of oxygen; second, the oxygen they contain is easier to displace than the one contained by oxides; third, the oxygen they contain can have lower oxidation degrees, rendering it more reactive [ZHO 15b]. The salts used in pyrotechnics as strong oxidizers are essentially the following: nitrates (NO_3^-), chlorates (ClO_3^-) and perchlorates (ClO_4^-), bromates (BrO_3^-), iodates (IO_3^-) and periodates (IO_4^-), peroxides (O_2^{2-}) and superoxides (O_2^-), permanganates (MnO_4^-) and persulfates ($S_2O_8^{2-}$). Other salts considered less reactive, such as the carbonates (CO_3^{2-}), oxalates ($C_2O_4^{2-}$), sulfates (SO_4^{2-}) or phosphates (PO_4^{3-}) have oxidizing properties that become manifest at high temperature. This is why they have to be mixed with highly reducing substances such as magnesium, aluminum or silicon: this is the "negative explosive" concept proposed by Shimizu [SHI 86].

One of the first studies on the combustion of mixtures of aluminum nanoparticles and ammonium perchlorate was done by Armstrong *et al.* The combustion velocity of these compositions, measured in pressure cell, increases exponentially when the diameter of aluminum nanoparticles passes from 200 to 40 nm. The combustion velocity of a composition containing aluminum of intermediate granulometry (≈120 nm) increases considerably with pressure, passing from around 25 mm/s at 0.14 MPa to 600 mm/s at 15 MPa [ARM 03].

In pyrotechnics, potassium permanganate ($KMnO_4$) is a very strong oxidizing agent, whose properties are similar to those of chlorates. Its use could enable the development of high-performance and low-toxicity energetic compositions, provided it is associated with inorganic reducing substances having good chemical stability. Prakash *et al.* have thus shown that the rates of pressurization generated in pressure cell by the mixtures of nanoparticles of $KMnO_4$ (~250 nm) and aluminum (~40 nm) were practically two orders of magnitude higher than those of CuO/Al and MoO_3/Al nanothermites [PRA 05a]. The drawback of these compositions resides in their instability in the presence of humid air, organic or acidic materials. In the presence of humidity, potassium permanganate progressively transforms into manganese dioxide, which leads to the ageing of the composition, characterized by lower oxygen balance and degradation of performances. When an energetic composition containing permanganate is accidentally brought into contact with certain organic substances, such as glycerin or oxalic acid, a spontaneous self-ignition reaction may occur. In the presence of concentrated sulfuric acid, permanganate is transformed into manganese heptoxide (Mn_2O_7), a sensitive explosive that reacts violently when in contact with organic matter. In order to render the $KMnO_4$/Al nanothermites more stable, Prakash *et al.* have conceived a very elegant technique, consisting of coating the permanganate, a strong oxidizer, with ferric oxide, a weak oxidizer. Prakash *et al.* resort to a three-stage atomization method to produce composite nanoparticles, formed of a permanganate core (~150 nm) covered by an iron oxide shell whose thickness is around 4 nm. When ferric oxide is in the proportion (> 14 vol.%) required to fully coat the permanganate, reactivity can be adjusted in the range between ~0.117 kPa/µs (Al/Fe_2O_3) and ~827 kPa/µs ($Al/\{KMnO/Fe_2O_3\}$) simply by varying the oxide content of the nanocomposite material [PRA 05b].

Using the same preparation method, Wu *et al.* have encapsulated perchlorates in metal oxides, namely iron oxide (Fe_2O_3) and copper oxide (CuO). The first material ($KClO_4/Fe_2O_3$) has an ideal core–shell structure, with excellent segregation of the two phases, such that perchlorate fills the core and the oxide forms the shell. In the second material ($KClO_4$/CuO), the oxide is to be found mainly in the shell, but also mixed with perchlorate, whose crystallization it disrupts. Finally, the third material (NH_4ClO_4/Fe_2O_3) is formed of hollow particles whose shell is a homogeneous mixture of the two constituents. When mixed with nanosized aluminum (~50 nm), these composite materials yield energetic compositions with definitely better performances than those of Al/Fe_2O_3 and Al/CuO, the homologue thermites. Perchlorate thermites react at lower temperature and generate in pressure cell rates of pressurization and maximum pressures significantly higher than those produced by the aluminothermic mixtures containing only the oxide. Wu *et al.* attribute this increase in reactivity to the fact that the thermolysis of potassium perchlorate releases oxygen more rapidly and at a lower temperature than the oxides Fe_2O_3 and CuO [WU 12].

The coating of perchlorates with metal oxides renders the former less hygroscopic and limits their diffusion in the surrounding environment. Nevertheless, the perchlorate anion is an endocrine disruptor that is active at very low doses. A report of the National Academy of Sciences in the United States recommends an acceptable daily intake level of 0.7 µg/kg. The toxicity of the perchlorate anion arises from the fact that compared to the iodine, it gets more easily fixed on a glycoprotein whose role is to carry iodine across the basolateral membrane of the thyroid cells. This action mechanism inhibits the iodine fixation by the thyroid, and subsequently the production of thyroid hormone [LEU 10]. The most recent studies on iodates [SUL 10a] and periodates [JIA 13a] have probably been motivated by the need to find substitutes for perchlorates.

Sullivan *et al.* studied the reaction of micro- and nanostructured silver iodate particles ($AgIO_3$) mixed with nanosized aluminum (50 and 80 nm). The reactivity of these compositions, measured in pressure cell, is higher than that of Al/Fe_2O_3 and Al/CuO nanothermites characterized under the same experimental conditions. Even though the nanosized $Al/AgIO_3$ composition generates a higher rate of pressurization than the Al/CuO nanothermite (57 psi/µs vs. 9.0 psi/µs), the combustion times of these two formulations, determined by the measurement of optical intensity, are comparable (172 µs vs. 192 µs). Sullivan *et al.* attribute the higher initial pressurization to the fact that $AgIO_3$ releases more gas (O_2, O, I) than

CuO. The kinetics of the reaction's further development is limited by the oxidation of aluminum. The mechanism of reaction of $AgIO_3$/Al compositions proposed by these authors is presented in Figure 3.13.

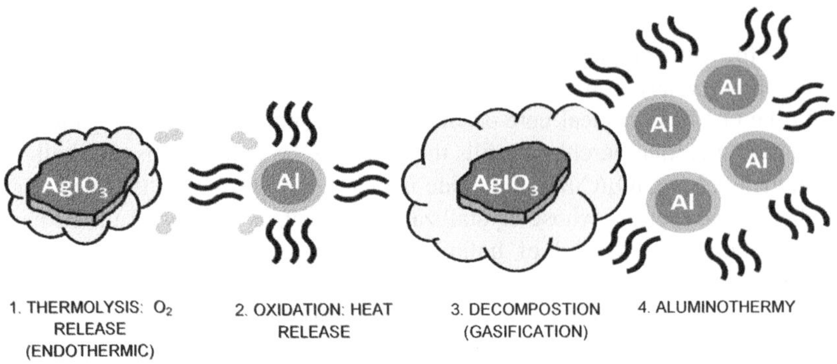

Figure 3.13. *Diagram of the mechanism of reaction of $AgIO_3$/Al nanothermites*

Sullivan *et al.* also note that silver iodate decomposes in a different manner when it is slowly heated (5 K/min) compared to when it is rapidly heated (5×10^5 K/s). In the first case, $AgIO_3$ melts (692 K) and then decomposes into silver iodide (AgI) and oxygen (740 K). When the temperature continues to rise, AgI melts in its turn (827 K) and evaporates. In the second case, the release of O_2, O and I gaseous species is observed at a significantly higher temperature (1,150 K) than that of the thermolysis generated by slow heating (740 K). Moreover, AgI and Ag are not detected in gaseous state. Finally, Sullivan *et al.* note that the ignition induced by a rapid heating of the Al/$AgIO_3$ nanothermite occurs at a temperature of $1,215 \pm 40$ K, which is slightly higher than that required to gasify $AgIO_3$. This led the researchers to conjecture that oxygen partial pressure plays an essential role in the ignition of aluminum [SUL 10a].

Wang *et al.* prepared aluminothermic compositions starting with other nanostructured metal iodates (Bi^{III}, Cu^{II}, Fe^{III}). These salts decompose into oxide, iodine and oxygen prior to the aluminothermic reaction taking place. The reaction of nanothermites based on metal iodates generates an adiabatic flame temperature (~4,050 K), a maximum pressure and a rate of pressurization that are higher than those of the nanothermites formulated starting from the corresponding metal oxides [WAN 15].

According to Jian *et al.* the nanostructured periodates have even more interesting pyrotechnic properties than perchlorates and a lower toxicity. Sodium periodates ($NaIO_4$) and potassium periodates (KIO_4) have been prepared in finely divided state

(50–300 nm) by an aerosol drying method. Combustion tests, conducted in a pressure cell, of the Al/KIO_4 and $Al/NaIO_4$ compositions show that the rates of pressurization generated by these formulations (2.4–2.6 MPa/µs) are significantly higher than those obtained with a traditional Al/CuO nanothermite (0.06 MPa/µs). Moreover, Jian *et al.* highlight the importance of the nanostructuration of periodate for reactivity. The maximum pressure and rate of pressurization generated by the aluminothermic reaction are thus multiplied by over 60 and 1,000, respectively, when periodates are used in the form of nanoparticles rather than as powders with micrometric granulometry (~100 µm). The temperature of ignition in air of $Al/NaIO_4$ (880 K) and Al/KIO_4 (950 K) compositions is lower than that of an Al/CuO nanothermite (1,040 K). Jian *et al.* demonstrate that the thermolysis of potassium periodate takes always place in two stages, irrespective of whether it is induced by slow (10 K/min) or rapid (5×10^5 K/s) heating. The first phase of the decomposition is exothermic. It corresponds to the reduction of periodate (KIO_4) into iodate (KIO_3), accompanied by a first release of oxygen. A second release of oxygen is observed at higher temperature; iodate is then reduced into potassium iodide (KI). The rapid heating (10^6 K/s) of an Al/KIO_4 composition placed under high vacuum in the chamber of a scanning electron microscope does not activate the aluminothermic reaction. The analysis of residues confirms that, under these operating conditions, potassium periodate decomposes without reacting with aluminum. This result seems all the more surprising as periodate nanothermites react strongly, both under oxidizing atmosphere (air) and under neutral atmosphere (argon) when they are thermally activated. Based on these results, Jian *et al.* conclude that oxygen plays a critical role in the ignition and combustion of these energetic nanomaterials [JIA 13a].

Comet *et al.* have described the properties of aluminothermic compositions, in which various metal sulfates with micro- or nanometric granulometry are used as oxidizers. The aluminum powder serving as fuel is formed of nanoparticles (~100 nm) and contains 74% by mass of metallic Al. The composition of sulfated nanothermites has been optimized by a series of measurements in bomb calorimeter, aimed at obtaining the highest possible heat of reaction. The aluminum mass content required to attain the energetic optimum obviously depends on the sulfate nature; but for most of the compositions, it ranges between 45 and 55%. The values of the heat of reaction of sulfated nanothermites are higher (4–6 kJ/g) than those of traditional aluminum/metal oxide mixtures (1.5–4.8 kJ/g). The propagation velocities of the combustion front, measured in tubes, range between 200 and 850 m/s. They depend on the sulfate chemical nature and morphology, and also on the porosity of the composition. The water of crystallization contained in metal sulfates has a significant influence on the ignition of sulfated aluminothermic compositions, and also on their threshold of sensitivity to shock and electrostatic discharge. Finally, contrary to most of the nanostructured aluminothermic compositions, sulfated nanothermites are not very sensitive to friction, which renders their preparation and handling particularly safe. Comet *et al.* also report that the salts of oxoacids of sulfur, such as sulfites (SO_3^{2-}), thiosulfates ($S_2O_3^{2-}$) and peroxodisulfates ($S_2O_8^{2-}$), when mixed with nanosized aluminum, give pyrotechnic reactions comparable to those having been observed with sulfates.

More recently still, Zhou *et al.* have compared the reactivity of nanothermites obtained by mixing aluminum nanoparticles (50 nm) and various oxidizing salts with coarser granulometry (≈0.4 μm). The potassium sulfate (K_2SO_4), iodate (KIO_4) and persulfate ($K_2S_2O_8$) that are used for this purpose are previously treated in divided state by drying atomized aqueous solutions. Tests conducted in pressure cell have shown that the composition based on persulfate was the one generating the highest maximum pressure and rate of pressurization. The study of the thermolysis of potassium persulfate under the effect of slow heating (10°C/min) or, conversely, very rapid heating ($4 \times 10^{5°C}$/s), shows that this salt always decomposes in three stages. The first transformation induced by heat is the deoxygenation of persulfate into pyrosulfate ($K_2S_2O_7$). When the temperature continues to rise, this salt decomposes in turn into potassium sulfate (K_2SO_4), dioxygen (O_2) and sulfur dioxide (SO_2). Finally, at even higher temperature, sulfate itself decomposes into potassium (K), O_2 and SO_2. Zhou *et al.* show that the ignition of the $K_2S_2O_8$/Al composition is, in reality, induced by the reaction of the oxygen released by the thermolysis of persulfate with aluminum. On the contrary, the potassium sulfate reacts with aluminum in condensed phase, as most metal oxides [ZHO 15b].

In conclusion, the nanostructured aluminothermic compositions produced by using oxygenated salts as oxidizers have higher performances than those elaborated starting from metal oxides. The excellent reactivity of these compositions is attributed to the rapid release of a significant quantity of oxygen by the oxidizing phase, from the very beginning of the reaction. With regard to these considerations, inorganic peroxides and superoxides should be oxidizers of choice when formulating reactive nanothermites, not only because they contain more oxygen than simple oxides but also because they more readily release it. Although Zhou *et al.* have recently suggested the use of peroxidized compounds as oxidizers for nanothermites [ZHO 15b], it seems that this type of composition has not yet been the subject of research published in scientific literature.

The barium peroxide and aluminum compositions with micrometric granulometry are more sensitive to ignition than the aluminothermic mixtures based on metal oxides. For this reason, they are often used as ignition substances for thermites [COM 06a]. The peroxides that react with water, such as Na_2O_2 and K_2O_2, should never be mixed in pure state with aluminum powder, since their hydrolysis yields a basic solution of concentrated hydrogen peroxide (H_2O_2). When in contact with it, aluminum oxidizes; the attack on metal yields hydrogen and a significant amount of heat. The rise in temperature triggers the decomposition of hydrogen peroxide, which releases oxygen. The series of these reactions leads to the apparently spontaneous ignition of said compositions.

The particles of alkaline peroxides or superoxides (see Table 3.8) could be stabilized by coating them with a layer of fluoropolymer such as Teflon, which has a triple advantage of being hydrophobic, oxidation resistant and able to react by detonation with aluminum [DOL 05].

Peroxide	Density (g/cm^3)	$T_{Fus.}$ (°C)	Oxygen content ($g/g–g/cm^3$)	Water effects
Li_2O_2	2.31	195*	0.697–1.611	Soluble
Na_2O_2	2.805	675	0.410–1.151	Hygroscopic reaction
NaO_2	2.2	552	0.582–1.280	Reaction
K_2O_2	2.14*	490	0.290–0.621	Reaction
KO_2	2.16	380	0.450–0.972	Hygroscopic reaction
MgO_2	≈ 3	100 decomposition	0.568–1.705	Insoluble
CaO_2	2.9	≈200 decomposition	0.444–1.287	Hygroscopic little soluble
ZnO_2	1.57	>150	0.329–0.516	Insoluble
BaO_2	4.96	450 decomposition	0.189–0.937	Little soluble*

Table 3.8. *Characteristics of peroxides and superoxides of interest for the formulation of highly reactive nanothermites. Sources: densities, melting temperature $T_{Fus.}$, effects of water [LID 05], except for the values marked by asterisks*

3.4. Methods for the characterization of nanothermites

Since nanothermites are energetic materials, their characterization should include both their reactivity and their morphology. In a short literature review, Pantoya *et al.*

had listed the main techniques that can be used for the characterization of nanothermites [PAN 05]. Some of these techniques, as well as other methods developed during this past decade, are described in this section. Despite the abundance of characterization means, one cannot fail to notice that few studies focus on the correlations between the reactive properties of nanothermites and their morphology. In this field, the most significant progress is still to come; it will be made at the confluence between pyrotechnics and materials science.

3.4.1. *Reactive characterization*

Since nanothermites are first and foremost energetic materials, the methods for the characterization of their reactive properties are of utmost importance. A nanothermite whose pyrotechnic properties are unknown to the experimenter should be first of all subjected to the flame test. The quantity used in this experiment should not exceed 10 mg. This simple test provides the pyrotechnician with information on the flammability and the mode of reaction of the nanothermite he is handling and allows for taking the appropriate precautions to ensure his safety. Some nanothermites are easily ignited and burn vividly (Bi_2O_3/B); others are more difficult to activate by flame and produce an explosion comparable to that of a primary explosive (CuO/Al; Bi_2O_3/Al); others, still, react either by vivid combustion or by explosion after having been heated until red-hot ($CaSO_4.2H_2O$/Al). Second, thresholds of sensitivity to temperature, shock, friction and electrostatic discharge should be determined by techniques that are described in Chapter 4 of this book. After having carried out these experimental stages that are essential for ensuring one's safety, the pyrotechnician can finally proceed to more advanced methods for the study of the reactivity of the nanothermite he has synthesized.

With the exception of thermal analysis devices, all the systems for the characterization of nanothermites have been developed in laboratories active in this field, in response to their specific needs. The techniques applied for the characterization of nanothermites combine high technology elements: high-speed video camera, lasers, firing enclosures, transducers for rapid measurement of pressure, optic fibers and photodiodes, rapid electron microscopy, time of flight mass spectrometry, etc.

A constantly growing number of studies relate to the development and use of microtests for the characterization of the reactivity of nanothermites. These techniques, which require small quantities of matter, are less restrictive from the point of view of pyrotechnic safety, but are not ideal as means of analysis, as they ignore the scale effects that significantly influence the reactivity of nanothermites.

3.4.1.1. *Thermal analysis*

The two main thermal analysis methods are TG and DSC. These experimental techniques are specifically designed for the study of gradual transformations undergone by matter subjected to slow heating (0.5–50 K/min). This is the reason why they are exclusively used to determine the thermal sensitivity of nanothermites or to study the physical and chemical stability of the phases they are composed of. The use of these techniques to study or compare the reactive performances of nanothermites makes no sense.

TGA consists of monitoring the evolution of the mass of a sample placed in a gas flow of defined composition and exposed to a rise in temperature. TGA is used, for example, to determine the metal content of certain nanoparticles (e.g. Al, Zn, Ti), or to study the thermolysis of oxides or metal salts. It can also be used to determine the water content adsorbed at the surface of nanoparticles or to determine the proportion of an organic substance introduced in a nanothermite in order to modify its properties.

DSC consists of measuring the difference in heat flow between the sample and a reference. This technique allows one to observe any physical or chemical transformation that involves heat release or absorption. It is essentially used to determine the temperature at which a nanothermite is activated. DSC proves to be particularly useful for the study of hybrid materials composed of a nanothermite and a reactive organic substance, and to determine the compatibility between these reactive systems. There are reports on some other more specific uses, such as the measurement of the decrease in melting temperature of metal nanoparticles or the study of the slow reaction mechanism of nanothermites. DSC cannot be used to determine the combustion heat of nanothermites because the slow transformation of these compositions does not yield the same products as their vivid combustion.

Sun *et al.* used thermal analysis methods to study the oxidation of aluminum powders by oxygen diluted in argon (O_2/Ar: 25/75) or in contact with molybdenum trioxide. These experiments have shown that aluminum becomes oxidized in the solid phase, before melting. The proportion of metal that reacts before melting is significantly higher when the aluminum is nanostructured. On the other hand, aluminum nanoparticles are more reactive than micron-sized particles of this metal. Their reactivity strongly depends on their granulometric distribution. Sun *et al.* also observed, without any explanation, that the amount of heat released by the oxidation in solid phase of nanostructured aluminum was smaller than that yielded by massive aluminum [SUN 06]. The reaction of slow oxidation of solid aluminum occurs at the interface between the alumina shell and the aluminum core. This particular environment may explain the lower energetic efficiency of the metal oxidation reaction and, consequently, the lower level of oxidation heat.

By using thermal analysis techniques, Pantoya *et al.* have shown that the slow reaction of mixtures of molybdenum trioxide and aluminum occurs at a temperature that is all the higher the larger the aluminum particles are [PAN 06].

Perry *et al.* used TGA in order to develop the synthesis of tungsten dioxide nanoparticles (WO_2) by the reduction in hydrogen of nanostructured tungsten trioxide (WO_3). In this case, TGA allows to determine the complete reaction, characterized by 6.9% loss of mass, required to treat WO_3 at a temperature of 650°C [PER 04]. In another study, Perry *et al.* use thermal analysis to characterize the dehydration of tungsten oxide, an endothermic reaction that consumes 80–100 J/g of hydrated oxide ($WO_3 \cdot H_2O$) [PER 07].

Farley *et al.* used TGA coupled with DSC in order to study the slow heating (10°C/min) induced decomposition into argon of iodine oxide (I_2O_5) either in pure state or in mixture with nanosized alumina (40 nm). The low proportion of iodic acid (HIO_3) contained in the oxide dehydrates at 206.7°C, yielding again oxide. The pure oxide I_2O_5 melts and decomposes in only one stage, at 390.6°C. Based on the interpretation of the curves resulting from the thermal analysis of the I_2O_5/Al_2O_3 mixture, Farley *et al.* show that the presence of alumina modifies the iodine oxide decomposition mechanism. The iodide ions formed by oxide thermolysis are fixed by sorption on the alumina. When the temperature rises, the initially amorphous alumina undergoes a phase change and transforms into gamma alumina. This modification triggers desorption of the iodine fixed by alumina. Subsequently, Farley *et al.* show that the mode of decomposition of Al/I_2O_5 thermites depends on the granulometry of the aluminum used for their formulation. In the thermite containing micron-sized metal (15 µm), the iodine oxide decomposes without actually interacting with aluminum. On the other hand, the mixture containing nanosized aluminum (80 nm) gives a complex reaction, consisting of several stages. The endotherm of iodine oxide decomposition at 390°C is concealed by the exotherm generated by the reaction of oxygen with aluminum. At a higher temperature, the endotherm produced by the release of the iodine adsorbed on alumina is partially concealed by an exothermic reaction of aluminum, iodine and oxygen. The endotherm corresponding to the melting of aluminum that was not consumed in the reaction is observed at 654°C [FAR 10].

Shin *et al.* used thermal analysis methods to compare the reaction of compositions in which iron oxide is either physically mixed with aluminum nanoparticles or formed around them by the sol-gel method. The thermogravimetric curves show that the materials synthesized by the sol-gel technique contain a significant proportion of impurities (26.4–47.1 wt%). Nevertheless, these impurities decompose before the temperature reaches 400°C; thus they do not interact directly with the reactions of nanosized aluminum, which occur at higher temperatures (450–900°C). The reaction of aluminum with the surrounding xerogel produces a unique decomposition exotherm, while that of the Al/Fe_2O_3 physical mixture is characterized by two distinct exotherms separated by the melting of aluminum [SHI 12]. The reaction of the aluminum dispersed in the xerogel is mainly due to the oxidation of the metal in solid phase, which is rendered possible by the excellent contact with the iron oxide coating the aluminum nanoparticles. In the case of a physical mixture, the reaction in the solid phase is restricted by the lack of contact between phases, and it is due to its melting that aluminum comes into contact with the oxide. According to the DSC measurements conducted by Shin *et al.*, the energies of reaction are close to 1 kJ/g. This value is far lower than the heat released by the aluminothermic reaction of iron oxide (3.96 kJ/g) [FIS 96], which shows that the classical DSC technique is not adequate for the characterization of the real reaction of nanothermites, triggered by rapid heating [COM 13a]. The heat of reaction produced by nanothermites can be measured precisely using a bomb calorimeter.

Li *et al.* used DSC to show that the nitrocellulose contained by Al/NC and Al/Fe_2O_3/NC nanocomposite fibers decomposes at lower temperature than pure NC. Oddly, these authors observe that thermal sensitization of NC is stronger in the powdery material, prepared by simple mixture of NC and nanothermite, than in the fibrous material [LI 13].

Slocik *et al.* used DTA to show that the energy released by the reaction of iron oxyhydroxide contained in ferritin protein cages increases with the number of ferritin layers covering the surface of aluminum nanoparticles (80 nm) [SLO 13].

DSC has been used by Comet *et al.* to clarify the mechanism of reaction of the mixtures of nanosized boron and bismuth oxide (Bi_2O_3), subjected to slow heating (4 K/min). For this purpose, thermal cycling experiments have been conducted by interrupting the sample heating at various temperatures and then cooling it to ambient temperature, before heating it once again. These thermal studies have shown that the metaboric acid (HBO_2) covering the boron nanoparticles reacted in an exothermic and irreversible manner with bismuth oxide. The slow borothermic reaction takes place at a higher temperature; it is characterized by two successive exothermic signals. The reduction of bismuth oxide takes place at the surface of boron nanoparticles. It yields metallic bismuth in liquid state, which migrates by capillarity to the surface of nanoparticles. This phenomenon slows down the borothermic reaction by limiting the diffusion of bismuth oxide toward boron. This mechanism explains why the slow reaction of Bi_2O_3/B nanothermites evolves in two successive stages [COM 14b].

3.4.1.2. *Ignition by laser radiation*

Lasers are used as high technology lighters to activate the combustion of nanothermites, either compacted or in the form of loose powders. They allow for a clean ignition, which can take place as their radiation goes through windows that are transparent to it. The interaction between laser radiation and the surface of the energetic material results in rapid heating, and this leads to ignition. The main drawback of lasers is the cumbersome character of the power and cooling systems that are required for their operation.

Nanothermites are compressed without great difficulty, by uniaxial pressing of loose powders in a metallic die. The punch and the die body should allow for enough room to avoid the jamming of parts when pellets are ejected. Given the granulometry of the phases it is composed of, the compacted powder inevitably diffuses in the interstices of the compression die. Despite the high sensitivity to friction of nanothermites, there is low probability of accidental ignition, as the nanothermite powder, distributed in fine layer on the smooth surfaces of the die, does not heat up, and acts as a lubricant. It is essential to carefully clean the metallic surfaces in between two successive pressing operations in order to avoid the accumulation of matter and the welding of the die parts in case of ignition of the nanothermite interstitial layer.

The pressing of nanothermites allows to considerably reduce their porosity, but their densification beyond 60% of the TMD requires significant compression forces, because of the crushing strength of the phases they are composed of. The compression of nanothermites that are exclusively composed of nanoparticles usually yields pellets with good cohesion. Nevertheless, the mechanical strength of pellets strongly depends on the morphology and relative size of the mixed particles. Thus, pressing a mixture of submicron-sized particles of bismuth oxide of ovoid form and spherical aluminum nanoparticles yields highly brittle pellets. On the other hand, the addition of carbonaceous fibers [SIE 10] or fibrous polymers improves the mechanical strength of nanothermite pellets.

Granier *et al.* used a continuous wave carbon dioxide laser ($\lambda = 10.6$ µm), emitting 50 W, to activate the combustion of MoO_3/Al thermite pellets. The maximum power density provided at the surface of the sample is around 100 W/cm^2. The diameter of the light spot (4 mm) is comparable to that of the thermite (4.5 mm). An electromechanical shutter is used to either obturate or allow the passage of the laser beam. Combustion is observed with a high-speed camera, which records at a rate of 32,000 frames per second, positioned perpendicular to the thermite pellet [GRA 04]. Siegert *et al.* have used a comparable device, whose design was inspired by the model developed by Granier *et al.* in order to study the combustion of nanothermites obtained by mixing aluminum nanoparticles with hollow carbon nanofibers filled with manganese oxide [SIE 10]. Comet *et al.* have also used this system to study the combustion of pellets of $CuO/Fe_2O_3/P$ ternary nanothermites [COM 11b] and of pellets of Cr_2O_3/Al nanothermites [COM 11c].

Stacy *et al.* measured the ignition delay time of nanothermite loose powders, exposed to the radiation emitted by an industrial Nd:YAG laser ($\lambda = 1.064$ nm). Being applied for 10 ms in the form of a 3 mm diameter spot, the pulse provides energy of around 1.5 J. The nanothermite powder (6–20 mg) is placed in a cylindrical cavity, with a depth of 6 mm and a diameter of 4.7 mm, arranged in a steel sample tray. The apparent density of all the tested compositions is below 30% of their TMD. The ignition has been obtained either by direct interaction between laser and nanothermite or through a glass slide confining the composition. These tests have shown that the ranking of the studied compositions according to their ignition delay time differed depending on whether they were confined or not. The deposit of an absorbing carbon layer on the surface of the glass slide in contact with the compositions has finally allowed Stacy *et al.* to establish an ascending ranking of the studied nanothermites depending on their absorbance:

$$Al/CuO < Al/C_2F_4 < Al/Fe_2O_3 < Al/I_2O_5 < Al/MoO_3$$

The absorbance properties of the oxidizer have a significant impact on the ignition delay time of nanothermites that are ignited by means of laser beam [STA 13].

Qin *et al.* used a CO_2 laser that emits radiation with 10.6 µm wavelength and 30 W output power in the form of a light point with a diameter of 5 mm to ignite Al@ZnO and Al@SnO_2 nanothermites. The samples are placed in ceramic microcrucibles (~0.03 mL). The reaction is monitored by measuring the evolution of light intensity by means of an optical device associated to an oscilloscope and by observation with conventional camera (30 frames per second) [QIN 13].

3.4.1.3. *Open tray burn tests*

The open tray tests consist of studying the combustion of a nanothermite powder placed in a linear channel fabricated in a metal sheet. The advantage of this method is its simplicity, as it allows to test the strong reaction of the material in open environment. The velocities measured by this technique are lower than those determined under confinement, as the propagation by convection of hot gases in the porosity occurs less efficiently under isobaric conditions.

The open system developed by Puszynski *et al.* for measuring the combustion velocity of several aluminothermic compositions has a channeled metal tray. The open-tray cavity has a length of 43 mm, width of 18 mm and depth of 8 mm. It is fitted with two (1 mm) ports 20 mm apart, allowing to transmit the light signal accompanying the reaction to photodiodes, through optical fibers. The tested matter (~100 mg) is spread on the base of the cavity and it only fills a small part of its volume (~6.2 cm^3) [PUS 07]. This system has been improved, being fitted with perforated baffles, whose role is to minimize the powder displacement induced by the reaction blast, and to avoid the premature recording of a light signal [VAL 05].

Perry *et al.* measured the flame propagation velocity in WO_3/Al nanothermites with various molar compositions. For this purpose, the characterized powder is placed in an open channel, being leveled with a notched blade. The velocity is measured by means of two optical fibers, 20 mm apart, connected to optical detectors [PER 04]. Malchi *et al.* used the same device to study the influence of alumina, added in various proportions to a CuO/Al nanothermite, on the combustion velocity of the latter [MAL 08].

Plantier *et al.* tested the air combustion of nanothermites based on iron oxide and aluminum, in a square cross-section channel (3.175 × 3.175 mm) with a length of 5 cm. The measured velocities (0.02–120 m/s) depend on the nature of the iron oxide, but also on the proportion of oxide and aluminum. The measurement is carried out by high-speed video camera [PLA 05].

Prentice *et al.* measured the combustion velocities of several Al-Fe_2O_3-SiO_2 ternary compositions in a square cross-section channel (3.175 × 3.175 mm) with a length of 10 cm, made in a transparent acrylic block. The amount of nanothermite tested (150 mg) is ignited by a spark [PRE 05].

3.4.1.4. *Burn tube experiments*

Burn tube experiments consist of studying the evolution of the reaction of a nanothermite confined in a transparent tube. The materials most commonly used to fabricate tubular confinements are borosilicate glass, poly(methyl methacrylate) or polycarbonate. The tubes should be long enough to allow the combustion to reach a stable combustion velocity, and this means a length ranging from 10 to 15 cm when the nanothermite is loaded in the form of loose powder. The combustion of nanothermites is easily transmitted through capillaries [SON 07] or layers [SON 09] with a thickness of several hundred micrometers. But on the other hand, as propagation velocity increases proportionally with the inverse of the tube diameter [SON 07], the value of the latter should be fixed so that it does not significantly limit velocity. In most cases the optimal diameter ranges between 3 and 5 mm. The use of tubes with higher diameters should be experimentally justified, since the mass of nanothermite required to fill the tube becomes then very significant. The tubes are filled either with loose powder or with powder compacted by tapping or vibration. The ignition requires powerful activation, such as electrical discharge or combustion of a pyrotechnic composition.

Combustion propagation is often observed with high-speed camera whose recording rate, defined as a function of the propagation rapidity, usually ranges between 2×10^4 and 2×10^5 frames per second. The tubes used for simple characterizations with high-speed camera can be maintained, by means of a weighty mass, in a V-shaped channel fabricated in a parallelepipedic metal part. The tube end opposite to where ignition takes place is positioned against a stop to prevent horizontal propulsion of the tube (see Figure 3.14). The opening at the tube outlet is capped by adhesive film, which preserves the composition and facilitates tube handling.

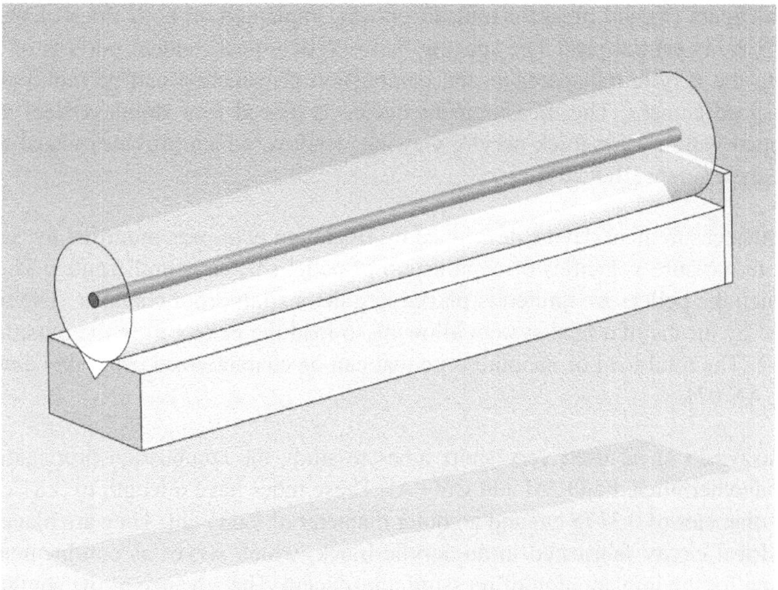

Figure 3.14. *Diagram of a simple system allowing to support a tube used for the nanothermite combustion characterization; the weighty mass that sticks the tube to the backing is not represented*

The tubes are sometimes instrumented, being fitted with optical fibers and pressure transducers for measuring the velocity of propagation in the material of the flame front and shock, respectively [BOC 05]. For this purpose, the tubes are inserted in a massive block of transparent polymer, which is used as backing for the measurement probes and also as confinement, to prevent tube explosion.

Firing nanothermites in tubular systems requires non-negligible masses of compositions (0.1–5 g). They should therefore be necessarily conducted in a detonation-proof chamber, which can resist the detonation of an equivalent mass of high explosive.

In 2005, Bockmon *et al.* described the first device for testing nanothermite combustion in tube. This system consists of an acrylic tube, whose inner diameter (0.3175 cm) is half the outer diameter (0.635 cm). The cylindrical cavity has a length of 10.16 cm and a volume of 0.8044 cm^3. The tube is loaded either with loose powder or with powder compressed by short (5 s) pulse or permanent vibration. The tube loaded with nanothermite is inserted in an acrylic block fitted with a series of optical fibers (6) and pressure transducers (6), implanted on both sides of the tube (180°), in its central part. The spacing between two measurement ports is of 1 cm. Due to the acrylic transparence, the combustion propagation can be monitored by high-speed camera. The measurement device is placed in a stainless steel vessel, equipped with a 3-cm-thick acrylic viewing window, which provides visual access to the experiment [BOC 05].

The measurement device developed by Bockmon *et al.* was modified by Sanders *et al.* to measure velocities of combustion of pellet-shaped nanothermites. The tube in which the pellets are ignited is placed in a detonation-proof chamber. The tube is loaded by means of a feed system allowing to load the pellets from the outside, one by one. The total load of nanothermite that can be characterized using this device is 5 g [SAN 07].

Puszynski *et al.* used very short tubes to study the combustion propagation in two nanothermites, Fe_2O_3/Al and CuO/Al. These tubes have a length of 3.81 cm, an inner diameter of 0.3175 cm and an outer diameter of 0.635 cm. They are placed in a cylindrical cavity fabricated in an acrylic block, which serves as confinement and backing for the implantation of pressure transducers. The advance of the combustion front is monitored by high-speed video camera. By using this device, Puszynski *et al.* noticed that the combustion of the Fe_2O_3/Al nanothermite propagated in a different manner, depending on whether the end of the tube, opposite from where the composition was ignited, was open or closed. In the open system, the depressurization gradient established in the cool composition drives the combustion, whose propagation accelerates. On the other hand, in the half-confined system, gases can only escape through the rear, and the pressure in the composition is nearly constant. In this case, the propagation velocity is stable [PUS 06].

The device that Martirosyan *et al.* used to measure the velocity of combustion propagation in an I_2O_5/Al nanothermite consists of a blind cylindrical cavity in a stainless steel block. This obstructed channel has an inner diameter of 2 mm and a length of 40 mm. The end where the composition is ignited leads into a higher volume chamber that allows for the expansion and exhaust of the gases resulted from the reaction. Velocity is measured by means of two photodiodes that were implanted perpendicular to the channel containing the nanothermite, 20 mm apart from each other. This system allows to test a low mass sample (~0.1 g) [MAR 09b].

Plantier *et al.* used a high-speed video camera (32,000 frames per second) to measure the velocity of flame propagation in several mixtures of nanostructured iron oxide and aluminum in Lexan tubes that are 10 cm long and have an inner diameter of 3.175 mm or 6.35 mm. The mass of matter to be tested (250 mg) is loaded at a bulk density of 315 kg/m^3, which represents 5–8% of the TMD. The tubes filled with powdery nanothermite are submitted to short-period vibration (5 s), in order to obtain a homogeneous density of the powder column, without inducing phase segregation. The experiments conducted in these tubes have shown that the distance needed for the combustion to transition to a stable regime is of around 6 cm, independently of the tube diameter and the nature of the iron oxide. But on the other hand, the velocity of propagation of the combustion wave varies from several meters per second to over 900 m/s, depending on the nature of the iron oxide composing the nanothermite. The combustion front is not flat, having a spearhead form, characteristic to propagation by gas convection. The presence of gas is also proven by the ejection of burning matter [PLA 05].

Malchi *et al.* studied the influence that alumina additions to an Al/CuO composition may have on its combustion by conducting experiments in instrumented tube. The nanothermite (250 mg) is loaded in an acrylic tube with a length of 8.9 cm, inner diameter of 0.32 cm and outer diameter of 0.64 cm. The powder is compacted by vibration in order to obtain uniform density (0.36 g/cm^3), reaching nearly 6% of the TMD. The tube is placed in a polycarbonate block on each side of which six pressure transducers and six optical fiber ports are located at 1 cm intervals. The reaction is activated by electric explosion of wire; its propagation is monitored with a high-speed camera, with a rate of 110,000 frames per second [MAL 08].

Weismiller *et al.* used a similar device to characterize the effect of the pressure of various gases (Ar, He, N_2) on the flame propagation in a CuO/Al nanothermite. The instrumented tube is placed in a large volume (23 L) chamber, fitted with optical windows for the observation of combustion with high-speed camera. The nanothermite powder (300 mg) is loaded in a tube at a density of nearly 6.6% of the TMD [WEI 09].

Dikici *et al.* studied the influence of preheating MoO_3/Al nanocomposite thermites on the propagation velocity of the flame resulted from their combustion. For this purpose, quartz tubes with a length of 100 mm and an inner diameter of 3 mm are used, loaded with thermites in the form of loose powders, at a density of 8% of TMD. A K type thermocouple, placed in the middle of the tube, allows to measure the thermite temperature. The tube is placed in a steel block that has a viewing window. A coiled heater is inserted around this device, then removed in order to conduct the firing [DIK 10].

Shende *et al.* used a Lexan tube, with an inner diameter of 3.2 mm and a length of 10 cm, open at both ends, inserted in an aluminum block. The combustion velocity is optically measured, by means of four optical fibers implanted at regular spacing (1 cm), and connected to photodetectors. The ignition is initiated by electric spark [SHE 08].

Apperson *et al.* designed a device that allows to measure, in one unique experiment, the propagation velocity of the combustion wave and the celerity of the shock wave generated by the reaction of nanothermites. This system is composed of polymethyl methacrylate (PMMA) tube, with an inner diameter of 3.175 mm, which is loaded with the nanothermite to be characterized and whose ignition is produced by electric spark. The velocity of propagation of the combustion front is measured by means of optical fibers, implanted in the PMMA tube and connected to photodiodes. This first section is coaxially assembled with two other tubular elements (2 × 10 cm) containing a column of air confined between an aluminum diaphragm (E ~100 μm) placed in contact with the nanothermite, and a metal plate fixed at the end of the device. The celerity of the shock is determined with pressure transducers, implanted in the tubular section filled with air. Pressure sensors are also placed in the PPMA tube in order to measure the pressure produced by the reaction of the nanothermite [APP 07].

Densmore *et al.* partially filled acrylic tubes, having an inner diameter of 3 mm and a length of 12 cm, with a CuO (50 nm)/Al (3.5 μm) thermite, varying the height of the composition column between 2 and 10 cm. The thermite powder is compacted by tapping to 14% of its TMD. The reaction is monitored by a high-speed camera, which is at the same time used to monitor the propagation of the luminous front and to measure the temperature. The velocity of the combustion wave in the section of the tube filled with thermite depends on the height of the composition column. It increases significantly when it goes from the loaded part of the tube to the section filled only with air, where its velocity approaches an asymptotic value (~1,000 m/s).

On the other hand, the temperature of the condensed phases projected in the unfilled space of the tube sustains a high value, which means that the reaction continues in the incandescent aerosol. Based on these observations, Densmore *et al.* concluded that burn tube experiments allow to measure rather the expansion velocity than the flame velocity [DEN 14]. In nanothermites, the porous volume facilitates the flow of gas and hot particles. On the other hand, matter constitutes an obstacle that causes a loss of load, which is all the stronger as the nanothermite is densified. In other words, a low-density composition offers less resistance to combustion propagation. The highest velocity that a nanothermite reaction can produce is then reached more easily and more rapidly. For this reason, it is preferable to use nanothermites in the form of loose powders when conducting burn tube experiments. The quantities of tested matter are also smaller, which offers numerous experimental advantages, especially in terms of pyrotechnic safety.

3.4.1.5. *Radial tests*

Radial tests consist of initiating the reaction of a thermite layer laid out in a circular confinement, then monitoring the combustion propagation from the center of the circle to its periphery.

Durães *et al.* used this method to study the slow combustion of a thermite composed of micron-sized particles of ferric oxide and aluminum. The thermite is compacted by cold pressing (200 MPa; 5 min) to 50–70% of its TMD, in the form of a disk of 1–2 mm thickness, in a cylindrical cavity with a diameter of 50 mm and a depth of 10 mm, fabricated in stainless steel. A PMMA lid serves as transparent confinement; it is pierced at its center by a cylindrical channel with a 3 mm diameter, filled with thermite in the form of loose powder, which serves to transmit the ignition to the underlying thermite disk. The ignition is initiated by a Nichrome resistive wire, heated by a 100 V capacitor discharge. The overall advance of the combustion front is monitored by a camera with a recording rate of 50 frames per second. The insertion of tungsten/rhenium thermocouples at 10 and 20 mm from the center allows a local measurement of velocity. A regular propagation of the reaction is characterized by an evolution of the combustion front, in the form of a succession of concentric circular profiles. On the other hand, a distortion of circularity indicates that combustion has been rendered irregular. This may be caused by: heterogeneity of the composition, asymmetrical ignition or a variation in the thickness of the thermite layer or still, combustion disruptions due to surface effects [DUR 06]. Durães *et al.* subsequently defined an asymmetry parameter to describe the irregularity of the combustion propagation. The nature of the confinement, which is stainless steel or poly(vinyl chloride), has a more significant influence on the variability of the asymmetry parameter than the composition of the Fe_2O_3/Al mixture. However, these two experimental factors are subordinate causes of the combustion irregularity [DUR 09].

Son *et al.* designed a system for the study of the radial combustion of a nanothermite based on molybdenum trioxide and aluminum. This device consists of a transparent polycarbonate cell, formed by fitting two parts. The space separating the male part from the base of the cavity in which it is inserted can be very precisely defined by machining. Loading this system is significantly easier than a slot or a capillary. Son *et al.* presented the results obtained with nanothermite layers that are only 380 µm thick. They have observed that the reaction of nanothermites whose composition is close to stoichiometry propagated radially and formed a continuous front; but on the other hand, combustion was irregularly and non-contiguously transmitted in the mixtures that are too poor or too rich in aluminum [SON 09].

3.4.1.6. *Bomb tests*

Tests in bomb are used to determine the performances of energetic materials reacting under confinement, in isochoric conditions. For example, this type of measurements is used to determine the heat of combustion of propellant powders or to study their combustion law. While the study of nanothermites by these characterization methods is possible, it is rendered difficult by the formation of reaction products in condensed phase, which can damage the metal chambers and deteriorate the transducers that are in contact with incandescent substances. As the combustion of nanothermites yields less gas than that of rocket propellants or organic explosives, the systems designed for the characterization of these energetic materials can be used to study nanothermites, without overpressure hazard.

3.4.1.6.1. Bomb calorimeter tests

The measurements conducted in a bomb calorimeter serve to determine the amount of heat generated by the combustion of nanothermites. It is worth stressing that it is the only correct manner to evaluate this essential quantity. DTA and DSC, which are sometimes used for this purpose, measure only the slow transformation of nanothermites, whose driving mechanism differs from that of combustion.

The measurement principle consists of activating the reaction of the nanothermite in a bomb calorimeter immersed in a water bath. The amount of heat released by combustion is calculated based on the rise in water temperature. The ignition of the nanothermite is ensured by means of a powder grain ignited by the heating of a metal wire. This ignition method has the advantage of being at the same time reliable and effective.

The firing of nanothermites in a bomb calorimeter does not pose any particular difficulty, other than the very significant clogging of the combustion chamber with metal residues. The morphology and adherence to walls of the combustion products depend on the nature of the nanothermite, but also on the proportion of its constituents. Thus, burning a WO_3/Al composition generates massive slags that do not adhere to the bomb's walls. On the other hand, the reaction of a Bi_2O_3/Al mixture covers the surface of the chamber with a layer of bismuth and alumina. The copper projected by the combustion of a CuO/Al nanothermite gets welded to the walls and in certain points it can even cut the steel. The combustion of borothermic compositions in the vicinity of their thermal optimum yields abundant metal splashes (see Figure 3.15). This effect disappears for compositions that are richer in boron, whose reaction yields voluminous residues with low adherence.

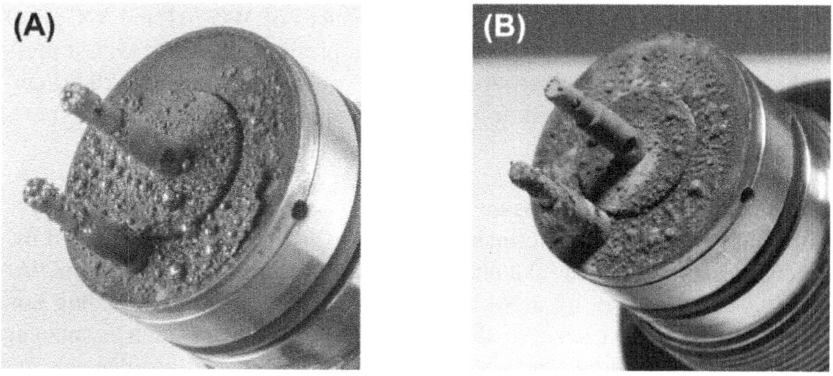

Figure 3.15. *Metal splashes produced by the combustion of Bi_2O_3/B a) and CuO/B b) nanothermites in a bomb calorimeter*

Comet *et al.* studied the evolution of the heat of reaction of Bi_2O_3/B and CuO/B nanothermites depending on their boron content, and they have then used the data obtained to calculate the reactive power of these materials [COM 14b]. The combustion heat values for mixtures of nanosized aluminum and metal sulfates are high (4–6 kJ/g); they vary quadratically with the aluminum content of these compositions [COM 15a].

The characterization in a bomb calorimeter of materials susceptible to the deflagration-to-detonation transition, such as the hybrid compositions of nanothermite and explosive, is strongly disadvised.

3.4.1.6.2. Pressure cell tests

Measurements conducted in pressure cells are used to study the reactivity of nanothermites under isochoric conditions. The experimental devices used for this type of experiments are developed by research teams, depending on their specific needs, by instrumenting laboratory pressure chambers. Ignition is provided by a hot wire or laser pulse. The reaction is monitored by measuring the pressure evolution in time and sometimes the light emission generated by combustion.

The analysis of the pressure signal allows to determine the ignition delay time, the rate of pressurization and the maximum value of pressure. Obviously, these values depend on the combustion chamber load rate, which needs to be constant to allow for a comparison of the measurements conducted on various nanothermites. Monitoring the light intensity provides supplementary information on the evolution of the reaction, and in particular on certain exothermic reactions that have little influence on the pressure.

Perry *et al.* compared the reactive performances of WO_3/Al and $WO_3.H_2O$/Al nanothermites by means of tests conducted in a pressure cell fitted with a pressure transducer and an optical fiber. A laser pulse of 30 ns, generating 20 mJ/cm^2 of energy per unit area is used to ignite the compositions and trigger pressure measurement [PER 07].

Malchi *et al.* used a pressure cell with a volume of 13 cm^3 to study the modification of energetic performances of an Al/CuO nanothermite induced by the addition of nanostructured alumina in various proportions. The Al/CuO/Al_2O_3 compositions are ignited by a Nd:YAG laser (1,064 nm) pulse, providing around 9 mJ of energy. With each test, the mass of Al/CuO composition is maintained constant (17.5 mg); the total mass of the characterized sample thus depends on the proportion of alumina it contains [MAL 08].

The results of the characterization in pressure cell of various nanothermites tested under similar conditions are presented in Table 3.9. The comparative analysis of these values shows that the performances reached with oxidizing salts are far better than those obtained with simple metal oxides. It also reveals the vital importance of gas generation for the reactivity of nanothermites.

Wang *et al.* studied the influence of several parameters on the reactivity of Bi_2O_3/Al nanothermites in a large volume commercial pressure cell (342 cm^3). The masses of tested composition are relatively significant (0.1–0.5 g) [WAN 11].

Patel *et al.* used a small volume pressure cell (0.15 cm^3), loaded with 30 mg of various nanostructured CuO/Al compositions, in order to study the performances of nanothermites that contain cupric oxide prepared in the form of nanorods by the structuring action of aloe gel [PAT 13]. Thiruvengadathan *et al.* used an even smaller cell (0.06 cm^3) to characterize the performances of nanocomposite materials produced by assembling aluminum nanoparticles and bismuth oxide with GO. The ignition is ensured by heating a nickel alloy wire with a diameter of 0.13 mm [THI 15].

Compositions	P_{max} (kPa)	dP/dt (kPa/μs)	D_c (μs)	Reference
Ag_2O/Al	68.9	0.047	1,381	[SUL 13]
$AgIO_3$/Al	2,041	393.0	172	[SUL 10a]
Bi_2O_3/Al	1,000	54	240	[WAN 15b]
$Bi(IO_3)_3$/Al	2,300	770	150	[WAN 15b]
$Bi(IO_3)_3$/Al@nitrocellulose	4,500	3,816	235	[WAN 15b]
$Bi(IO_3)_3$/Al/nitrocellulose	730	53	298	[WAN 15b]
CuO/Al	–	30.7	–	[PRA 05a]
CuO/Al	800	62.0	192	[SUL 10a]
CuO/Al	676	64.8	–	[WU 12]
CuO/Al	800	61.5	192	[FEN 13]
CuO/Al	1,100	100	220	[WAN 15b]
CuO/Al/B	–	~0–113	–	[SUL 09]
CuO/Al@nitrocellulose	300–1,800	0.095–255	247–2,250	[WAN 14]
$Cu(IO_3)_2$/Al	1,800	225	170	[WAN 15b]
$Cu(IO_3)_2$/Al@nitrocellulose	4,900	3,966	238	[WAN 15b]
$Cu(IO_3)_2$/Al/nitrocellulose	1,400	0.07	2,162	[WAN 15b]
Fe_2O_3/Al	–	0.12	–	[PRA 05a]
Fe_2O_3/Al	92.4	0.117	936	[SUL 10a]
Fe_2O_3/Al	89.6	0.117	–	[WU 12]
Fe_2O_3/Al	92.4	0.116	936	[FEN 13]
Fe_2O_3/Al	60	0.02	3,330	[WAN 15b]
$Fe(IO_3)_3$/Al	1,900	590	170	[WAN 15b]
$Fe(IO_3)_3$/Al@nitrocellulose	4,000	3,186	161	[WAN 15b]
$Fe(IO_3)_3$/Al/nitrocellulose	170	0.10	3,667	[WAN 15b]
I_2O_5@Fe_2O_3/Al	108–1,262	0.0363–45.1	183–4,280	[FEN 13]
$KClO_4$@CuO/Al	1,924–3,682	772–2,455	–	[WU 12]

$KClO_4$@Fe_2O_3/Al	1,655–3,413	669–1,420	–	[WU 12]
KIO_4/Al	3,800	2,400	124	[JIA 13a]
KIO_4/Al	909 ± 185	91 ± 19	130 ± 35	[ZHO 15b]
K_2SO_4/Al	104 ± 14	0.06 ± 0.02	2,800 ± 400	[ZHO 15b]
$K_2S_2O_8/Al$	1,206 ± 208	151 ±26	205 ± 38	[ZHO 15b]
$KMnO_4/Al$	–	1,999	–	[PRA 05a]
$KMnO_4/Al$	–	152–1,999	–	[PRA 05b]
$KMnO_4$@Fe_2O_3/Al	–	~0–841	–	[PRA 05b]
$NaIO_4/Al$	4,000	2,600	124	[JIA 13a]
NH_4ClO_4@Fe_2O_3/Al	2,682	807	–	[WU 12]

Table 3.9. *Values of maximum pressure (P_{max}), rate of pressurization (dP/dt) and combustion duration (D_c) measured for various nanothermites, in a 13 cm^3 pressure cell loaded with 25 mg of composition. The @ sign indicates that the compound in the prefix is coated with the compound in the suffix; the / sign indicates a simple mixture*

3.4.1.7. *Combustion microtests*

Combustion microtests consist of testing very small quantities of nanothermites, which from the pyrotechnic safety perspective offers a definite advantage. From a scientific point of view, these techniques have allowed for significant advances in comprehending the mechanism of reaction of nanothermites. Nevertheless, these methods cannot replace the larger scale tests, which provide more satisfactory information on the "macroscopic" pyrotechnic properties of nanothermites.

3.4.1.7.1. "T-jump" technique

The "T-jump" technique was developed by Zhou *et al.* for the characterization of energetic materials [ZHO 09b]; it has since been used to characterize numerous nanothermites based on oxides or metal salts. This experimental method consists of the rapid heating (10^5–10^6 K/s) of a nanothermite or of one of its constituents, while measuring the temperature and analyzing the composition of the reaction products. The synchronicity of measurements allows to correlate temperature and a chemical transformation and thus to determine the mechanism of reaction of a given composition. For this purpose, the nanothermite is deposited starting from a suspension on the central area (5 mm) of a platinum wire with a diameter of 76 μm and a length of around 10 mm. The passage of electric current provides the heating.

The combustion of the nanothermite layer is initiated at both ends; the two combustion fronts propagate toward the center of the platinum filament where they meet. The ignited matter is ejected but continues to react. A complex device permits to gather the ionized species produced by the combustion, which are characterized by time of flight mass spectrometry [ZHO 10].

By using this analysis method, Zhou *et al.* showed that the decomposition of copper oxide (CuO) and the accompanying release of oxygen occur at significantly lower temperature (~1,150 K) than for iron oxide (Fe_2O_3; ~1,450 K). Moreover, the oxygen contained by copper oxide is more fully and rapidly released than that contained by the iron oxide. The decomposition temperature of nanostructured zinc oxide (ZnO) (~1,900 K) is higher than the adiabatic temperature of the reaction of ZnO/Al composition (~1,822 K). This explains why this aluminothermic reaction does not occur [ZHO 10] under the experimental conditions described by these authors.

3.4.1.7.2. On-chip measurements

Gangopadhyay *et al.* invented a miniaturized device that allows for the ignition of energetic nanomaterials and the measurement of their combustion velocity, on a centimeter-sized glass sheet (2.5 × 7.5 cm). A photolithographic method is used to define a template composed of two patterns, whose respective functions are the ignition of the nanothermite and the measurement of its combustion velocity. These two elements are fabricated by first depositing a thin film of titanium (20 nm) then a thicker layer (100–200 nm) of platinum by cathode sputtering. The surface of the substrate is then spin coated with a molecular binding agent such as Poly(4-vinylpyridine) in order to facilitate subsequent fixation of the energetic nanomaterial. The morphology of the deposit of nanothermite is defined by means of a template formed by photolithography. The energetic material, in a pasty form, is finally deposited by spin coating. After the evaporation of the solvent, the excess nanothermite is eliminated by an adequate washing method. Using the chips fabricated by this method, Gangopadhyay *et al.* were able to measure the combustion velocities of various nanostructured compositions based on copper oxide (CuO) and aluminum (300–2,400 m/s), by using only several milligrams of these highly reactive substances [GAN 07]. Based on this method, Apperson *et al.* fabricated a system, by depositing two platinum films on a glass blade. The first platinum-coated area, which is heated by the passage of current, serves to ignite the nanothermite. The passage of combustion to the second platinum film generates a variation of the electrical resistance of the latter. The measurement of the duration of this phenomenon allows to determine the time needed for the combustion front to cover the length of the "film detector" (32 mm) and to deduce from it the average combustion velocity. The high sampling rate of 1.25×10^6 recordings per second allows to measure rapid combustions over a quite short distance. Apperson *et al.* observed that the combustion of a thin layer ($E < 5$ μm) of CuO/Al nanothermite generates only one signal (350 m/s), while that of a thicker layer generates two, corresponding to velocities of 1,100 m/s and 350 m/s. They also note that the thickening of the deposit of nanothermite leads to the intensification of the first signal (1,100 m/s) and advance the hypothesis that the top part of the deposit, which is in contact with air, burns more slowly than the underlying stratum. As a consequence, Apperson *et al.* recommended the use of deposits that are thick enough to allow to measure the velocity of the self-confined nanothermite. The values of velocities determined this way are very close (±5%) to those measured in tube by high-speed video. Apperson *et al.* used this miniaturized characterization device to measure the ignition delay time of CuO/Al nanothermites, and show that this time decreases with the increase in the electric power used for the ignition or with the decrease in the ignition film surface [APP 06]. Bhattacharya *et al.* used the same system to study the combustion velocity of nanothermites composed of aluminum (80 nm) and copper oxide (CuO, 9 nm) or bismuth oxide (Bi_2O_3, 150 nm), and have shown that the reaction of the former (442 ± 55.4 m/s) is more rapid than that of the latter (147 ± 6.5 m/s) [BHA 06]. The combustion velocity of the Bi_2O_3/Al composition is surprisingly low compared to the one determined by Martirosyan *et al.* (~2,500 m/s) for similar compositions [MAR 09a]. These divergent results may be explained by how these velocities are measured, in the form of unconfined trail in one case, and in a tube in the other case. According to the thermochemical data published by Fischer and Grubelich [FIS 96], the reaction of one gram of a Bi_2O_3/Al composition yields in effect 2.6 times more gas and 1.9 times less heat than the reaction of the same mass of CuO/Al. The validity of the on-chip measurement of velocity may depend on the nature of the aluminothermic composition studied.

3.4.1.7.3. Measurements in microchannels

Son *et al.* studied the combustion of nanothermites composed of nanostructured aluminum and molybdenum trioxide in microchannels. These tests have been conducted either in tubes with small inner diameter (0.48, 1.01, 1.85 and 3.63 mm) or in electrical discharge fabricated steel slots, with submillimeter hydraulic diameters (121–769 μm) and a length of 2.54 cm. The steel sheets containing the microchannels are placed in a stand, sandwiched between two sheets of PMMA, one of which is used as viewing window. For both experimental configurations, ignition is triggered by a spark. The flame front propagation is monitored by high-speed video camera. Using these experimental devices, Son *et al.* were able to show that the combustion of an Al/MoO_3 nanothermite propagates through very small cross section channels, but the combustion front velocity *(V)* decreases linearly with the inverse of tube diameter *(d)*, according to equation [3.10]:

$$V = V_{\infty} - \frac{k}{d} \qquad [3.10]$$

In light of these results, Son *et al.* proposed the quantity (ξ), calculated according to equation [3.11], to compare the performances of nanothermites:

$$\xi = \frac{k}{V_{\infty}} \qquad [3.11]$$

The nanothermites that have a smaller value of (ξ) have a smaller failure diameter and a better combustion efficiency [SON 07].

Malchi *et al.* tested the combustion in microchannel of three CuO/Al nanothermites, by placing 2 mg of sample in a 2.54 cm long channel. The first material is pure nanothermite, composed uniquely of cupric oxide and aluminum, mixed by simple dispersion in hexane. The second material is prepared by an assembly method; this composition contains a certain proportion of organic matter, brought by the ligands grafted on the nanoparticles of oxide and aluminum for their assembly. The third material is a reference composition, produced in a manner that is comparable to the second material, but with molecules that do not allow the final assembly of nanoparticles. Ignition is produced by an electric spark. A small quantity of pure CuO/Al nanothermite is used as relay to facilitate the ignition of modified compositions. The flame propagation velocity in pure nanothermite is 285 ± 48 m/s. The nanothermite produced by assembly reacts significantly more slowly, and two combustion regimes can be observed. At first, combustion propagates with a celerity of 10 m/s, over a distance of around 0.5 cm. Then velocity stabilizes at 0.25 m/s, up to the end of the microchannel. Tested under the same conditions, the reference nanothermite does not ignite. Based on a comparison between this latter nanothermite and the one formulated by assembly, Malchi *et al.* concluded that the organization of the fuel and oxidizer facilitates the aluminothermic reaction [MAL 09].

3.4.1.7.4. Projection measurement

Zhou *et al.* reported on an original method for measuring the projection velocity of a polyimide film in an exploding foil initiator device. The reference system consists of a 1 µm thick copper deposit, in the form of a bridge (450 µm × 450 µm) connecting the tops of two triangular zones. In the modified system, the copper film is covered by a foliated thermite layer (2 µm), formed of alternating layers of aluminum (250 nm) and cupric oxide (500 nm). A capacitor discharge in this device triggers an electric explosion of the copper film, which initiates the combustion of the foliated thermite. The observation of the burst by high-speed video camera shows that in the presence of the thermite, the duration of the light emission goes from 300 µs to 600 µs and the ignited particles are projected at a distance of 6 mm. Each of these reactive systems is covered by a polyimide film (25 µm), then by a sleeve of defined height (0.54 mm) on top of which a polyvinylidene fluoride (PVDF) film is finally placed. The explosion of the reactive layer, composed of copper or copper coated with thermite, projects polyimide against the PVDF film whose compression generates a high-intensity current, due to piezoelectric effect. The current emergence marks the moment when the projectile arrives. The time of flight is calculated by subtracting from this value the moment when the polyimide is ejected, which corresponds to the emergence of the voltage peak that accompanies the capacitive discharge in the copper film. This ingenious measurement device enabled Zhou *et al.* to show that the explosion of the copper wire by itself projects the polyimide slapper at a higher velocity than that obtained for thermite coated copper (381 m/s vs. 326 m/s) [ZHO 11].

3.4.1.8. *Impact tests*

Since nanothermites have low sensitivity to impact, powerful shocks are required to activate their reaction. The sensitization of nanothermites to shock can be achieved by associating them to organic substances. A further method would be to use the energy provided by the shock to generate friction forces or an electrical discharge, forms of stress to which nanothermites are highly sensitive. In the first case, the conversion could be induced by the addition of particles of a hard material, for example, silicon carbide [GIB 11c]. Mischmetal particles could be used to convert friction into thermal sparks. In the second case, fragments of piezoelectric ceramics could serve as generators of electric sparks, which are activated by shock.

Bouma *et al.* studied the ignition of MoO_3/Al compositions compressed under the action of a shear stress, either generated by a striker activated by a drop hammer or by the impact against a steel plate. The first experimental device is an instrumented form of drop hammer impact device. Thermites to be tested have the form of dense disks, fabricated by compression, with a diameter of 5 mm and a height of 1.2 mm. The thermite pellet is placed on sandpaper, on which it is squeezed under the action of the striker. The experiments conducted with this device have shown that the time of activation of the reaction of MoO_3/Al thermites is shorter when molybdenum oxide is nanostructured. On the other hand, the sensitivity to initiation by mechanical deformation diminishes with the increase in porosity. Finally, the proportion of aluminum and molybdenum oxide does not influence the mechanical sensitivity of MoO_3/Al compositions. Bouma *et al.* placed a thermite composed of nanosized aluminum, micron-sized molybdenum oxide and Viton (10 wt%), compressed to 95% of its TMD, in a 30 mm projectile ogive. The activation threshold of the reaction by the impact against a steel plate ranges between 1.26 and 1.54 km/s [BOU 07].

Cheng *et al.* determined the sensitivity to impact of Fe_2O_3/Al thermites, compressed to around 70 ± 3% of their maximum density. The composition pellets are 2 mm thick and 3.23 mm in diameter. They are glued to 7.62 mm caliber copper projectiles. They are launched by compressed air gun, which accelerates the projectile to velocities between 200 and 500 m/s. They impact a hardened steel target, placed in a chamber maintained under primary vacuum (50–100 mTorr). Cheng *et al.* used this experimental device to show that the Fe_2O_3/Al nanothermites prepared by physical mixing or assembly had a higher threshold of sensitivity to impact than the Fe_2O_3/Al thermites composed of micron-sized particles. This result is attributed to a better distribution of the impact-induced stress in the nanostructured materials [CHE 10].

The activation of the thermite reaction by an impact at 1,000 m/s against metal plates was studied by Russel *et al.* The three tested compositions (Al/I_2O_5; Al/Ag_2O; Nd/I_2O_5) consist of reactive materials with micrometric granulometry. They are loaded in a cylindrical cavity (V = 2.4 cm^3) within a brass projectile with 12.5 mm caliber and a total mass of around twenty grams. The projectiles are launched by a gun into a metal chamber (D = 40 cm; L = 338 cm), which is in turn connected to a smaller instrumented chamber (D = 40 cm; L = 149 cm) where the impact takes place. The target is a mild steel plate (E = 3.25 mm), which is perforated upon impact. The projectile fragments are in the end stopped by a metal stop block placed behind the target. The measurements of temperature and pressure are performed by means of probes, thermocouples and pressure transducers implanted in the vicinity of the impact zone, as well as at a 38 cm distance, upstream and downstream of the target. The impact is monitored by high-speed camera, whose framing rate is fixed at 10,000 frames per second. Russell *et al.* used this experimental device to show that the target should be thick enough (E > 0.6 mm) for the reaction of the thermite to be activated. The impact triggers the ignition of the compositions, but the essential part of the reaction is observed behind the target. Based on the analysis of the experimental video recordings Russell *et al.* concluded that the propagation velocity of the reaction activated by the impact has the same order of magnitude as the projectile velocity (~1,000 m/s). The measurement of temperature shows that the Al/Ag_2O composition heats more strongly the environment in which it reacts than the other compositions tested. Russell *et al.* noted that the heating produced by the reaction of the Al/I_2O_5 thermite is stronger when the iodine oxide is micron-sized [RUS 11].

3.4.1.9. *Pyrometric measurements*

The measurement of the nanothermite combustion temperature is a difficult operation due to the extreme conditions in which it has to be conducted. First of all, the reaction of nanothermites generates very high temperature (1,500–4,000 K) and projects a large quantity of incandescent particles, which can damage the measurement systems that are in contact with the sample. On the other hand, the measurement of short duration dynamic phenomena requires extremely short response times, which are unattainable with classical thermocouples, designed to operate under conditions close to thermal equilibrium. The pyrometric optical methods, which allow to remotely measure temperature, with very short acquisition times, are more adequate.

Weismiller *et al.* used multiple wavelength pyrometry to measure the temperature generated by the combustion of three nanothermites: Al/CuO, Al/MoO_3 and Al/Fe_2O_3. A first method consists of recording an emission spectrum integrated over time, by monitoring the combustion of a small amount (~10 mg) of nanothermite. A second, more complex method allows to acquire a time-resolved emission spectrum, by monitoring the reaction of nanothermites confined in acrylic tubes. In both cases, the measurement device is equipped with a collimator lens, which focuses the light emitted by the reaction and transmits it by an optical fiber to the measurement devices. The average temperatures integrated over time, determined by using the first method, are equal to 2,390 ± 150 K (Al/CuO), 2,150 ± 100 K (Al/MoO_3) and 1,735 ± 50 K (Al/Fe_2O_3). Based on these values, Weismiller *et al.* deduced that the gas emitted by the combustion of nanothermites came essentially from the vaporization of the oxide (MoO_3) or from its decomposition (CuO and Fe_2O_3) [WEI 11b]. It is worth noting that these experimental values are significantly lower than those calculated by Fischer *et al.* for the same aluminothermic compositions: 2,843 K (Al/CuO), 3,253 K (Al/MoO_3) and 3,135 K (Al/Fe_2O_3) [FIS 96]. The presence of a non-negligible proportion of alumina in these compositions is one of the causes of this phenomenon.

Densmore *et al.* developed an original method for measuring the temperature of the products of pyrotechnic reactions with good time resolution. For this purpose, a high-speed color camera is used as optical pyrometer after complex calibration in order to conduct measurements of temperature within a given range. Densmore *et al.* initially employed this technique to monitor the evolution of the temperature generated by the explosion of a C-4 charge [DEN 11], then the combustion in tube of a CuO/Al thermite [DEN 14].

3.4.2. *Morphological characterizations*

Due to their pyrotechnic nature, nanothermites are more frequently studied as combustible compositions than as materials per se. The morphological study of nanothermites and of their combustion products is nevertheless very instructive, since it often offers the opportunity to correlate the energetic properties of nanothermites to their particular form.

Most of the techniques used for the characterization of classical materials can be employed for the study of nanothermites. It should, however, be kept in mind that accidental activation of a nanothermite, triggered by inappropriate handling, may damage an analysis device or even hurt an untrained or inattentive experimenter. As a measure of protection against the risks inherent to nanothermite handling, it is essential to be aware of their pyrotechnic behavior and, in particular, their thresholds of sensitivity to various forms of stress, before studying their morphological characteristics. Characterization experiments should always be conducted with caution on the minimum quantity required by the scientific quality of the measurement to be conducted. Several points of reference permit an estimation of the incurred risk depending on the quantity of nanothermite handled. Provided that the experimenter does not have direct contact with the nanothermite, the risk posed by handling quantities below 10 mg is insignificant. For quantities ranging between 10 and 100 mg the risk is low, if the safety distance between operator and nanothermite is maintained above 20 cm. Handling larger quantities (0.1–1.0 g) requires more specific protection means, such as face shields, Kevlar gloves and clamps for manipulating samples at adequate distance (>50 cm). When larger quantities are involved, the handling of nanothermites should exclusively take place within specific pyrotechnic laboratory infrastructures.

The most commonly used methods for morphological analysis are microscopy techniques, XRD and specific area measurements.

3.4.2.1. *Microscopy techniques*

Microscopy techniques are used for the study of the morphology of nanothermites, in view of optimizing the methods of formulation and understanding their reaction mechanisms.

Optical microscopy resolution is too low for the observation of the fine morphology of nanothermites. It can however be used for observing aggregates, deposits or micron-sized objects composed of nanothermites.

Scanning electron microscopy (SEM) offers a level of resolution that is high enough to observe the morphology of nanoparticles, measure their size and analyze the distribution of various phases present in the nanothermite. The inorganic nature of the constituents of nanothermites confer them a certain thermal stability, which facilitates their observation by SEM. It is however essential to metallize the samples of nanothermites by cathode sputtering of gold or platinum, to facilitate the migration of electrons and improve the quality of images.

TEM provides information on the internal structure of matter and the distribution of elements at the scale of individual nanoparticles. Special care should be taken during preparation of samples to avoid their denaturing by dispersion in the liquid used to deposit them on the grids on which they are observed. The TEM technique is ideal for the study of nanothermites prepared by coating [COM 12] or of those having complex fine structures [YAN 13b], but it is not adequate for the observation of hybrid materials containing a nanothermite associated to heat sensitive or energetic organic substances, susceptible to degradation when subjected to the heat transferred to the sample by the electron beam.

AFM can be used to observe nanothermites containing organic phases. For this purpose, the powder should be first compacted in order to form a surface having the least possible roughness. The pressing may be soft in order to avoid altering the fine structure of the organic nanoparticles. The differences in hardness among the mineral constituents of the nanothermite and the organic compound are taken advantage of when establishing the distribution of phases in the material at nanoscale.

Blobaum *et al.* used XRD, TEM and Auger electron spectroscopy to characterize the layered structure of a CuO_x/Al thermite prepared by cathode sputtering. The successive layers of aluminum and copper oxide have thicknesses of 0.3 and 0.7 µm, respectively. The deposited copper oxide is paramelaconite (Cu_4O_3), therefore a decrease in oxygen of the oxide of origin (CuO) occurs during cathode sputtering. The Auger depth profiling indicates that the oxygen is exclusively distributed in the areas containing copper and that aluminum layers are free of copper and oxygen. The interface between layers is composed of domains of amorphous alumina with nanocrystalline sections in places. It is thicker (~10 nm) than the native alumina layer formed on the aluminum exposed to air (2–3 nm). This effect would presumably be caused either by the in-depth penetration of oxygen ions resulted from CuO dissociation during sputtering or by the fact that the CuO_x oxide layer facilitates the growth of alumina [BLO 03].

Shin *et al.* used SEM to study the morphology of a composite nanomaterial, obtained by trapping aluminum nanoparticles in an iron oxide xerogel prepared according to the gelling method using epoxides, described by Gash *et al.* [GAS 01]. The morphology depends on the Al/Fe molar proportion; when this is below two, the aluminum nanoparticles associate in clusters in between which xerogel blocks get inserted, forming a very heterogeneous mixture. When the aluminum proportion increases (Al/Fe > 2), the xerogel gets distributed among the metal particles, similarly to a cement. The mixture then becomes homogeneous at submicron scale [SHI 12].

Gesner *et al.* used transmission microscopy to measure the thickness of the alumina layer (2.7–8 nm) formed at the surface of aluminum nanoparticles (95 nm) depending on the duration of an isothermal (480°C) oxidation treatment [GES 12].

Li *et al.* used AFM to measure the modulus of elasticity of pure or Al/Fe_2O_3 nanothermite-doped NC fibers. Having sub-micrometer diameters, the fibers are arranged above a channel made in a silicon plate. Through the tip of the AFM, a force is applied in the middle of the fiber segment overhanging the channel. The measurement of the deflection produced allows to calculate the modulus of elasticity. Li *et al.* thus showed that the stiffness of the NC fibers (71 GPa) is enhanced by the incorporation of 5% by mass of Al/Fe_2O_3 nanothermite (96 GPa), but that it then strongly drops when this content exceeds 10% (48 GPa) [LI 13].

Dynamic transmission electron microscopy (DTEM) has been developed at Lawrence Livermore National Laboratory (United States) by LaGrange *et al.* for monitoring very rapid transformations of nanometric matter, such as the changes in crystalline phases or intermetallic reactions. This system reaches a spatial resolution of 10 nm and can capture images by steps of 15 ns. For this purpose, a Nd:YLF laser operating in the ultraviolet spectrum emits a 10 ns pulse toward the photocathode of the transmission electron microscope, which in turn emits an electron flow (10^8–10^9) for around 15 ns. The sample is heated using a Nd:YAG laser emitting in infrared, which generates 10 ns pulses [LAG 08]. Egan *et al.* used this microscopy technique to clarify the sintering mechanism affecting aluminum nanoparticles submitted to rapid heating (10^{11} K/s). These scientists have also used a heating substrate formed of an integrated chip coated with silicon nitride to produce less rapid heating (~10^6 K/s) [EGA 14].

Qin *et al.* used TEM to measure the thickness and to characterize the morphology of zinc oxide (ZnO) or tin oxide (SnO_2) shells formed at the surface of aluminum nanoparticles by successive deposits of molecular layers. The thickness of oxide coatings is regular, except for the areas of contact between particles, where the oxide films are thinner. Qin *et al.* analyzed the composition of the surface of these composite nanoparticles by XPS and their overall composition by energy dispersive X-ray spectrometry [QIN 13].

Dong *et al.* studied the morphology of $WO_{2,9}$ oxide nanowires coated with aluminum and the very thin alumina layer formed at the interface between these phases by high-resolution TEM [DON 13].

Comet *et al.* used AFM to observe the behavior of a RDX@Cr_2O_3 composite material, formed of porous chromium oxide (Cr_2O_3) loaded with hexogen (RDX), subjected to heating. These experiments show that the hexogen nanostructured in the matrix of chromium oxide started to decompose at a much lower temperature (~130°C) than the massive explosive (~205°C), and that the gases generated by this reaction triggered the expansion of the porous Cr_2O_3 matrix [COM 08b]. Based on these observations, Comet *et al.* proposed a mechanism that explains the combustion of hybrid RDX@Cr_2O_3/Al nanothermites, which consists of a synergy between the combustion of the explosive and the aluminothermic reaction [COM 13a].

3.4.2.2. *X-ray diffraction*

The XRD is a valuable tool that is essentially used to determine the chemical nature of the crystallized phases that are part of the composition of nanothermites or result from their reaction. It requires a much larger quantity of matter than microscopy (100–500 mg).

Siegert *et al.* drew interesting conclusions from the XRD analysis of residues of the combustion of nanothermites composed of hollow carbon nanofibers filled with manganese oxide (MnO_x) and mixed with nanosized aluminum. The presence of aluminum carbide (Al_4C_3) shows that in this type of mixture a part of the aluminum reacts with the walls of the carbon tubes. The presence of the metallic manganese nevertheless proves that the aluminothermic reaction with the encapsulated manganese oxide actually takes place. The ashes contain also two manganese oxides, the akhtensite (ε-MnO_2) and the hausmannite (Mn_3O_4), which are stable at low (<250°C) and high (900–1,200°C) temperature, respectively. The akhtensite is formed by the oxidation of manganese during its cooling, while the hausmannite is produced by the heating of the non-reduced fraction of encapsulated oxide (MnO_x) [SIE 10].

Comet *et al.* used XRD to clarify the composition of oxidizing $Al_xMo_yO_z$ phases [COM 06b] or to determine the nature of the products of the reaction in bomb calorimeter of sulfate-based aluminothermic compositions [COM 15a]. XRD can sometimes be used to calculate, by means of the Scherrer method, the average core diameter of the nanoparticles of a fuel, provided that this latter is in a crystallized state and, at the same time, the size of its particles is small (<150 nm).

3.4.2.3. *Specific surface area measurements*

The specific surface area of a material is calculated by dividing its geometrical surface area by its mass. It is calculated by the Brunauer, Emmett and Teller method, starting from the nitrogen adsorption isotherm, measured at the temperature of liquid nitrogen. The experimental determination of the specific surface area of nanothermites should be conducted very cautiously, since it requires the use of relatively significant quantities of matter (0.05–1 g). The nanothermite sample is first degassed in order to eliminate the molecules adsorbed at the surface of particles. For this purpose, the nanothermite is placed under primary vacuum or dry helium flow, prior to being heated to a temperature whose value has been previously defined by TGA.

The specific surface area is a very important quantity, whose in-depth analysis provides valuable information on the sample. When the specific surface area (S_{BET}) of a nanothermite is equal to the sum of the surface areas of its constituents (S_i), weighted by their mass fraction in the mixture (μ_i), this means that the particles are only feebly aggregated, by surface forces (equation [3.12]). On the other hand, if the experimental specific surface area is smaller, this indicates that the particles are strongly aggregated.

$$S_{BET} = \sum_i \mu_i S_i \quad \sum_i \mu_i = 1 \qquad [3.12]$$

The specific surface area (S_{BET}) is also used to calculate the mean characteristic size of a population of particles, with known density (ρ) and defined morphology.

Equation [3.13] gives, for example, the mean diameter of spherical particles:

$$\Phi_s = \frac{6}{\rho \times S_{BET}} \qquad [3.13]$$

Equation [3.14] allows to calculate the length of the edge (A) or the longer diagonal (D) of cubic particles:

$$A = \frac{8}{\rho \times S_{BET}} \quad D = \frac{8\sqrt{3}}{\rho \times S_{BET}} \qquad [3.14]$$

Equation [3.15] gives the mean diameter (Φ_c) of filiform cylindrical particles:

$$\Phi_c \approx \frac{4}{\rho \times S_{BET}} \qquad [3.15]$$

Equation [3.16] is applicable to platelet-shaped particles, whose thickness *(E_p)* is negligible compared to their surface:

$$E_p \approx \frac{2}{\rho \times S_{BET}} \qquad [3.16]$$

These equations are valid for the simplest and most frequently encountered morphologies. But it is also possible to apply this approach to the analysis of materials with much more complex morphologies. The measurement of the specific surface area of aluminum nanoparticles is used for the calculation of the average thickness of the alumina layer covering their metallic core [PES 04]. The measurement of the specific surface area can also serve to evaluate the quality of a nanothermite produced by coating. In this case, the specific surface area is defined by the initial size of the coated particles, the thickness of the deposited layer, the density of the two associated materials and the degree of aggregation. The gap between the specific surface area of the ideal material, calculated starting from the three first parameters, and the experimentally measured area is related to the degree of aggregation of the particles. The closeness of these values is characteristic of an ideal coating, while a significant gap between them indicates strong aggregation [COM 13b].

Qin *et al.* used the measurements of specific surface area to study the degree of aggregation of thermite nanoparticles, produced by coating nanosized aluminum with zinc or tin oxides films. They have also shown that the composition of Al@ZnO consisted of more strongly aggregated particles than the Al@SnO_2 composition [QIN 13]. Nevertheless, it appears that the reasoning developed by Qin *et al.* does not take into account the increase in density that induces the coating of aluminum by the oxides and its influence on the specific surface area values.

Yang *et al.* used gas adsorption to measure the size distribution of the pores of a manganese oxide (α-MnO_2) formed of entangled nanowires and a nanothermite produced by the assembly of this oxide with aluminum nanoparticles. They have also shown that aluminum fills the interparticle porosity of the oxide [YAN 13a].

Gibot *et al.* measured the specific surface area (113 m^2/g) of a chromium oxide (Cr_2O_3) composed of spherical nanoparticles, and they then used this value to calculate the average diameter of the sample particles (~10 nm). The average size of crystallites, calculated according to the Scherrer method based on the X-ray diffractogram, was of 12 nm; Gibot *et al.* were in a position to conclude that the oxide particles were monocrystalline [GIB 11a].

Comet *et al.* used specific surface area measurements to study the distribution of an explosive, the hexogen (RDX), in a porous matrix of chromium oxide (Cr_2O_3). For this purpose, the thickness of the hexogen deposit at the oxide surface is supposed to vary linearly with the mass proportion of this explosive in the composite material. The coefficient of proportionality (α) is calculated for each composite material, based on its experimental specific surface area. By analyzing the evolution of the value of (α), Comet *et al.* showed that the distribution of hexogen depended on its proportion in the RDX@Cr_2O_3 composite material. For values ranging between 0 and 36.8 wt%, the explosive solidifies in the form of independent particles at the oxide surface. For higher content (36.8–73.8 wt%) the explosive forms a continuous layer that becomes thicker in direct proportion with the hexogen content; the initial hypothesis is thus verified. Finally, above a proportion of 73.8 wt%, the explosive crystallizes outside the porous matrix of the oxide. Comet *et al.* showed that the combustion and detonation velocities of these composite materials, and the properties of Al/RDX@Cr_2O_3 nanothermites formulated starting from them, depended on the distribution of explosive in the porous matrix of oxide [COM 08b].

Comet *et al.* used the measurements of specific surface area to optimize the formulation of hybrid gels, composed of agar and ammonium paramolybdate, serving as precursors for the synthesis of nanostructured $Al_xMo_yO_z$ mixed oxidizing phases. The hybrid gels consist of interconnected filaments with nanometric diameter (20–90 nm), in which the paramolybdate takes the form of spheres, defining a rosary-like morphology. The modeling of this structure has allowed to show that above a certain value of the content in gel (~80 wt%), the mineral phase crystallizes in the form of micron-sized particles, outside the agar [COM 06b].

3.5. Conclusion: performance of nanothermites and their enhancement

There has been significant progress in the research conducted in the field of nanothermites during the past two decades, in line with the development of knowledge in this domain. The first studies were related to the fabrication of nanosized aluminum and particularly to the delicate issue of its stabilization by the formation of an alumina layer. The mode of reaction of the aluminum core–shell structure has been the subject of a lot of research [STA 15] and controversies. The properties of the aluminothermic compositions prepared starting from various metal oxides (Cr_2O_3, Fe_2O_3, WO_3, MoO_3, CuO, Bi_2O_3) have then been tested in relation with numerous experimental parameters. In recent years, a new trend has been observed: oxides are more and more frequently replaced by metal salts, which more readily release the oxygen they contain and consequently yield nanothermites that are significantly more reactive. At the same time, there is an incipient emergence of new fuels, such as boron or phosphorus. One of the most recent developments consists of compensating for the feeble gas generation by the nanothermites by the addition of substances that can be gasified by heating.

The evolution of the state of the art clearly indicates the directions to be followed for further enhancing the performances of nanothermites.

First, future research should focus on new nanostructured metal oxidizers containing a significant proportion of chemically available oxygen, such as peroxygenated salts. The morphology of the oxidizing particles has a significant influence on the performance of nanothermites [SHE 08]. The use of oxide particles with tubular morphology [PAT 02] could not only enable a better contact between oxidizer and fuel, but also facilitate gas convection. The hollow oxide fibers would also play a role of reactive reinforcement, allowing for an enhanced cohesion of the nanothermites formulated by pressing.

Several substances that have the potential of being used as nanostructured fuels in nanothermites have been mentioned in this chapter. Among them, few have been studied until present. The main difficulty that remains to be solved in this field is the passivation of metal nanoparticles, which is an essential requirement for their long-term stability. The substances used to protect metal against oxidation should have good chemical compatibility with the latter, high insensitivity to atmospheric conditions and ideally oxidizing properties at high temperature. Based on these criteria, the fluorinated polymers are among the substances with the most promising properties.

The exceptional pyrotechnic properties of nanothermites are attributed to the significant porous volume they contain. Future research should devise the means to set the porosity of the powders of nanothermites in order to confer good physical stability to these materials, while preserving their remarkable reactivity. Stabilized porosity facilitates convection phenomena but also limits to a large extent the phenomenon of coalescence of nanoparticles described by Sullivan *et al.* [SUL 12c].

When reacting, most of the thermites produce a proportion of gas that is generally quite low [FIS 96]. Some aluminothermic compositions, such as the mixtures based on iodine oxide (Al/I_2O_5) or bismuth oxide (Al/Bi_2O_3) generate a very significant quantity of gas [MAR 11]. This approach is quite restrictive, as it restrains the field of possibilities to several compositions. The second method is more general, since it consists of introducing a gasifiable substance in the nanothermite. The heat generated by the combustion of the thermite enables the thermolysis of the gas generating agent. Thiruvengadathan *et al.* produced hybrid materials, by associating Al/CuO nanothermites to various secondary explosives, such as ammonium nitrate, hexogen (RDX) or hexanitrohexaazaisowurtzitane (CL-20). The reactivity of these materials is related either to that of the nanothermite or to that of the explosive depending on their composition [THI 11]. Qiao *et al.* coated micron-sized hexogen particles with thin (~1–10 μm) or coarser (~50 μm) granulometry with a Fe_2O_3/Al nanothermite. These hybrid materials have better reactivity than the two materials that compose them, taken separately [QIA 15]. The non-energetic organic substances can also be used to increase the quantity of gas produced by the reaction of nanothermites. Bezmelnitsyn *et al.*, for example, used a polymer deriving from cellulose to reinforce gas generation by a Fe_2O_3/Al nanothermite [BEZ 10]. Comet *et al.* employed azodicarbonamide to enhance the performance of a WO_3/Al nanothermite [COM 12].

More generally, future nanothermites will be compositions with multiple constituents, whose performances will be customized in order to meet specific needs.

The shift in the status of nanothermites from laboratory curiosities to elements of pyrotechnic systems will require significant efforts in the field of technological development in the decades to come. There have been reports of several methods that allow the transition from powder to object. Methods for the fabrication of hydrophobic membranes [YANG 13a] or mats [YAN 12] of nanothermites have already been reported, but these techniques are only in their infancy.

Numerous other aspects, more anecdotal but nevertheless highly interesting, related to the reaction of nanothermites are still to be studied. Some examples are as follows: the ignition of nanothermites under the action of a flash [OHK 11a], the activation of combustion by microwaves [MEI 12] or the emission of electromagnetic radiation during the reaction [KOR 05].

4

Nanothermites and Safety

4.1. Introduction

As we have seen in the previous chapters of this book, nanothermites are new materials in terms of both morphology and pyrotechnic properties. The preparation, manipulation and use of nanothermites for research or industrial development purposes pose safety problems. The security problems have this far been studied little and represent a broad field of fundamental and applied research to be explored by scientists in the coming decade. The transfer of nanothermite technology from the laboratory to industry will rapidly stress the demand for precise knowledge and information in this field.

This chapter focuses on some of the numerous facets of this novel subject, such as pyrotechnic safety, neutralization and toxicological risk.

4.2. Pyrotechnic safety

The only way to guarantee safety in pyrotechnics is to have or acquire in-depth experimental knowledge on the energetic material to be manipulated. There is no exception to this rule, and no regulation, no matter how binding, will ever be a substitute for it. This "first principle" of pyrotechnics is obviously applicable to nanothermites.

When the experimenter has no knowledge of a nanothermite, the first preparation should always involve a limited quantity (~0.1 g) taking similar precautions to those required when handling primary explosives. The first test consists of exposing several milligrams of nanothermite to a flame and observing the mode of reaction: propagation of incandescence, vivid combustion or explosion. Upon completion of this stage, the thresholds of sensitivity to friction and electrostatic discharge (ESD) should be determined, given that nanothermites are particularly sensitive to these types of stress. Finally, if an assessment is made that the reaction of the nanothermite is vivid and/or, one of the thresholds of sensitivity is particularly low, it is essential to develop a method for the desensitization of the material prior to preparing it in a more significant quantity.

The following paragraphs deal with the sensitivity of nanothermites: measurement techniques, specific effects of various forms of stress, desensitization methods, pyrotechnic risk assessment and, incidentally, regulatory aspects.

4.2.1. *Definition and measurement of sensitivity*

The sensitivity of a nanothermite lies in its tendency to decompose when subjected to various types of stress: heating, impact, friction or ESD [SIE 11]. The threshold of sensitivity is the highest intensity stress for which no reaction is observed during six consecutive tests. The lower the threshold, the higher the sensitivity of the material. The threshold values on which an energetic material can be categorized according to its sensitivity to impact and friction have been reported by Piercey *et al.* These authors also note that the electrostatic energy that the human body is able to store ranges from 5 to 20 mJ [PIE 10].

Thermal sensitivity is determined by differential scanning calorimetry (DSC). For this purpose, a sample (1–5 mg) is placed in a crucible for analysis and submitted to gradual heating (1–20 K/min). The pyrotechnic reaction is characterized by the emission of a significant heat flow. The temperature at which the exothermic reaction starts characterizes the sensitivity of the nanothermite to slow heating. This temperature obviously depends on the heating rate: the higher the rate, the higher the temperature at which the reaction starts. Therefore, when measuring thermal sensitivity it is preferable to resort to as low as possible heating rates (1–5 K/min). Most of the nanothermites react at high temperature (>400°C), and this is a characteristic of low thermal sensitivity. Nevertheless, organic substances added to nanothermites, in order to augment the proportion of gas in the combustion products, react at lower temperature than the aluminothermic composition to which they are associated. This results in a thermal sensitization of the material. When a nanothermite is characterized by DSC in an open crucible, the study should imperatively be conducted under a flow of analytical grade argon. However, the best work method consists of using a crucible sealed by compression in order to prevent the composition from coming into contact with a reactive atmosphere.

The sensitivity to impact of nanothermites is measured by letting a drop hammer, which is guided by vertical rails, fall on a steel cell containing the sample to be tested (see Figure 4.1). The energy, expressed in joules, is calculated as the product of weight multiplied by the height of fall. Upon impact, the hammer squeezes the powdery sample placed between two metal rods that can freely slide in a steel ring, serving as both guide and confinement. The volume of the tested material (40 µL) is constant, but the mass of the sample obviously depends on its intrinsic density and consistency. In contrast to explosives, the reaction of nanothermites generates a noise that is difficult to distinguish from the one produced by the hammer drop. There are, nevertheless, several criteria based on which we can assess whether the reaction has actually occurred. First, the formation of a thin disk with metallic brightness often signals that the nanothermite has not reacted, and this can be confirmed by submitting a fragment of this material to the flame test. On the other hand, the presence of colored residues adhering to the walls of the steel cell shows the reaction has taken place. Other more original criteria may also be considered such as the mephitic odor emitted by sulfur residues formed during the reaction of sulfated nanothermites [COM 15a]. Most of the nanothermites have quite a high threshold of sensitivity to impact, which means they are little sensitive to this form of stress. The addition of organic substances generally lowers the threshold of sensitivity to impact of nanothermites, thus rendering them more sensitive.

The sensitivity to friction of nanothermites can be measured using a Julius–Peters device (see Figure 4.1). A sample of a given volume (10 µL) is placed on a rough ceramic, whose surface is covered with parallel striae. Friction is applied perpendicular to the latter, by a porcelain peg, moving to and fro. A positive reaction is usually accompanied by flame, crackling, even explosion, matter projection and often change in color. The intensity of the force applied is measured in newtons. It is determined by means of weights of variable mass hung on a lever in various positions. The threshold of sensitivity to friction of aluminum and metal oxide based compositions is lower as the granulometry of the oxide is finer, provided that the two phases are properly mixed. This way, aluminothermic compositions prepared with micron-sized oxide particles or, with oxide nanoparticles assembled into large size aggregates, are insensitive to friction. For these compositions, the threshold of sensitivity to friction constitutes a sort of macroscopic gauge for the homogeneity of the mixture.

The sensitivity to ESD is measured by sending a spark through a nanothermite (5–20 mg) placed between two electrodes 1 mm apart (see Figure 4.1). The energy of the spark is defined by fixed capacitance and voltage. Nanothermites have extremely low thresholds of sensitivity to ESD. Puszynski has shown that the thresholds of sensitivity to ESD of Al/Bi_2O_3 compositions were extremely low (0.075–2 μJ) and they decreased with the granulometry of bismuth oxide [PUS 11].

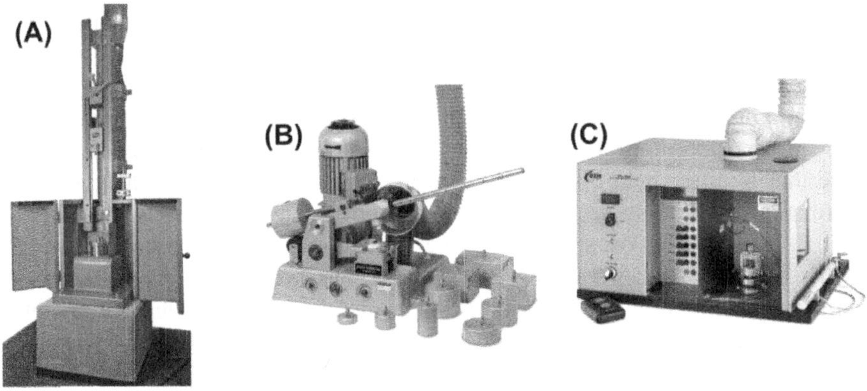

Figure 4.1. *Drop hammer impact device; a) Julius–Peters device; b) and electrostatic discharge test device; c) used for determining the thresholds of sensitivity of nanothermites to impact, friction and ESD*

4.2.2. *Techniques used for the desensitization of nanothermites*

As we have seen, the types of stress to which nanothermites are most sensitive are friction and ESD. A review of the research in this field shows that raising the threshold of sensitivity of nanothermites always induces a more or less marked decrease in their reactivity. Therefore, the challenge that pyrotechnicians face is to increase the threshold of sensitivity of nanothermites without significantly decreasing their reaction kinetics. In reality, sensitivity and performance are not opposed, rather they are two complementary aspects connected by a subtle equilibrium that the pyrotechnician can use in order to obtain the desired properties.

Puszynski *et al.* have studied the electrostatic charge that develops spontaneously on the surface of powders of aluminum and oxides of molybdenum (MoO_3) and bismuth (Bi_2O_3) kept in glass flasks. The overall charge of a given powder is proportional to its mass. The sign and the absolute value of the electrostatic charge, expressed in nanocoulomb per gram, depend on the nature and specific surface area of the material. The accumulation of charges within a material can lead to accidental ignition, all the more readily when the threshold of sensitivity to ESD is low. Puszynski *et al.* have determined the thresholds of sensitivity to ESD for three classical nanothermites, formulated from oxide nanoparticles with comparable granulometry: Al/Bi_2O_3 (0.125 µJ), Al/MoO_3 (50 µJ) and Al/Fe_2O_3 (1.25 mJ). The desensitization to ESD of the Al/Bi_2O_3 composition is difficult to achieve. Coating the aluminum with ammonium dihydrogenophosphate (3.3 wt%) has no effect. The encapsulation of aluminum in oleic acid (5 wt%) seems more promising, as it makes it possible to raise the threshold of sensitivity of the composition to 4.5 µJ. According to Puszynski *et al.* the guar gum is ultimately the most efficient desensitizing agent. The hemispherical particles (~1 mm) formed starting from solution droplets have a very high threshold of sensitivity to ESD (30.2 mJ). Nevertheless, the material becomes sensitive once more when the composite macroparticles are reduced to powder by milling [PUS 07].

Siegert *et al.* have used hollow carbon nanofibers (n-FC) to modify the sensitivity to friction, impact and ESD of nanothermites based on manganese oxide (MnO_x) and aluminum (see Table 4.1). This method proves to be particularly effective for raising the threshold of sensitivity to friction. The nanothermite produced with manganese oxide encapsulated in carbon tubes (MnO_x@n-FC/Al) is both less sensitive to impact and much more sensitive to ESD than the simple mixture of the three phases (MnO_2/n-FC/Al) [SIE 10]. This result highlights the significant influence that phase layout may have on the levels of sensitivity of these nanothermites. The increase in the threshold of sensitivity to impact is attributed to the physical separation of oxide and aluminum by the carbonaceous wall. The gap between the thresholds of sensitivity to ESD of the two ternary compositions can be explained by the mechanism of dispersion of the energy provided to the sample by the spark. When the oxide is encapsulated, the current that travels through the carbon fibers efficiently heats the nanoparticles (MnO_x and Al) situated in close proximity to each other, on both sides of the carbonaceous membrane. On the other hand, when carbon fibers are dispersed in the aluminothermic composition, the heating of the surrounding composition is less effective, as the porosity that separates aluminum and oxide nanoparticles restricts the transfer of heat required to activate the ignition. The main drawback of the method of desensitization by encapsulation in carbon nanofibers is that it significantly slows down combustion, due the addition of too high a proportion of carbon to the composition.

Composition	Oxide/n-FC/Al (wt%)	S_F (N)	S_I (J)	S_{ESD} (mJ)
MnO_2/Al	55.1/0.0/44.9	<5	31.9	1.0
MnO_2/n-FC/Al	34.6/37.2/28.2	>360	29.4	1,800
MnO_x@n-FC/Al	35.1/37.2/26.7	>360	44.2	35

Table 4.1. *Thresholds of sensitivity to friction (S_F), impact (S_I) and electrostatic discharge (S_{ESD}) of nanothermites based on manganese oxides and aluminum depending on their mass composition; "@" indicates that the oxide MnO_x is encapsulated in hollow carbon nanofibers, whereas "/" indicates a simple mixture of phases [SIE 10]*

Bach *et al.* have used carbon black, either in pure state or activated by various treatments, to modify the thresholds of sensitivity of a WO_3/Al nanothermite. The proportion of carbonaceous additive introduced into the nanothermite is low (5 wt%). Bach *et al.* have noted that the thresholds of sensitivity to impact (>50 J) and ESD (<0.14 mJ) of pure nanothermite were not modified by the addition of carbon, while on the other hand, the threshold of sensitivity to friction could be modulated using this technique (64 N > 360 N). The addition of carbon in significant proportions (20–50 wt%) makes it possible, however, to raise the threshold of sensitivity to ESD, but this effect is obtained at the expense of pyrotechnic performances [BAC 15].

Pichot *et al.* have rendered Bi_2O_3/Al nanothermites less sensitive to friction and ESD by coating the bismuth oxide nanoparticles (~200 nm) with nanodiamonds (nDs, 3.7 nm) produced by the detonation of hexolite. nD restricts the contacts between oxide and aluminum surfaces, while ensuring interparticle lubrication. As a result, the threshold of sensitivity to friction increases steadily with the nD content of the coated oxide. The increase in the threshold of sensitivity to ESD occurs when nD is in sufficient proportion to fully cover the surface of the oxide. The formation of such a screen requires three to four layers of nanoparticles, which means a proportion of 1.2 wt% nD in the oxide (see Table 4.2). Under these conditions, the molten metal projected by the explosion of the core–shell structure of the aluminum nanoparticles [LEV 06] cannot reach the oxide [PIC 15].

nD (wt%)	0	0.5	1.2	1.8	4.5
S_F (N)	5	28	128	216	>360
S_{ESD} (mJ)	0.14	0.14	0.14	2.71	1,249

Table 4.2. *Thresholds of sensitivity to friction (S_F) and electrostatic discharge (S_{ESD}) of Bi_2O_3@nD/Al compositions depend on the mass proportion of nanodiamond (nD) in the coated oxide [PIC 15]*

Thiruvengadathan *et al.* have used graphene oxide (GO, 1–5 wt%) to prepare Bi_2O_3/Al nanothermites by assembly. They observed that the threshold of sensitivity to ESD, measured according to American military standards, increases with the proportion of GO. Thus, the threshold of sensitivity of an Al/Bi_2O_3 composition prepared by physical mixing is below 0.16 mJ, while that of a GO/Al/Bi_2O_3 nanocomposite material is above 1.2 mJ. The increase in the threshold of sensitivity of the nanothermite is attributed to the increase in its electrical conductivity enabled by the GO, but also to the decrease in its surface charge. In effect, the assembly of nanoparticles into macrostructures reduces the surface susceptible to carrying an electrical charge [THI 15].

Foley *et al.* have desensitized a CuO/AL nanothermite to ESD a CuO/Al nanothermite to ESD by adding to it a fluorinated polymer, Viton A® (see Table 4.3). Pressure cell tests conducted in order to evaluate the influence of Viton A® addition on performance have led Foley *et al.* to conclude that this additive strongly increases the ignition threshold, while it has no marked influence on the pressure peak [FOL 07]. The fact that performance is unchanged is most likely due to the reaction of aluminum with the fluorine contained by Viton A®.

Viton A® (wt%)	0	3	5	10
S_{ESD} (mJ)	<0.14	0.80	2.11	21.22

Table 4.3. *Threshold of sensitivity to electrostatic discharge (S_{ESD}) depends on the mass proportion of Viton A® in a CuO/Al nanothermite [FOL 07]*

Yang *et al.* have applied polydimethylsiloxane to a membrane of MnO_2/Al nanothermite in order to render it hydrophobic. The silicone layer is deposited by a vapor phase method (235°C); it evenly covers the aluminum and oxide nanoparticles. The siliconization of the surface of reactive agents enables the controlled increase of the threshold of sensitivity to friction of the nanothermite (see Table 4.4) [YAN 13a].

D_t (min)	0	30	60	120	T
E_s (nm)	0	~1	~2	~3	$-1.26.10^{-4}.t^2 + 4.05.10^{-2}.t - 2.73.10^{-2}$
S_F (N)	<5	~140	~170	~240	$1.12.t + 105$

Table 4.4. *Thickness of the silicone layer (E_S) and threshold of sensitivity to friction (S_F) of a membrane of MnO_2/Al nanothermite depend on its duration of exposure to silicone vapors (D_t) [YAN 13a]*

Weir *et al.* have studied the threshold of sensitivity to ESD of nanothermites based on molybdenum trioxide (44 nm) and aluminum of various granulometries (0.05–40 μm). The compositions containing micrometric aluminum (4 and 20 μm) prove to be insensitive to ESD (S_{ESD} > 100 mJ). On the other hand, when the granulometry of aluminum is below 2 μm, the threshold of sensitivity to

ESD decreases very rapidly with the size of aluminum particles. Weir *et al.* have noted that the threshold of sensitivity to ESD of MoO_3/Al compositions increased with their electrical resistance [WEI 13]. In reality, the alumina shell covering aluminum nanoparticles plays the role of a dielectric, which allows the material to accumulate electrical charges and then lose them by capacitive discharge.

Comet *et al.* have patented a technique that allows mixing the constituents of a nanothermite (fuel/oxidizer) in the very instant preceding its ignition. This method relies on setting into motion, under the action of a magnetic field, metal or magnetic particles that serve as microstirring bars. Since the activation of the composition takes place during the mixing of nanostructured phases, this method ensures full desensitization of the pyrotechnic system precisely until the moment when it needs to be used [COM 13d].

4.2.3. *Pyrotechnic risk assessment*

Pyrotechnic risk is associated with the rapid release of chemical energy stored in an energetic material under the effect of unintentional stress. Very in-depth and precise knowledge on the reactive behavior of the substance to be manipulated makes it possible to minimize risk, pushing it toward zero. The pyrotechnic risk specific to the manipulation of nanothermites is still little documented in the scientific literature. The vividness of the reaction of nanothermites and their thresholds of sensitivity to friction vary considerably depending on their constituents' chemical nature and morphology. In terms of vividness, the reaction is situated between the propagation of incandescence, observed, for example, with a Fe_2O_3/P mixture, and the detonation produced by a Bi_2O_3/Al composition. The "pyrotechnic response" of nanothermites depends also on the intensity of the stress deployed to activate them: the progressive heating applied to a sample during thermal analysis triggers a progressive decomposition that does not at all reflect the reactivity induced by a rougher ignition [COM 13a]

To help tackle this problem, the following formula has been proposed [COM 13b], which enables quantitative assessment of the risk inherent to the manipulation of a nanothermite in accordance with equation [4.1]:

$$Risk \propto \frac{\left(\frac{V_d}{V_R}\right)^2}{\min\left(\frac{S_i}{S_{R/i}}\right)} \qquad [4.1]$$

where V_d is the velocity of decomposition of the energetic material, V_R is the reference velocity, S_i is the threshold of sensitivity to a given stress (impact, friction, ESD) and $S_{R/i}$ is the reference value for the threshold of sensitivity S_i.

This formula reflects the fact that the risk inherent to the manipulation of a given energetic substance increases in direct proportion with the square of its velocity of decomposition (brisance) and in inverse proportion with its threshold of sensitivity.

First, the thresholds of sensitivity to impact, friction and ESD are determined experimentally according to the previously described techniques. The values are then normalized by taking as reference, for example, the thresholds of sensitivity to impact of 2 J and to friction of 80 N, which correspond to the minimum values recommended by the United Nations for an energetic material to be transportable on public roads. As far as the ESD is concerned, the critical energy is the energy that the human body can store. According to the authors, this threshold ranges from 5 to 20 mJ [PIE 10] and can even reach 156 mJ [SIE 11]. This level of energy can serve as a reference, since in the field of pyrotechnic safety it is advisable to always favor the realistic value that maximizes danger.

The normalization of the thresholds of sensitivity over the three forms of stress has the benefit of defining dimensionless quantities that can be compared. The lowest normalized value of the threshold of sensitivity indicates the nature of the stress to which the material is the most sensitive; this value is thus retained when calculating the risk.

The velocity of decomposition is measured by an appropriate technique, using experimental conditions that realistically reflect the accidental stress to which the material may be submitted. The velocity is fixed at 10 m/s, where any value above it gives a very powerful combustion.

The limit values are defined by choosing pyrotechnic materials with extreme and opposite characteristics. The Bi_2O_3/Al compositions are a typical example of nanothermites that are both very reactive and very sensitive. According to Puszynski, a mixture composed of bismuth oxide (40 nm) and aluminum (80 nm) nanoparticles has a sensitivity to ESD of only 0.075 µJ [PUS 11]. Moreover, the study conducted by Martirosyan on the reactivity of a chemically and morphologically analogous composition shows that its reaction rate reaches 2,500 m/s [WAN 11]. The risk associated with manipulating this nanothermite, calculated according to the above-proposed formula, is approximately 1.3×10^{11}. In the same way, it is possible to define the lowest level of risk starting from a pyrotechnic composition with low reactivity and sensitivity based on red phosphorus and iron oxide nanoparticles (Fe_2O_3), such as the one described by Comet [COM 10a]. The risk calculated starting from the velocity of combustion of this nanothermite (0.83 cm/s) and from its threshold of sensitivity to ESD (24.12 mJ) is only 4.5×10^{-6}. The corresponding logarithmic scale thus ranges from –5.4 to +11.1. A risk value of zero could be observed in the case of a material whose lowest velocity of reaction and threshold of sensitivity were equal to those taken as reference ($V_d = V_R$; $S_i = S_{R/i}$). A null risk is here defined as the frontier between the domain where the risk is so low that it does not pose a real danger and the domain in which the incurred risk is significant. The ranking of nanothermites based on this scale has the merit of enabling the objective comparison of various materials that have very different chemical compositions and morphologies based on a purely pyrotechnic criterion.

A complementary approach, which is perfectly adapted to the reactive characteristics of nanothermites, consists of determining their ejection rate and reactive power under comparable experimental conditions. This test is conducted on nanothermites in the state of loose powders, which give significantly more vivid reactions than the compacted state. The reaction is triggered by an ESD whose energy is higher than the minimum energy required for activating the nanothermite. One discharge of 10–100 mJ triggers a strong ignition of most nanothermites, without causing a notable dispersion of the sample through the thermal effect of dissipation of electrostatic plasma. A device for measuring the ESD threshold such as the one presented in Figure 4.1, associated with a high-speed camera with acquisition rate above 50,000 frames per second, allows for measuring the duration of the main stage of the reaction (d_r). The ejected mass (m_e), whose value is needed for the calculation of the ejection rate (T_e) according to equation [4.2], is determined by differential weighing. The reactive power (P_r) is assessed from the heat of explosion (Q_{ex}) measured in bomb calorimeter, according to equation [4.3]:

$$T_e = \frac{m_e}{d_r} \qquad [4.2]$$

$$P_r = \frac{Q_{ex}}{d_r} \qquad [4.3]$$

This method has been used by Comet to show that Bi_2O_3/B and CuO/B compositions formulated from nanometric boron are two orders of magnitude less reactive than Bi_2O_3/Al and CuO/Al aluminothermic compositions [COM 14b].

4.2.4. *Regulatory aspects*

In France, research on energetic materials is governed by regulations that were developed in the late 1970s, with the explicit intention of providing a legal framework for the industrial activities related to the fabrication, transformation, packaging and use of explosives. For lack of more specific legislation, this decree has also been applied to research [COM 13c]. The resulting technical constraints have acted as a powerful brake on innovation, and have seriously impaired the French industry. To be able to successfully carry out their research, pyrotechnicians need a legal framework within which they are guaranteed a certain degree of freedom. The short-term effect of more flexible regulations would be to instill new energy in innovation, while on a longer timescale this would ensure the sustainability of French and European pyrotechnic industries, enabling them to compete with players from emerging countries. Future regulations should exempt research from the constraints imposed to industry, drawing inspiration from the European regulation REACH, which stipulates that chemical substances used in limited quantities for research purposes are exempted from registration, authorization and limitation obligations.

To ensure pyrotechnic safety, regulations maximize the risk, instead of measuring it. Such reasoning is justifiable when the energetic substance is manipulated in large quantities by an operator who has limited knowledge of the incurred risk, but is not applicable when small quantities are manipulated for research purposes by an expert in the field. The classes of explosives are defined by the regulation group substances with very different reactive properties, which leads to a simplification that is certainly useful for the industry, but meaningless for the researcher and, above all, harmful to their creativity.

Given their particular pyrotechnic properties, nanothermites are difficult to categorize. In terms of effects produced by their reaction, they do not belong to the category of propellant powders more than to that of primary or secondary explosives, though they do exhibit some characteristics of each of these families of energetic materials. In reality, the development of these atypical nanostructured compositions offers an extraordinary opportunity to relax the pyrotechnic regulations, to Europeanize legislation and to adapt it to the scientific and technical breakthroughs in the field. Future regulations in the pyrotechnics field should take scientific fact more determinedly into account than they do at present.

4.3. Neutralization of nanothermites

The neutralization of nanothermites consists of removing their pyrotechnic properties by the application of a specific treatment. Despite its importance, this subject is never approached by publications dealing with nanothermites. Neutralization seems to be wrongly considered by pyrotechnicians a routine operation, something that can bring no real added value to scientific activity. It is, nevertheless, essential to have knowledge on how to dispose of nanothermite residues in the laboratory, or how to destroy a non-complying batch of such materials when they are fabricated at industrial scale.

The methods for destroying organic explosives are not applicable to nanothermites. The neutralization of secondary explosives is generally carried out by burning. When they are burnt in small, unconfined quantities, substances such as hexogen (RDX) or hexanitrostilbene (HNS) are decomposed by slow self-propagating combustion. This technique is also applicable to certain primary explosives that can readily transition to detonation, such as triacetone triperoxide, by first dissolving them in a combustible liquid and the solution then being burned in the form of an aerosol. Because of their deflagrable behavior, it is not possible to destroy nanothermites by mass burning, since it would be classified as a blasting operation, and forbidden by pyrotechnic regulations. The destruction by progressive heating of certain nanothermites is a laboratory method, adequate for the destruction of small quantities of material of the order of several hundred milligrams at the most, under very specific operating conditions. It consists of using the mechanism of slow degradation of nanothermites, which is observed by thermal analysis [COM 13a]. When an aluminothermic composition is subjected to a rise in temperature, of the order of several degrees per minute, it undergoes an exothermic transformation that is not accompanied by pyrotechnic effects. In the presence of air, nanometric aluminum oxidizes first in solid phase by the diffusion of oxygen through the alumina shell that covers it. Aluminum melting is accompanied by the cracking of the alumina layer: liquid aluminum flows through the formed cracks and reacts either with the surrounding gaseous species or with the oxide nanoparticles it contacts. The slow reaction of a WO_3/Al composition produces a mixture of alumina and aluminum tungstate, while the vivid combustion of the same formulation yields alumina and tungsten.

The decomposition at high temperature of aluminothermic compositions makes it possible to thermally destroy the gas-generating agents that they contain, while the reaction of the latter does not initiate the brutal reaction of the nanothermite. This behavior has been observed with a substance generating inert gas such as the azodicarbonamide, incorporated in the form of micrometric particles in a WO_3/Al nanothermite [COM 13a]. Similar results were also obtained with aluminothermic compositions formulated starting from macroporous chromium(III) oxide (Cr_2O_3), containing nanometric particles of hexogen [COM 08b].

The neutralization of nanothermites in larger quantities should be carried out using a method in solution. The general principle implemented for abolishing the pyrotechnic properties of an energetic composition consists of separating the oxidizer from the reducing agent, either by physical treatment or by chemical destruction of one of the phases.

For example, an ignition composition such as ZPP, composed of potassium perchlorate ($KClO_4$) and zirconium, can be neutralized by a physical method consisting of ZPP dispersion in water, then separation of the perchlorate solution from the zirconium powder by decantation or filtration. This method can be successfully applied to compositions that are similar to nanothermites, in which the oxidizing phase is a soluble oxygenated salt. Referring strictly to nanothermites, the physical neutralization methods seem difficult to implement, since on the one hand metal and metal oxides powders are fairly insoluble, and on the other hand, nanoparticles are difficult to settle and they easily cross through or clog the filtration membranes. The application of a physical method in the field of neutralization of nanothermites is limited to specific cases, in which organic substances are incorporated into nanothermites to modify their properties. Coating waxes, gas generating substances and organic explosives incorporated in hybrid nanothermites can be extracted by washing carried out with a Soxhlet device, using an appropriate organic solvent.

ZPP can also be destroyed chemically by attacking, for example, zirconium with *aqua regia* or hydrofluoric acid. Chemical destruction methods are particularly well adapted to neutralize nanothermites. Oxidation of the fuel element, which is the most reactive component, should be systematically favored. Neutralization operations have to be conducted in vessels of relatively significant size in order to limit heating, avoid confinement of reactive gaseous species and overflowing of foam produced at the surface of the solution. The nanothermite should be added in small successive fractions to the solution maintained under moderate stirring. To avoid stagnation of very reactive gases and formation of an explosive gaseous atmosphere, the work should be carried out under a properly ventilated fume hood. When gas emission is observed, it is advisable to wait for it to diminish before resuming operations. Particular attention should be paid to making sure the composition is not flammable in contact with water, especially if the gaseous products of the fuel oxidation reaction are combustible. Thus, a mixture of sodium peroxide and aluminum ignite when in contact with water. Pouring a composition of sodium persulfate and nanometric aluminum into a caustic potash solution leads to a gas explosion at the surface of the solution due to the mixture of hydrogen and oxygen resulting from the reduction of water by aluminum and from its oxidation by the persulfate anion in the alkaline medium respectively.

When in contact with water, nanometric aluminum oxidizes. This reaction is, nevertheless, too slow to be used for neutralization purposes. Mixing adequate proportions of nanometric aluminum with water yields a combustible energetic composition [RIS 07], in which water plays the role of oxidizer. From a practical perspective, the operations of neutralization of aluminum powders or nanothermites should be always conducted in a volume of aqueous solution that is largely in excess compared to the quantity of treated metal, in order to avoid the formation of a pyrotechnic composition.

The aluminum nanoparticles contained in nanothermites can be destroyed by dispersion in a concentrated solution of sodium hydroxide (NaOH) or potassium hydroxide (KOH). The reaction is characterized by a strong release of hydrogen and is accompanied by marked heating of the solution. The attack on aluminum is slow as long as the alkaline solution stays cold, but it increases in virulence when its temperature rises. Since the dissolution of sodium or potassium hydroxides in water is exothermic, it can effectively be used for the initial heating of the solution, enabling the reaction of aluminum oxidizing to take place, while avoiding it racing unchecked.

It is not recommended to use a solution of concentrated hydrochloric acid for the destruction of aluminum powders, since it takes a very long time to induce the reaction of aluminum oxidation by the acid. The reaction occurs brutally and races unchecked due to the combined effect of metal oxidation by the hydronium ions and subsequent complexation by the chloride anions. This uncontrollable reaction is not observed in the case of zinc nanoparticles, which instantly react when they come in contact with the hydrogen chloride solution. Moreover, at ambient temperature, concentrated hydrochloric acid (37 wt%) readily dissolves the submicron-sized particles of quite a number of metal oxides used to formulate nanothermites, such as Bi_2O_3, CuO, Fe_2O_3, MnO_2 and NiO.

In general, mineral acids containing explosophore groups [FUZ 00], such as nitric or perchloric acid should not be used to neutralize fuels, since they form an energetic composition when in contact with the latter. Fuming nitric acid (HNO_3) is characterized by very marked oxidizing properties, thanks to which it is an oxidizer of choice and a constituent of many hypergolic mixtures. It is therefore advisable to resort to this acid as little as possible, always taking care to avoid bringing it into contact with organic substances. Tepid nitric acid, diluted by at least half in water, is very useful for destroying certain metalloids by oxidation, such as red phosphorus or boron, which are part of the composition of certain nanothermites.

Sulfuric acid (H_2SO_4) and orthophosphoric acid (H_3PO_4) are less adapted than the previous acids to the neutralization of nanothermites. When concentrated, these acids have a syrupy consistency, which indicates high viscosity. This property renders the diffusion of ionic species more difficult and considerably slows down the attack of powdery solids dispersed in these acids. Nevertheless, their viscosity strongly diminishes when they are diluted in water or heated. When in aqueous solution and at ambient temperature, the oxidation power of sulfate and phosphate anions is very low; the only active oxidizing species is the hydronium ion. At high temperature, the oxygenated anions SO_4^{2-} and PO_4^{3-} have marked oxidizing properties that should be carefully considered during the neutralization operations. Thus, the mixture of a nanometric aluminum powder with these mineral acids leads to pastes that deflagrate vividly when exposed to flame. A solution of concentrated phosphoric acid (85 wt%) diluted in an equivalent volume of water can easily dissolve a nanometric cupric oxide powder, but it has very limited dissolving action on nanoparticles of Bi_2O_3, Co_3O_4, Fe_2O_3, NiO and WO_3 at ambient temperature. The nanoparticles of Fe_2O_3 and NiO nevertheless dissolve when the solution is brought to boiling point. After a certain induction time, the previous solution of phosphoric acid slowly attacks the nanometric aluminum powder dispersed in it. The reaction takes place far more rapidly under heat. At ambient temperature, a solution of concentrated sulfuric acid (95 wt%) diluted in an equivalent volume of water does not dissolve the cupric oxide. When under heat, the aluminum and iron triad oxides (Co_3O_4, Fe_2O_3, NiO) are attacked and fully dissolved. When attacked under heat, bismuth oxide produces a colloid suspension that gelifies while cooling, but no real solution is observed.

4.4. Toxicological risk

The toxicological risk related to manipulating nanothermites and their constituents, as well as to fumes resulting from their combustion, has hardly been studied at all. It is however an essential point, especially for the industrial development of pyrotechnic systems containing nanothermites. From a thermochemical perspective, the use of certain toxic, and even radioactive, metals or metal oxides such as beryllium (Be), thorium (Th), oxides of cobalt (Co_3O_4), mercury (HgO), lead (Pb_3O_4 and PbO_2), osmium (OsO_4), or uranium (U_3O_8) is possible, but not advisable.

In the absence of specific toxicological data, the toxicity of nanothermites and their reaction products can be considered comparable to that of the oxides of the metal elements it contains. This approximation is justified by the fact that the surface of most of the nanostructured phases composing nanothermites, or resulting from their combustion, consists of metal oxides.

The various ways in which nanoparticles penetrate into the body, and their effects on various systems and internal organs of the human body, have been discussed by Buzea *et al.* in a very detailed literature review [BUZ 07].

4.4.1. *Toxicity of the constituents and reaction products of nanothermites*

The toxicity of some nanostructured substances that are representative for the constituents and reaction products of nanothermites has been described in recent scientific literature. It is worth noting that it is often studied on quite different biological systems, and that little research has focused on the effects of nanoparticles on macroscopic living beings. The toxicity of nanoparticles for the latter may be lower than the one measured on isolated cells, due to protective mechanisms that restrict the penetration of nanomaterials into the body and also to the defense mechanisms of the immune system. A wide range of nanoparticles are naturally present in the environment. They can be found in water in the form of organic colloidal substances and silica nanoparticles. Carbonaceous nanoparticles are produced by the combustion of organic materials; nanostructured mineral substances are emitted in significant quantities by volcanic eruptions. Viruses, whose sizes range from 10 to 400 nm, are an example of very complex biological nanoparticles. It is therefore plausible that the body knows how to defend itself against a nanometric aggression as long as the exposure threshold is low, since nanothermites are no exception to the Paracelsus principle, linking the toxic effect to the dose.

In a literature review, Niazi *et al.* describe various aspects of the toxicity of metal nanoparticles for microorganisms. The bacteria present in the environment are the first target for man-made nanoparticles. Therefore, the study of the interactions between these unicellular beings and the finely divided matter is vitally important, since it permits evaluation of the potential consequences of disseminating nanoparticles in the natural environment. Metal nanoparticles act mainly on the membrane of aerobic bacteria, which they disrupt and perforate. The release of reactive oxygenated species triggers the peroxidation of membrane lipids. They can also have a deleterious effect on other organic molecules that are essential to the smooth functioning of the cell: carbohydrates, proteins and DNA. Anaerobic bacteria are less sensitive to the toxicity of dissolved metals, since when presented with metal ions in excess, their external membrane reacts by setting electrochemical conditions favorable to the precipitation of the dissolved species in the form of metal or oxide nanoparticles [NIA 09]. This biological mechanism enables the local reduction of toxicity, which means that nanoparticles are less toxic than the metal ions they originate from.

According to Djurišić *et al.* the main mechanisms that could explain the toxicity of metal oxide nanoparticles for bacterial cells are as follows: (1) the production of reactive oxygenated species leading to oxidative stress, (2) the release of metal ions by the dissolution of oxide, (3) the accumulation of nanoparticles at the surface of the membrane and (4) the penetration of nanoparticles into the cell. It is worth noting that nanoparticles carrying a positive surface charge are the most toxic, as they can easily bind by electrostatic interaction to the negatively charged membrane of bacterial cells [DJU 15].

Huang *et al.* have published a graphic representation of the viability of human bronchial epithelial cells (BEAS-2B) as a function of the concentration of nanostructured metal oxides to which they are exposed. Particles of chromium oxide (Cr_2O_3), iron oxide (Fe_2O_3) and titanium oxide (TiO_2) are the least toxic. Oxides of manganese (Mn_2O_3) and nickel (NiO) have average toxicity, while oxides of copper (CuO), cobalt (CoO) and zinc (ZnO) have the highest toxicity. According to Huang *et al.* oxidative stress may be caused by the metallic or semiconductor character of nanoparticles. DNA deterioration by reactive oxygenated species is carried out through various mechanisms and can lead to cell necrosis or apoptosis. Oxidative stress also disrupts an intracellular calcium homeostasis and triggers an inflammatory response [HUA 10].

More recently, Sabella *et al.* have proposed a universal mechanism that explains the intracellular toxicity of nanoparticles containing metals. Nanoparticles penetrate the cells either by endocytosis or by a mechanism that does not require any input of energy. In the first case, nanoparticles cross the plasma by invagination and vacuolation. They are then released in the cytosol in an encapsulated form and able to interact with the endosomes, cellular organelles whose role is to sort out the substances that enter the cell. The particles are finally transferred to the lysosomes, which are in charge with breaking down the wastes resulted from cell operation. The internal acidity of lysosomes ($3.5 < pH < 5$) causes the dissolution of nanoparticles in the form of toxic metal ions, which then disperse in the cytosol and carry out their deleterious action on cellular organelles. In other terms, instead of protecting the cell, the action of lysosomes contributes to its destruction. The nanoparticles penetrating the cell by a mechanism that differs from endocytosis disperse in the cytosol ($pH = 7$) without releasing a significant amount of metal. Their toxic effect is therefore limited. For example, nanodiamond and silica nanoparticles, which do not dissolve in acid environment, are biocompatible. The modification of the surface of Fe_3O_4 nanoparticles by silica grafting lowers their cellular toxicity by preventing the passage of iron into solution; the particles thus treated nevertheless pressure their size and surface charge. Relying on the mechanism they have elucidated, Sabella *et al.* propose to diminish the intracellular toxicity of metal nanoparticles by using a coating that would allow either avoiding endocytosis or facilitating rapid expulsion of the nanoparticle out of the lysosome, or resistance to the internal acidity of this cellular organelle [SAB 14].

Braydich-Stolle *et al.* have studied the toxicity of nanoparticles of silver (15 nm), molybdenum trioxide (30 nm) and aluminum (30 nm) on a line of stem cells (C18-4) selected for the *in vitro* evaluation of nanotoxicity to the line of germ cells. The nanoparticles are placed in a culture medium with concentrations between 5 and 100 µg/mL. Incubation is carried out at 33°C, for 48 h, in an atmosphere containing 5% CO_2. The tested nanoparticles cause an impairment of the plasma membrane at very low median concentrations: 2.5 µg/mL (Ag), 5 µg/mL (MoO_3) and 4.7 µg/mL (Al). Silver and molybdenum trioxide nanoparticles also affect the mitochondrial function and metabolic activity of the cell, as well as, to a more significant extent, silver carbonate and ammonium molybdate. The research conducted by Braydich-Stolle *et al.* shows that the studied nanoparticles are cytotoxic ($Ag > Al > MoO_3$) and that their deleterious effects are manifest at low concentration and are dose dependent [BRA 05].

Puzyn *et al.* have developed a model based on the quantitative structure activity relationship to predict the cytotoxicity of metal oxides. The proposed equation links the median effective concentration of oxide nanoparticles (CE50), for which the viability of *Escherichia coli* bacteria is halved, to a descriptor. The latter corresponds to the enthalpy of formation in gas state of the metal cation at its degree of oxidation in the oxide. The predictions of this model have been experimentally verified on a series of 17 metal oxides [PUZ 11].

Chen *et al.* have studied *in vitro* the effect of alumina nanoparticles (8–12 nm) on the human endothelial cells from brain capillaries. Nanometric alumina causes mitochondrial dysfunction in the cells, which leads to oxidative stress that impairs the proteins that ensure membrane junction. The intercellular space then becomes permeable, which allows for the transfer of substances or cells contained in the blood to the brain [CHE 08]. In other terms, the cerebral vascular toxicity of alumina is characterized by the blood–brain barrier breakdown. The effects of nanometric alumina are all the more serious as the reaction products of most nanothermites contain this substance in finely divided state. According to Chen *et al.* nanometric alumina is absorbed by the pulmonary or digestive systems, and perhaps by the skin. It is probably excreted in the form of aluminum citrate. The alumina nanoparticles penetrate the cells and accumulate in the cellular organelles. The accumulation determines the many-sided toxic action of the alumina on the brain. Nevertheless, the tests conducted *in vitro* show that the presence of glutathione prevents oxidative stress in cells exposed to alumina nanoparticles [CHE 08].

Park *et al.* have orally administered aluminum nanoparticles (35.0 ± 18.8 nm) to mice, in daily doses of 15, 30 or 60 mg/kg, for 28 days. Upon completion of this experiment, the blood count of the animals exposed to the highest dose (60 mg/kg) is altered, the number of white blood cells being lower and the number of platelets higher. This effect has not been observed for a lower dose of aluminum. The analysis of the aluminum content of various organs has shown that the metal accumulates in the thymus, lungs and brain, but not in the liver or spleen. Histopathological analysis of brains, kidneys, livers and lungs of the mice exposed to the strongest dose of aluminum (60 mg/kg) did not allow determination of a difference in comparison with the organs of mice that did not ingest aluminum [PAR 11].

Yang *et al.* have studied the effects produced by a unique dose of nanometric alumina (50 or 500 mg/kg) administered to mice by tube feeding. Organs of animals sacrificed 1 day after exposure do not contain statistically higher quantities of aluminum than those of animals that did not ingest alumina. On the other hand, the ingestion of alumina has a very limited incidence on the quantity of trace elements (Fe, Cu, Zn) normally present in the organs. Biochemical analyses of blood serum of the mice exposed reveal, however, an increase in the concentration of lactate dehydrogenase enzyme, which is an indicator of cell suffering. The dosage of other biological markers shows that alumina causes hepatic and renal toxicity at low dose (50 mg/kg), although heart and immune activity are not affected. The analyses carried out over longer timescales, 7 and 14 days after the intake of alumina, show that the toxic effects disappear over time [YAN 12].

Rajsekhar *et al.* have studied the effect on mice of repeated inhalation of fine (α-Al_2O_3, 150 nm) and ultrafine (γ-Al_2O_3, 10 nm) alumina nanoparticles, dispersed in the air at concentrations ranging from 0 to 160 mg/m^3 and from 0 to 500 mg/m^3, respectively. Biochemical analysis of bronchoalveolar lavage fluid shows that the two varieties of alumina have a harmful effect on the lungs, which is characterized by marked inflammation. Histopathological examination shows that the less fine alumina causes an emphysema, which means loss of elasticity of pulmonary alveoli. The lesions observed depend on the concentration of alumina to which mice are exposed. The finer alumina accumulates in the macrophages, without causing emphysema. The quantity of nanometric alumina deposited in the lungs is proportional with the concentration of the inhaled aerosol. The accumulation is attributed to the macrophages being unable to eliminate alumina. The biological half-time of fine alumina is longer than the one of ultrafine alumina; consequently, the ultrafine alumina is more easily eliminated [RAJ 12].

Radziun *et al.* have studied *in vitro* the cytotoxicity of alumina nanoparticles (γ-Al_2O_3; 50–80 nm) for human (BJ) and murine (L929) fibroblasts. Fibroblasts are connective tissue cells that synthesize collagen, a protein that reinforces tissues and confers them mechanical strength. The cells are incubated for 24 h in alumina suspensions whose concentration varies between 10 and 400 μg/mL. The cytotoxic effect is manifest at the highest concentration (400 μg/mL) and is then slightly more marked for murine fibroblasts. The lower alumina concentrations (10–200 μg/mL) do not induce any significant cytotoxic effect. The alumina nanoparticles penetrate the two types of cells studied, without causing apoptosis. The degree of penetration of the aluminum increases with the concentration of the suspension to which fibroblasts are exposed. From these results Radziun *et al.* conclude that the alumina nanoparticles have no cytotoxic effect on the mammal cells they have studied [RAD 11].

Strigul *et al.* studied the toxicity of nanoparticles of boron (10–20 nm), aluminum (100 nm) and titanium dioxide (6 nm) dispersed in water for aquatic organisms. Acute toxicity was tested on daphnias (*Daphnia magna*), crustaceans of millimetric size. These tests have shown that boron is more toxic to daphnies than aluminum, which is itself more toxic than titanium dioxide. Strigul *et al.* noted that nanoparticles (B, Al, TiO_2) accumulated in the digestive tract of daphnias, and that their exposure for 24 h to high concentrations of TiO_2 (80 and 250 mg/L), induced a slowdown in their movements. Strigul *et al.* also observed that the dispersion of boron nanoparticles in water led to a slight increase in pH. According to them, these two effects do not explain the acute toxicity of nanoparticles, which is instead linked to their chemical properties [STR 09]. It is worth noting here that exposing boron to substances containing oxygen causes superficial oxidation and formation of boric acid (H_3BO_3), a substance antibacterial and insecticidal properties that probably explain the observed effect. Strigul *et al.* have also shown that boron nanoparticles had a toxic action on *Vibrio fischeri* bacteria, contrary to aluminum and titanium oxide.

Sadiq *et al.* have shown that the exposure for 72 h of microalgae, *Scenedesmus* sp. and *Chlorella* sp., to alumina suspensions (3–192 mg/L) inhibits their growth and lowers their chlorophyll content. These effects increase as the alumina concentration rises. Nanometric alumina (9–172 nm) has a more manifest toxic action than micrometric alumina (<5 μm). This result is explained by the aggregation of alumina at the surface of cell walls. The screen that this forms prevents light from penetrating the cell and causes a decrease in photosynthesis. The finely divided alumina, which more effectively covers the microalgae cell walls, therefore exhibits a higher toxicity [SAD 11].

Kennedy *et al.* have studied the properties of cupric oxide (CuO) nanoparticles suspended in water in the form of spheres or nanorods, as well as of combustion residues of aluminothermic compositions produced with these two oxide morphologies. The sedimentation kinetics of spherical CuO nanoparticles is slower than that of nanorods. Moreover, the measurement of the concentration of copper resulting from the dissolution of oxide shows that CuO nanorods rapidly release a significant quantity of copper when dispersed in water, while the spheres lose the metal they contain in a more gradual manner. The combustion residues of CuO/Al nanothermites sediment more rapidly and more readily release the copper they contain than the oxide of origin. Toxicity tests conducted by Kennedy *et al.* on *Ceriodaphnia dubia* have brought them to conclude that copper is less toxic for macroorganisms when it is in the form of nanoparticles than in the form of ions. The toxicity of nanoparticles containing copper is linked to the release of metal in the solution; consequently, it is more significant for nanoparticles with higher specific surface area. The addition of a polydentate ligand, ethylenediaminetetracetic acid in the appropriate concentration, reduces the toxicity of dissolved copper by complexing it. In the environment, natural organic molecules may play this role and thus limit the toxicity of nanoparticles containing copper [KEN 13].

Park *et al.* have studied *in vitro* the toxicity of nanoparticles of silver (150 nm), aluminum (100 nm), zinc (100 nm) and nickel (100 nm) nanoparticles on a human alveolar epithelial cells line (A549). These metal nanoparticles do not induce significant changes in cell morphology, but they have a cytotoxic effect when their concentration exceeds 50 µg/mL. Park *et al.* have classified these nanoparticles according to their toxicity: n-Zn > n-Ni > n-Al > n-Ag [PAR 07].

Hussain *et al.* have studied the toxicity of several nanostructured metals (Ag; Al) and oxides (MoO_3, Fe_3O_4 and TiO_2) on a line of rat hepatic cells (BRL 3A). This research revealed that silver nanoparticles are the most toxic, while molybdenum trioxide is moderately toxic. Iron and titanium oxides are little, if at all, toxic at the concentrations used in these studies [HUS 05]. Turkez *et al.* studied the toxicity of tungsten trioxide nanoparticles (100 nm) on rat hepatocytes. When at low concentrations (<150 ppm), the WO_3 nanoparticles have no toxic effects. At higher concentrations (300–1,000 ppm), they reduce the viability of cells and trigger the formation of reactive oxygen species that damage DNA. WO_3 nanoparticles thus have an indirect genotoxic action resulting from the formation of DNA adducts [TUR 14]. Hasegawa *et al.* have shown that nanometric tungsten trioxide (100 nm) had a mutagenic action on two strains of *Salmonella typhimurium* (TA1537 and TA98), while the oxide with micrometric granulometry (10.5 µm) has no such action [HAS 11].

The effects of nanoparticles of alumina (30 and 40 nm) and aluminum (50, 80 and 120 nm) on rat pulmonary alveolar macrophages have been studied by

Wagner *et al.* At high concentrations (100 and 250 μg/mL), the alumina has a delayed toxic action, which only appears after a sufficiently long exposure time (96 and 144 h). Aluminum toxicity at the same concentrations manifests more rapidly (24 h). The exposure of macrophages to non-toxic levels of aluminum (25 μg/mL) reduces their phagocytic activity. None of the tested nanoparticles included an inflammatory response in the studied macrophages [WAG 06].

In light of the research described above, it appears that nanometric aluminum and alumina have relatively low toxicity. The toxicity of metal oxides that are of interest for the preparation of nanothermites depends on their nature, but also on their preparation method. Oxides of bismuth (Bi_2O_3), iron (Fe_2O_3 and Fe_3O_4) and titanium (TiO_2) have low toxicity. Oxides of cobalt (Co_3O_4), nickel (NiO) or copper (CuO) have *a priori* much higher toxicity. The biogenic synthesis of oxide nanoparticles may in certain cases enable reduction of their toxicity [SEA 15].

4.4.2. *Analyses of specific risks and good practices*

Though the toxicological risks inherent to the fabrication and manipulation of the most studied nanothermites are *a priori* quite low, it is nevertheless essential to take particular precautions to protect persons exposed to nanothermites on a regular basis while exercising their profession. Ultrafine particles are capable of crossing the respiratory epithelium. By passing into the bloodstream, they can reach internal organs such as the liver, spleen or brain. They can also have damaging effects on blood coagulation, cardiac functioning or the central nervous system [KRE 13]. Until present, there has been no codification of good practices; they simply result from the analysis of experimental observations conducted in laboratories operating in this field. Observation by scanning electron microscopy of numerous powders composed of nanoparticles shows that the elementary particles of nanometric or submicrometric size manifest a strong propensity to assemble in micrometric aggregates under the action of surface forces (see Figure 4.2). The proportion of free particles susceptible to disperse in the atmosphere is in reality very low. The interactions that keep nanoparticles together are so strong that, in order to obtain homogeneously mixed nanothermites, clusters of oxide and fuel nanoparticles have to be dissociated by dispersing them in a liquid and applying powerful ultrasonication.

Respiratory tract penetration by particles depends on their size. A report by the Swiss Federal Commission for Air Hygiene on the fine powders in Switzerland mentions that micrometric particles (10–30 μm) deposit in the nose, throat and trachea; particles with diameters below 10 μm reach the bronchi and the bronchioles, while the smallest particles (<2–3 μm) deposit in the pulmonary alveoli [CFH 07].

Figure 4.2. *Scanning electron microscope images at × 1000 and × 100,000 magnification of aluminum; a) and tungsten trioxide; b) nanoparticles, illustrating the phenomenon of nanoparticle aggregation into micron-sized structures*

Most elementary particles used in the formulation of nanothermites have diameters ranging from 0.02 to 2 μm. Therefore, they are theoretically capable of reaching the pulmonary alveoli and depositing there, crossing the protection barrier constituted by the pharynx, trachea and bronchi. Nevertheless, considering that aerosols mainly consist of particle clusters of the above-mentioned types observed by SEM (see Figure 4.2), it is safe to state that the majority of the inhaled mass deposits in the higher respiratory tract. Under the action of body fluids (saliva and mucus), the aggregates can then dissociate and interact with the tissues. At this stage, toxicity depends on the chemical nature, solubility and form of particles. The intrinsic chemical toxicity of a material is generally increased by nanostructuration, which facilitates its penetration into the organism through biological membranes. The high specific surface area of nanoparticles facilitates their dissolution, which shortens their lifetime as particles in the body, but increases their chemical toxicity by the rapid formation of ionic species in high quantities, which are capable of interference with biological mechanisms. The particles with high elongation factors, such as carbon nanotubes, have higher toxicity than the more compact particles. Though most constituents of nanothermites do not have filiform morphology, certain nanostructured fibrous substances are sometimes added to raise the threshold of sensitivity to ESD, and to improve the cohesion of elements shaped by compression [SIE 10]. The addition of fibrous substances that are harmful in themselves may, however, reduce the dispersibility of the nanoparticle mixture during operations involving manipulation of energetic material by the formation of entanglements that facilitate the formation of large size clusters.

At the macroscopic scale, oxides of transition metals are considered relatively insoluble in an aqueous environment, except for strongly acidic or alkaline environments, depending on the metal nature. The body liquids that are present in the respiratory tract have pH values close to those of neutral substances, and therefore do not facilitate the dissolution of oxide particles. Inhaled metal oxides therefore have a certain physical and chemical stability, which limits their deleterious action but also prevents their elimination by the body unless they are mechanically disposed of. In the long term, gradual dissolution of certain finely divided oxides by the complexing action of anionic species contained in body liquids remains a possibility.

Experiments conducted on animals have shown that several metals pass from the nasal cavity to the olfactory bulb via olfactory receptor neurons [SUN 01]. Some of these metals can then migrate deeper into the brain via secondary olfactory neurons. Metals used to formulate nanothermites or those resulting from their reaction (Al, Bi, Cu, Mn) accumulate in the olfactory bulb. Sunderman further reports that experiments conducted on rabbits and rats have shown that aluminum in the form of soluble salts crosses the nasal barrier and reaches various locations in the brain [SUN 01]. Analysis of the results compiled by Sunderman shows that nasal absorption of metals is studied with metal salts soluble in an aqueous environment. The formation of a cation, subsequently complexed by carnosine, is supposed to be the origin of this unusual metal absorption mode. Due to the low solubility of metal oxides used to formulate nanothermites, it seems unlikely that the metal they contain can penetrate in significant quantities via the nasal tract into the brain.

Nanothermites in which nanometric aluminum powders are used as fuel are among the most reactive. This is the reason why most of the research published thus far in the scientific literature has focused on aluminothermic formulations. The risk inherent to the use of nanometric aluminum is therefore an important aspect of the fabrication and implementation of nanothermites in research laboratories and also in industry. As noted previously, the nanometric aluminum particle has a specific structure, in which the aluminum core is covered by an alumina shell. This system is metastable, since from a thermodynamic point of view oxidation of the aluminum core is possible, but does not occur, because the alumina layer limits the diffusion of gaseous oxygen to the extent where the oxidation rate tends to zero at ambient temperature. In the presence of liquid water, the alumina layer does not efficiently protect aluminum against corrosion. According to Puszynski, the hydration reaction continues until full oxidation of the metal is reached. It starts at the surface of the aluminum and advances according to a diffusion controlled mechanism, described by a core collapse model, without modification of the overall form of the hydrating particle [PUS 06]. In an advanced stage, when the aluminum proportion is below 20% of the initial content, the hydration products form a distinct, separate phase. The "digestion" observed *in vitro* by slow hydrolysis of aluminum nanoparticles in pure water is probably accelerated *in vivo* in the presence of chloride anions contained in the bodily liquids. It may be assumed that the essential part of the gelatinous mass of aluminum hydroxide thus formed is then eliminated by mucus or expectorated.

The reddening of absorbent papers in contact with nanometric aluminum powder is an interesting experimental observation. It signals that the reducing character of aluminum nanoparticles is marked enough to induce the dehydroxylation of the polysaccharides in the paper fibers. The effect of this chemical carbonization is comparable to that observed when exposing the paper to high temperature. It indicates a strong irritating action of the aluminum nanoparticles when they come into contact with the tissues, and in particular with the mucosae.

Reaction of a nanothermite produces an incandescent aerosol that disperses in the surrounding atmosphere and rapidly extinguishes. During experiments of firing under confinement, fumes condense in massive deposits (see Figure 4.3), whose macro- and microscopic morphologies depend on the nature of the composition. The combustion of a nanothermite based on tungsten trioxide yields slags in which the initial nanostructuration is completely abolished (see Figure 4.3). Residues of the combustion of a composition formulated from cupric oxide are instead composed of clusters of nanoparticles assembled by sintering (see Figure 4.3). When in contact with air, metal particles released by combustion are covered by an oxide layer. From a toxicological point of view, their short-term behavior can be equated to that of the metal oxide particles. In the longer term, the behavior of these particles depends on the stability of the oxide layer and on its capacity to limit the dissolution of the metal core it coats.

Exposure to certain vapors or metal nanoparticles emitted by the combustion of nanothermites may trigger symptoms comparable to those experienced by personnel working with incandescent metals, such as welders and foundry workers. Though benign, the main acute pathology is metal fume fever, causing flu-like syndromes that disappear in 1 day or 2 days. The chronic pathologies that can be caused by recurring exposure can be allergic (asthma) or inflammatory (bronchitis). In the most extreme cases, inhaling metals or oxides can lead to pneumoconioses, which irreversibly damages lungs.

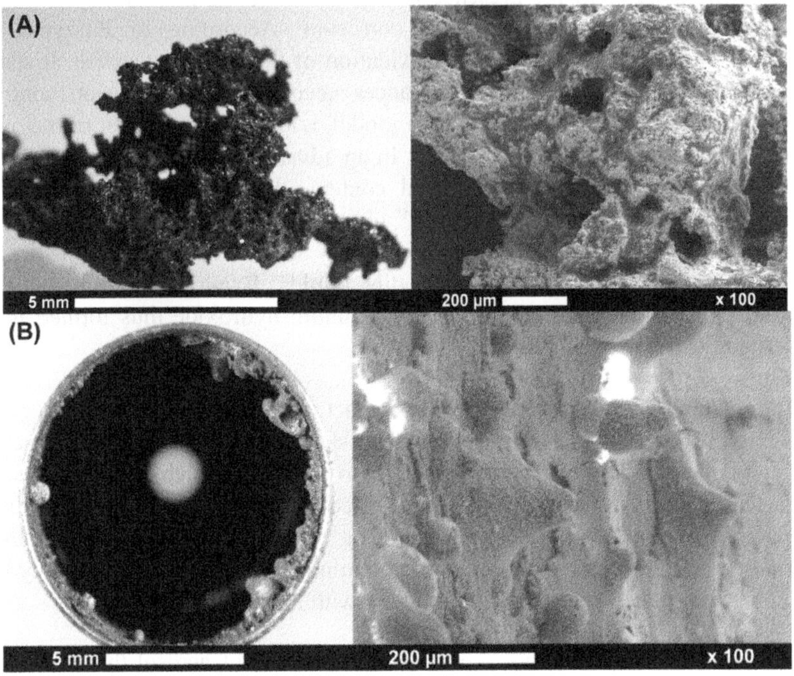

Figure 4.3. *Macro- and microscopic morphology of the residues produced by the reaction in a pressure cell of WO_3/Al a) and CuO/Al b) nanothermites*

Fortunately, there are several means to efficiently reduce exposure to the constituents and products of reaction of nanothermites.

First, the unitary operations of sampling, weighing and mixing should be systematically conducted under a laboratory hood, and as far as possible in confined volumes. Even though the nanoparticles dispersed in the atmosphere are locally animated by random movements related to Brownian motion, on the whole they follow the movement of the mass of air in which they are situated. In the absence of an efficient ventilation system, the nanostructured powders should be produced, manipulated and stored under hermetically sealed confinement in order to avoid all dispersion in the atmosphere. At the industrial scale, continuous elaboration by methods such as the flash evaporation technique developed by Risse *et al.* is the safest and most efficient way to prepare significant quantities of nanostructured energetic powders [RIS 12].

Second, it is advisable to avoid all dispersion, particularly under the action of electrostatic forces. In this respect, wearing insulating gloves is not recommended, since it facilitates dispersion of nanoparticles by static electricity and by the mechanical effects induced by glove removal. Applying hand cream is an effective way to protect oneself from skin retention of nanoparticles due to skin porosity. In case a significant quantity of nanothermite powder is spilled on the epidermis, the procedure to be followed consists of first disposing of the powder that does not adhere to the skin by softly shaking it off, under a laboratory hood. As a second step, the skin is washed under a water jet, avoiding friction. The final step in the decontamination consists of thoroughly washing with plenty of water, using a tensioactive liquid, whose use is essential to eliminating most of the particles embedded in the epidermis. Skin is a barrier impervious to nanoparticles, with the exception of hair follicles. Nanoparticles can penetrate this barrier when the skin is damaged or subjected to mechanical stress [BUZ 07].

Finally, the use of a particle-filtering mask offers a high level of protection, significantly reducing the proportion of fine particles or the air inhaled from (99%). The efficiency of aerosol capture depends on the size of its component particles; particles with sizes ranging from 30 to 1,000 nm are only partially retained by the filtering devices. However, the most penetrating particles (200–300 nm) are also those whose overall deposit rate in the respiratory tract is lowest [CFH 07].

Manipulation of nanothermites or of their constituents for experimental purposes is inevitably accompanied by small quantity spillage of these materials on dedicated work surfaces. When contaminated, surfaces should be cleaned with moist absorbent paper, which allows efficient collection of particles, and enables their fixation by the paper fibers to avoid their dispersion. To destroy the nanoparticles, a treatment specific to the soiled paper can be implemented, depending on the nature of the absorbed matter.

One can minimize exposure to aerosols produced by the combustion of nanothermites by carrying out the works in armored chambers, which are proven to resist the effects of explosions, and are maintained under dynamic low pressure conditions by an air aspiration device. This type of infrastructure does not provide absolute protection, and it is therefore essential to wear a respirator mask, similar to those used when elaborating nanothermites, for the entire duration of experiments that involve dispersion of nanostructured matter by vivid reactions. When accidentally exposed to combustion fumes, it is recommended to thoroughly wash the contaminated work suit and especially hair, which is a real trap for nanoparticles. Due to the electrostatic charges accumulation and the presence of sebum, hair surface facilitates the capture and retention of nanoparticles. Obviously, the surface area of a person's hair is highly variable, since it depends on hair diameter, length and number. Elementary calculation gives, nevertheless, a surface area ranging from 10 to 60 m^2, with an average from 1 to 10 m^2. In comparison with the average surface area of the human body (1.7 m^2), these values are high and they show the crucial importance of protecting one's hair by wearing headwear, and maintaining special care with regards to hair hygiene for the personnel in contact with nanoparticle aerosols resulting from combustion of nanothermites.

Hair has been exposed to an aerosol produced by the combustion of a WO_3/Al nanothermite. Scanning electron microscope images have shown that hair was covered with a continuous layer of submicron-sized particles and also, in places, by spherical aggregates (~1-10 μm) of these particles (see Figure 4.4).

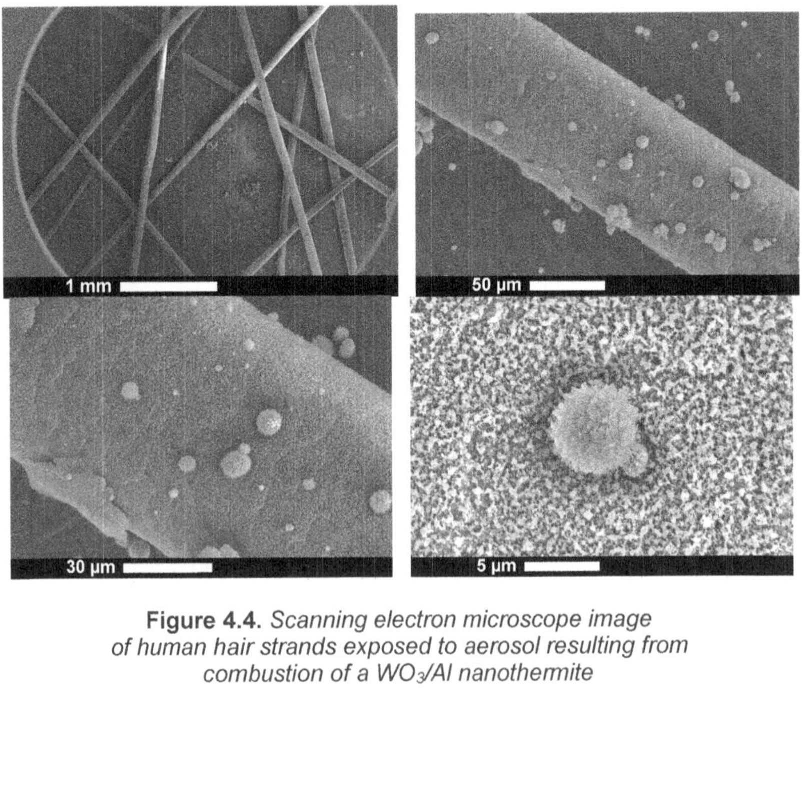

Figure 4.4. *Scanning electron microscope image of human hair strands exposed to aerosol resulting from combustion of a WO_3/Al nanothermite*

Exposure to constituents of nanothermites and their reaction products could be measured quantitatively by setting up the metallic profile [GOU 10] of personnel working in this field, starting with blood sample collection performed by occupational health care services.

4.5. Conclusions and perspectives

The pyrotechnic and toxicological risks inherent to the preparation, manipulation and use of nanothermites at the laboratory scale are quite limited due to the small quantities studied and the appropriate protective measures taken by the scientists working in this field. Significant research efforts are still underway in order to render nanothermites less sensitive to friction and ESD without significantly degrading their reactivity. With regard to pyrotechnic safety, developing methods by which very small quantities of matter are used to establish a ranking of nanothermites according to the risks they pose to the experimenter will be of high importance.

The integration of nanothermites into complex pyrotechnic systems produced at the industrial scale will require more precise knowledge of the toxicological risks incurred by operators who are exposed to nanothermites and products of their reaction on a daily basis. Methods of fabrication yet to be conceived will have to allow for maintaining the nanostructured matter under confinement until its final integration, in a stabilized form, in the fabricated object. This approach will enable a strong limitation of nanoparticle dispersion in the atmosphere. The emission of metal or oxide nanoparticles in the air is certainly not a new problem, as numerous metallurgical industries have released these types of aerosols [BER 13] since the beginning of the industrial age. The protection of operators responsible for manipulating nanostructured matter on a recurrent basis requires protective means that are adequate for the specific circumstances of the exposure: the nature of products, their concentration in the air, the frequency and duration of the operation. There should be systematic measurement of nanoparticle content in the workplace atmosphere in all industrial activities that use finely divided matter. For example, the National Research and Safety Institute (Institut National de Recherche et de Sécurité) has issued recommendations concerning the characterization of potential emission and occupational exposure to aerosols during operations that involve nanomaterials [WIT 12].

Conclusion

Nanothermites are atypical energetic materials with exceptional pyrotechnic properties. Their vivid combustion is activated by a low energy albeit high-power input, such as an open flame [COM 15], a wire heated by the Joule effect [ZHO 09b], a laser beam [GRA 04], an electrostatic discharge [PUS 07] or a flash [OHK 11a]. The reaction of nanothermites releases a significant amount of heat, but it generally yields little gas. It propagates at highly variable velocities, ranging from several centimeters [COM 11c] to several kilometers per second [WAN 11]. The addition of gasifiable energetic substances reinforces the reactivity of nanothermites [THI 11].

In the coming years, research on nanothermites will focus on other nanostructured fuels besides aluminum. It will also engage in the development of oxidizing substances that are stronger than simple metal oxides. The development of methods for the preparation of organic nanoparticles such as the flash-evaporation technique [RIS 12] provides interesting perspectives to the preparation of high-performance hybrid materials.

The development of highly reactive nanothermites will require significant research efforts to control the sensitivity of these materials to impact, friction and electrostatic discharge. In this field, the challenge is to maximize performance, which means to develop materials that are at the same time reactive and not too sensitive.

In terms of morphology, nanothermites are characterized by highly developed interparticle porosity, and this is the "secret" behind their impressive reactivity. The large porous volume they contain facilitates the rapid propagation of combustion by hot gas convection. One of the major challenges to be met in view of the development of numerous applications is the stabilization of the porosity of nanothermites.

Nanothermites will very likely be the most frequently used energetic materials in future pyrotechnic systems, since it is possible to customize their properties in view of a defined purpose. The reaction of nanothermites propagates at velocities ranging from incandescence propagation velocity (mm/s) to detonation velocity (km/s). The pyrotechnic effect can be modulated by modifying the chemical nature of the nanothermite, varying the size and form of its constituent particles, its porosity, the presence of additives allowing to moderate or reinforce its reaction, its macroscopic form and its confinement.

Nanothermites have applications in numerous pyrotechnic systems. Puszynski *et al.* patented the use of nanothermites as main component in the loading of percussion primers [PUS 10]. Martirosyan *et al.* used Al/Bi_2O_3 and Al/I_2O_5 nanocompositions to fabricate matrices for micropropulsion systems [MAR 12]. Gas-generating nanothermites, composed of chromium oxide impregnated with an explosive mixed with aluminum, have been used for guiding 40 mm projectiles [CIS 12]. The addition of thermites, including the nanostructured ones, to explosive loads has been patented by Jones, with the purpose of forming energy dense explosives [JON 04]. Crane *et al.* developed a spray gun for discharging incandescent aerosol resulted from the combustion of a micro- or nanostructured Al/Fe_2O_3 composition at a distance of 26.6 mm on a temperature sensor or a silicon substrate [CRA 11]. This type of device could be used to cut up metal parts. Kim *et al.* prepared nanothermites with very particular morphology, which are able to ignite in water when exposed to a flash [KIM 15d]. Finally, nanothermites can be used to transmit a pyrotechnic signal with a given delay to ignite propellant powders [BAS 14]. Hybrid compositions combining a nanothermite with a high explosive have detonating properties, which were used by Comet *et al.* to transmit the detonation to a secondary charge of PETN [COM 15b]. This phenomenon was called “pyrotechnic transduction”, because it permits to transform the low-power pyrotechnic signal produced by a flame or a hot wire into detonation. Hybrid nanothermites are promising materials for replacing lead-based primary explosives, which are both sensitive and toxic. Hybrid nanothermites of various natures have been prepared by Thiruvengadathan *et al.* [THI 11] and, more recently, Qiao *et al.* [QIA 15].

Clark *et al.* showed that iodine vapors resulting from the combustion of a nanostructured Al/I_2O_5 composition were effectively destroying spores of *Bacillus atrophaeus* [CLA 10]. Russell *et al.* confirmed the biocidal potential of iodine oxide based compositions applied to spores of *Bacillus subtilis* [RUS 12]. It would be interesting to test the action of these compositions on highly pathogen bacteria such as anthrax (*Bacillus anthracis*) or plague (*Yersinia pestis*). The thermal effect of the combustion of nanothermites could be taken advantage of for destroying the stockpiles of chemical weapons containing organophosphate substances. Oxygenated salts such as iodates of silver [SUL 10a], iron, copper, bismuth

[WAN 15b] or potassium persulfate [ZHO 15b] have been used as oxidizers in nanostructured aluminothermic compositions susceptible to have biocidal properties.

To enable the integration of nanothermites into functional pyrotechnic systems, a final step has to be made, and this consists of passing from powder to object. For the time being, no study has been dedicated to this aspect. There is no doubt that this is the most difficult stage, since the development of nanothermites should be carried out without destroying their porosity. The resolution of this technological problem will open the way for the wide-scale use of nanothermites in pyrotechnic systems.

Bibliography

[ABD 95] ABDELLAOUI M., GAFFET E., "The physics of mechanical alloying in a planetary ball mill: Mathematical treatment", *Acta Metall. Mater.*, vol. 43, no. 3, pp. 1087–1098, 1995.

[ABD 96] ABDELLAOUI M., GAFFET E., "The physics of mechanical alloying in a modified horizontal rod mill: Mathematical treatment", *Acta Mater.*, vol. 44, no. 2, pp. 725–734, 1996.

[ABD 08] ABDOLVAND A., KHAN S.Z. YUAN Y. *et al.*, "Generation of titanium-oxide nanoparticles in liquid using a high-power, high-brightness continuous-wave fiber laser", *Appl. Phys., A*, no. 91, pp. 365–368, 2008.

[ABR 46] ABRAMS R., POTTS A.M., BEILMAN C.E. *et al.*, Production and Analysis of Radioactive Aerosol, Report, U.S. Atomic Energy Commission, 1946.

[ABU 05] ABU MUKH-QASEM R., GEDANKEN A., "Sonochemical synthesis of stable hydrosol of Fe_3O_4 nanoparticles", *J. Colloid Interface Sci.*, no. 284, pp. 489–494, 2005.

[ABU 14] ABULIZI A.,YANG G.H., OKITSU K. *et al.*, "Synthesis of MnO^2 nanoparticles from sonochemical reduction of MnO_4 in water under different pH conditions", *Ultrason. Sonochem.*, no. 21, pp. 1629–1634, 2014.

[ACA 10] ACACIA N., BARRECA F., BARLETTA E. *et al.*, "Laser ablation synthesis of indium oxide nanoparticles in water", *Appl. Surf. Sci.*, no. 256, pp. 6918–6922, 2010.

[ADS 92a] ADSCHIRI T., KANAZAWA K., ARAI K., "Rapid and continuous hydrothermal crystallization of metal oxide particles in supercritical water", *J. Am. Ceram. Soc.*, vol. 75, no. 4, pp. 1019–1022, 1992.

[ADS 92b] ADSCHIRI T., KANAZAWA K., ARAI K., "Rapid and continuous hydrothermal synthesis of boehmite particles in subcritical and supercritical water", *J. Am. Ceram. Soc.*, vol. 75, no. 9, pp. 2615–2618, 1992.

[AGU 04] AGUIRRE C.M., MORAN C.E., YOUNG J.F. *et al.*, "Laser-induced reshaping of metallodielectric nanoshells under femtosecond and nanosecond plasmon resonant illumination", *J. Phys. Chem., B*, vol. 108, no. 22, pp. 7040–7045, 2004.

[AGU 98] AGUILERA J.A., ARAGON C., PENALBA F., "Plasma shielding effect in laser ablation of metallic samples and its influence on LIBS analysis", *Appl. Surf. Sci.*, nos. 127–129, pp. 309–314, 1998.

[ALN 13] ALNASSAR S., ADEL K.M., FADHIL Z., "Study the effect of laser fluences on the production of alumina nanoparticles (Al_2O_3) synthesized by pulsed laser ablation techniques in aqeous solutions", *Mach. Technol. Mater.*, no. 8, pp. 47–50, 2013.

[ALO 07] ALONSO E., MONTEQUI I., LUCAS S. *et al.*, "Synthesis of titanium oxide particles in supercritical CO_2: Effect of operational variables in the characteristics of the final product", *J. Supercrit. Fluids*, no. 39, pp. 453–461, 2007.

[AMA 12] AMADOSS A. KIM S.J., "Synthesis and characterization of HfO_2 nanoparticles by sonochemical approach", *J. Alloys Compd.*, no. 544, pp. 115–119, 2012.

[AME 09] AMENDOLA V., MENEGHETTI M., "Laser ablation synthesis in solution and size manipulation of noble metal nanoparticles", *Phys. Chem. Chem. Phys.*, no. 11, pp. 3805–3821, 2009.

[AME 95] AMEYAMA K., OKADA O., HIRAI K. *et al.*, "Microstructure of a Ti-45 mol%Al mechanical alloyed powder and its $\alpha \rightarrow \gamma$ massive transformation during consolidation", *Mater. Trans., JIM*, vol. 32, no. 2, pp. 269–275, 1995.

[AMO 99] AMORUSO S., BRUZZESE R., SPINELLI N. *et al.*, "Characterisation of laser-ablation plasmas", *J. Phys., B*, no. 32, pp. R131–R172, 1999.

[AMU 05a] AMORUSO S., AUSANIO G., DE LISIO C. *et al.*, "Synthesis of nickel nanoparticles and nanoparticles magnetic films by femtosecond laser ablation in vacuum", *Appl. Surf. Sci.*, no. 247, pp. 71–75, 2005.

[AMU 05b] AMORUSO S., AUSANIO G., BRUZZESE R. *et al.*, "Femtosecond laser pulse irradiation of solid targets as a general route to nanoparticle formation in a vacuum", *Phys. Rev., B*, no. 71, pp. 033406(1)–033406(4), 2005.

[AMU 07] AMORUSO S., AUSANIO G., BARONE A.C. *et al.*, "Nanoparticles size modifications during femtosecond laser ablation of nickel in vacuum", *Appl. Surf. Sci.*, no. 254, pp. 1012–1016, 2007.

[AND 06] ANDANSON J.-M., BOPP P.A., SOETENS J.-C., "Relation between hydrogen bonding and intramolecular motions in liquid and supercritical methanol", *J. Mol. Liq.*, no. 129, pp. 101–107, 2006.

[AND 97a] ANDRIANAINARIVELO M., CORRIU R.J.P., LECLERCQ D. *et al.*, "Non hydroliytic sol-gel process: aluminium titanate gels", *Chem. Mater.*, no. 9, pp. 1098–1102, 1997.

[AND 97b] ANDRIANAINARIVELO M., CORRIU R.J.P., LECLERCQ D. *et al.*, "Non hydroliytic sol-gel process: zirconium titanate gels", *Chem. Mater.*, no. 9, pp. 279–1284, 1997.

[ANI 07] ANISICHKIN V.F., "Isotope studies of detonation mechanisms of TNT, RDX, and HMX", *Combust. Explos. Shock Waves*, vol. 43, no. 5, pp. 580–586, 2007.

[APP 06] APPERSON S., BHATTACHARYA S., GAO Y. *et al.*, "On-chip initiation and burn rate measurements of thermite energetic reactions", *Mater. Res. Soc. Sympp. P.*, vol. 896, pp. 81–86, 2006.

[APP 07] APPERSON S., SHENDE R.V., SUBRAMANIAN S. *et al.*, "Generation of fast propagating combustion and shock waves with copper oxide/aluminum nanothermite composites", *Appl. Phys. Lett.*, vol. 91, no. 243109, 2007.

[ARA 03] ARAUJO-ANDRADE C., ESPINOZA-BELTRAN F.J., JIMENEZ-SANDOVAL S. *et al.*, "Synthesis of nanocrystalline Si particles from a solid-state reaction during a ball-milling process", *Scripta Mater.*, no. 49, pp. 773–778, 2003.

[ARI 11] ARITA T., HITAKA H., MINAMI K. *et al.*, "Synthesis of iron nanoparticle: Challenge to determine the limit of hydrogen reduction in supercritical water", *J. Supercrit. Fluids*, no. 57, pp. 183–189, 2011.

[ARM 03] ARMSTRONG R.W., BASCHUNG B., BOOTH D.W. *et al.*, "Enhanced propellant combustion with nanoparticles", *Nano Lett.*, vol. 3, no. 2, pp. 253–255, 2003.

[ARR 99] ARRIAGADA F.J., OSSEO-ASARE K., "Synthesis of nanosize silica in a nonionic water-in-oil microemulsion: Effects of the water/surfactant molar ratio and ammonia concentration", *J. Colloid Interface Sci.*, no. 211, pp. 210–220, 1999.

[ASA 04] ASAY B.W., SON S.F., BUSSE J.R. *et al.*, "Ignition characteristics of metastable intermolecular composites", *Propell. Explos. Pyrot.*, vol. 29, no. 4, pp. 216–219, 2004.

[ASH 08] ASHKARRAN A.A., ZAD A.I., AHADIAN M.M. *et al.*, "Synthesis and photocatalytic activity of WO_3 nanoparticles prepared by the arc discharge method in deionized water", *Nanotechnology*, no. 19, pp. 195709-1–195709-7, 2008.

[ASH 09] ASHKARRAN A.A., ZAD A.I., MAHDAVI S.M. *et al.*, "ZnO nanoparticles prepared by electrical arc discharge method in water", *Mater. Chem. Phys.*, no. 118, pp. 6–8, 2009.

[ASH 10a] ASHKARRAn A.A., "A novel method for synthesis of colloidal silver nanoparticles by arc discharge in liquid", *Curr. Appl. Phys.*, no. 10, pp. 1442–1447, 2010.

[ASH 10b] ASHKARRAN A.A., AHMADI AFSHAR S.A., AGHIGH S.M. *et al.*, "Photocatalytic activity of ZrO_2 nanoparticles prepared by electrical arc discharge method in water", *Polyhedron*, no. 29, pp. 1370–1374, 2010.

[ASH 10c] ASHKARRAN A.A., ZAD A.I., MAHDAVI S.M. *et al.*, "Photocatalytic activity of ZnO nanoparticles prepared via submerged arc discharge method", *Appl Phys., A*, no. 100, pp. 1097–1102, 2010.

[ASH 10d] ASHKARRAN A.A., KAVIANIPOUR M., AGHIGH S.M. *et al.*, "On the formation of TiO_2 nanoparticles via submerged arc discharge technique: synthesis, characterization and photocatalytic properties", *J. Clust. Sci.*, no. 21, pp. 753–766, 2010.

[ASL 11] ASLANI A., OROOJPOUR V., FALLAHI M., "Sonochemical synthesis, size controlling and gas sensing properties of NiO nanoparticles", *Appl. Surf. Sci.*, no. 257, pp. 4056–4061, 2011.

[AUM 95] AUMANN C.E., SKOFRONICK G.L., MARTIN J.A., "Oxidation behavior of aluminum nanopowders", *J. Vac. Sci. Technol., B*, vol. 13, no. 3, pp. 1178–1183, 1995.

[AUS 06] AUSANIO G., AMORUSO S., BARONE A.C. *et al.*, "Production of nanoparticles of different materials by means of ultrashort laser pulses", *Appl. Surf. Sci.*, no. 252, pp. 4678–4684, 2006.

[AUS 76] AUSTIN L.G., SHOJI K., LUCKIE P.T., "The effect of ball size on mill performance", *Powder Technol.*, no. 14, pp. 71–79, 1976.

[AVE 03] AVETTAND-FÉNOËL M.-N., TAILLARD R., DHERS J. *et al.*, "Effect of ball milling parameters on the microstructure of W–Y powders and sintered samples", *Int. J. Refract. Met. Hard Mater.*, no. 21, pp. 205–213, 2003.

[AVV 01] AVVAKUMOV E., SENNA M., KOSOVA N., *Soft Mechanochemical Synthesis, a Basis for new Chemical Technologies*", Kluwer Academic Publishers, Boston, 2001.

[AWA 03] AWATI P.S., AWATE S.V., SHAH P.P. *et al.*, "Photocatalytic decomposition of methylene blue using nanocrystalline anatase titania prepared by ultrasonic technique", *Catal. Commun.*, no. 4, pp. 393–400, 2003.

[AZI 15] AZIZIAN-KALANDARAGH Y., SEDAGHATDOUST-BODAGH F., HABIBI-YANGJEH A., "Ultrasound-assisted preparation and characterization of β-Bi_2O_3 nanostructures: Exploring the photocatalytic activity against rhodamine B", *Superlattices Microstruct.*, no. 81, pp. 151–160, 2015.

[BAB 47] BABAT G.I., "Electrodeless discharges and some allied problems", *J. Inst. Elec. Eng.*, vol. 94, no. 27, pp. 27–37, 1947.

[BAB 97] BABURAJ E.G., HUBERT K.T., FROES F.H., "Preparation of Ni powder by mechanochemical process", *J. Alloys Compd.*, no. 257, pp. 146–149, 1997.

[BAC 15] BACH A., GIBOT P., VIDAL L. *et al.*, "Modulation of the reactivity of a WO_3/Al energetic material with graphitized carbon black as additive", *J. Energ. Mater.*, vol. 33, no. 4, pp. 260–276, 2015.

[BAG 11] BAGAZEEV A.V., KOTOV Y.A., "Some characteristics of electric explosion of zinc wires", *Tech. Phys. Lett.*, vol. 37, no. 9 pp. 91–96, 2011.

[BAI 09] BAI L., FAN J., HU P. *et al.*, "RF plasma synthesis of nickel nanopowders via hydrogen reduction of nickel hydroxide/carbonate", *J. Alloys Compd.*, no. 481, pp. 563–567, 2009.

[BAL 07] BALAZ P., TAKACS L., GODOCIKOVA E. *et al.*, "Preparation of nanosized antimony by mechanochemical reduction of antimony sulphide Sb_2S_3", *J. Alloys Compd.*, nos. 434–435, pp. 773–775, 2007.

[BAM 14] BAMIDURO F., WARD M.B., BRYDSON R. *et al.*, "Hierarchical growth of ZnO particles by a hydrothermal route", *J. Am. Ceram.Soc.*, vol. 97, no. 5, pp. 1619–1624, 2014.

[BAN 00] BANDYOPADHYAYA R., KUMAR R., GANDHI K.S., "Simulation of precipitation reactions in reverse micelles", *Langmuir*, no. 16, pp. 7139–7149, 2000.

[BAN 97] BANDYOPADHYAYA R., KUMAR R., GANDHI K.S., "Modeling of precipitation in reverse micellar systems", *Langmuir*, no. 13, pp. 3610–3620, 1997.

[BAN 07] BANG J.H., SUSLICK K.S., "Sonochemical synthesis of nanosized hollow hematite", *J. Am. Chem. Soc.*, no. 129, pp. 2242–2243, 2007.

[BAP 07] BAPAT A., GATTI M., DING Y.-P. *et al.*, "A plasma process for the synthesis of cubic-shaped silicon nanocrystals for nanoelectronic devices", *J. Phys. D: Appl. Phys.*, no. 40, pp. 2247–2257, 2007.

[BAR 07] BARCIKOWSKI S., HAHN A., KABASHIN A.V. *et al.*, "Properties of nanoparticles generated during femtosecond laser machining in air and water", *Appl. Phys., A*, no. 87, pp. 47–55, 2007.

[BAR 06] BANERJEE I., KHOLLAM Y.B., BALASUBRAMANIAN C. *et al.*, "Preparation of γ-Fe_2O_3 naoparticles using DC thermal arc-plasma route, their characterization and magnetic properties", *Scripta Mater.*, no. 54, pp. 1235–1240, 2006.

[BAR 05] BARROSO F., DE DIOS M., TOJO C. *et al.*, "A computer simulation study on the influence of the critical nucleus on the mechanism of formation of nanoparticles in microemulsions", *Colloids Surf., A*, nos. 270–271, pp. 78–82, 2005.

[BAR 11] BARRECA F., ACACIA N., SPADARO S. *et al.*, "Tungsten trioxide (WO_{3-x}) nanoparticles prepared by pulsed laser ablation in water", *Mater. Chem. Phys.*, no. 127, pp. 197–202, 2011.

[BAS 14] BASCHUNG B., BOUCHAMA A., COMET M. *et al.*, "Experimental investigation of different ignition methods for LOVA gun propellant", *28th International Symposium on Ballistics*, Atlanta, Georgia, USA, September 2014.

[BAU 06] BAUMANN W., THEKEDAR B., PAUR H.R. *et al.*, "Characterization of nanoparticles synthesized in the microwave plasma discharge process by particle mass spectrometry and transmission electron microscopy", *AIChE Fall and Annual Meeting*, San Francisco, CA, USA, November 2006.

[BAZ 06] BAZYN T., KRIER H., GLUMAC N., "Combustion of nanoaluminum at elevated pressure and temperature behind reflected shock waves", *Combust. Flame*, vol. 145, no. 4, pp. 703–713, 2006.

[BAZ 10] BAZYN T., LYNCH P., KRIER H. *et al.*, "Combustion measurements of fuel-rich aluminum and molybdenum oxide nano-composite mixtures", *Propell. Explos. Pyrot.*, vol. 35, no. 2, pp. 93–99, 2010.

[BEC 51] BECHTOLD, M.F., SNYDER O.E., Chemical processes and composition, US Patent no. 2,574,902, December, 15, 1948.

[BEC 11] BECKER C.R., APPERSON S., MORRIS C.J. *et al.*, "Galvanic porous silicon composites for high-velocity nanoenergetics", *Nano. Lett.*, vol. 11, no. 2, pp. 803–807, 2011.

[BEG 97] BEGIN-COLIN S., WOLF F., LE CAËR G., "Mécanosynthèse d'oxydes nanocristallins", *J. Phys., III*, no. 7, pp. 473–482, 1997.

[BEL 14] BELMONTE T., HAMDAN A., KOSIOR F. *et al.*, "Interaction of discharges with electrode surfaces in dielectric liquids: application to nanoparticle synthesis", *J. Phys. D: Appl. Phys.*, no. 47, pp. 224016-1–224016-18, 2014.

[BEL 90] BELOSHAPKO A.G., BUKAEMSKII A.A., STAVER A.M., "Formation of ultradispersed compounds upon schock wave loading of porous aluminum. Study of particles obtained", *Combust. Explos. Shock Waves*, vol. 26, no. 4, pp. 457–461, 1990.

[BEN 70] BENJAMIN J.S., "Dispersion strengthened superalloys by mechanical alloying", *Metall. Trans.*, vol. 1, pp. 2943–2951, 1970.

[BEN 74] BENJAMIN J.S., VOLIN T.E., "The mechanism of mechanical alloying", *Metall. Trans.*, vol. 5, pp. 1929–1934, 1974.

[BER 13] BERGES M.G.M., "Exposure during production and handling of manufactured nanomaterials", in *Nanomaterials*, Wiley-VCH, Weinheim, 2013.

[BER 00] BERTHE L., SOLLIER A., PEYRE P. *et al.*, "The generation of laser shock waves in a water-confinement regime with 50 ns and 150 ns XeCl excimer laser pulses", *J. Phys. D: Appl. Phys.*, no. 33, pp. 2142–2145, 2000.

[BER 97] BERTHE L., FABBRO R., PEYRE P. *et al.*, "Shock waves from a water-confined laser-generated plasma", *J. Appl. Phys.*, vol. 82, no. 6, pp.2826–2832, 1997.

[BER 99] BERTHE L., FABBRO R., PEYRE P. *et al.*, "Wavelength dependent of laser shock-wave generation in the water-confinement regime", *J. Appl. Phys.*, vol. 85, no. 11, pp. 7552–7555, 1999.

[BES 06] BESNER S., KABASHIN A.V., MEUNIER M., "Fragmentation of colloidal nanoparticles by femtosecond laser-induced super continuum generation", *Appl. Phys. Lett.*, vol. 89, no. 23, pp. 233122(1)–233122(3), 2006.

[BES 09] BESNER S., KABASHIN A.V., WINNIK F.M. *et al.*, "Synthesis of size-tunable polymer-protected gold nanoparticles by femtosecond laser-based ablation and seed growth", *J. Phys. Chem., C*, vol. 113, no. 22, pp. 9526–9531, 2009.

[BES 10] BESNER S., MEUNIER M., "Femtosecond laser synthesis of AuAg nanoalloys: photo-induced oxidation and ions release", *J. Phys. Chem., C*, no. 114, pp. 10403–10409, 2010.

[BES 05] BESSON C., FINNEY E.E., FINKE R.G., "A mechanism for transition-metal nanoparticle self-assembly", *J. Am. Chem. Soc.*, no. 127, pp. 8179–8184, 2005.

[BEZ 10] BEZMELNITSYN A., THIRUVENGADATHAN R., BARIZUDDIN S. *et al.*, "Modified nanoenergetic composites with tunable combustion characteristics for propellant applications", *Propell. Explos. Pyrot.*, vol. 35, pp. 384–394, 2010.

[BHA 06] BHATTACHARYA S., GAO Y., APPERSON S. *et al.*, "A novel on-chip diagnostic method to measure burn rates of energetic materials", *J. Energ. Mater.*, vol. 24, no. 1, pp. 1–15, 2006.

[BIL 07] BILLIK P., PLESCH G., BREZOVA V. *et al.*, "Anatase TiO_2 nanocrystals prepared by mechanochemical synthesis and their photochemical activity studied by EPR spectroscopy", *J. Phys. Chem. Solids*, no. 68, pp. 1112–1116, 2007.

[BIL 32] BILLY M., "Titane", in PASCAL P., BAUD P. (ed.), *Traité de chimie minérale – Tome V*, Masson et Cie, Paris, 1932.

[BIR 84] BIRRINGER, R., GLEITER, H., KLEIN, H.P. *et al.*, "Nanocrystalline materials: an approach to a novel solid structure with gas – like disorder?", *Phys. Lett.*, no. 102A, pp. 365–369, 1984.

[BLE 14] BLESSI S., SONIA M.M.L., VIJAYALAKSHMI S. *et al.*, "Preparation and characterization of SnO_2 nanoparticles by hydrothermal method", *Int. J. Chem. Tech. Res.*, vol. 6, no. 3, pp. 2153–2155, 2014.

[BLO 03] BLOBAUM K.J., REISS M.E., PLITZKO LAWRENCE J.M. *et al.*, "Deposition and characterization of a self-propagating CuO_x/Al thermite reaction in a multilayer foil geometry", *J. Appl. Phys.*, vol. 94, no. 5, pp. 2915–2922, 2003.

[BOC 05] BOCKMON B.S., PANTOYA M.L., SON S.F. *et al.*, "Combustion velocities and propagation mechanisms of metastable interstitial composites", *Appl. Phys. Lett.*, vol. 98, no. 064903, 2005.

[BOG 05] BOGAERTS, A., CHEN Z.Y., "Effect of laser parameters on laser ablation and laser-induced plasma formation: A numerical modeling investigation", *Spectrochim. Acta Part B*, vol. 60, nos. 9–10, pp. 1280–1307, 2005.

[BOG 91] BOGUSH G.H., ZUKOSKI IV C.F., "Studies of the kinetics of the precipitation of uniform silica particles through the hydrolysis and condensation of silicon alkoxides", *J. Colloid Interface Sci.*, vol. 142, no. 1, pp. 1–18, 1991.

[BON 84] BONAFEDE J. D., "Explosive isopropanol", *J. Chem. Educ.*, vol. 61, no. 7, pp. 652, 1984.

[BON 09] BONNARD N., BRONDEAU M.-T., FALCY M. *et al.*, Nickel et ses oxydes, Fiches Toxicologiques, Institut National de Recherche et Sécurité (INRS), FT 68, 2009.

[BOR 97] BORNER I., ECKERT J., "Nanostructure formation and steady-state grain size of ball-milled iron powders", *Mater. Sci. Eng., A*, nos. 226–228, pp. 541–545, 1997.

[BOU 95] BOULOS M.I., "The inductively coupled R.F. (radio frequency) plasma", *Pure Appl. Chem.*, vol. 57, no. 9, pp. 1321–1352, 1985.

[BOU 07] BOUMA R.H.B., MEUKEN D., VERBEEK R. *et al.*, "Shear initiation of Al/MoO_3-based reactive materials", *Propell. Explos. Pyrot.*, vol. 32, no. 6, pp. 447–453, 2007.

[BOU 82] BOUTONNET M., KIZLING J., STENIUS P., "The preparation of monodisperse colloidal metal particles from microemulsion", *Colloids Surf.*, no. 5, pp. 209–225, 1982.

[BOW 81] BOWLES R.S., KOLSTAD J.J., CALO J.M. *et al.*, "Generation of molecular clusters of controlled size", *Surf. Sci.*, no. 106, pp. 117–124, 1981.

[BRA 11] BRATESCU M.A., CHO S.P., TAKAI O. *et al.*, "Size-controlled gold nanoparticles synthesized in solution plasma", *J. Phys. Chem., C*, no. 115, pp. 24569–24576, 2011.

[BRA 05] BRAYDICH-STOLLE L., HUSSAIN S., SCHLAGER J.J. *et al.*, "In vitro cytotoxicity of nanoparticles in mammalian germline stem cells", *Toxicol. Sci.*, vol. 88, no. 2, pp. 412–419, 2005.

[BÜC 00] BÜCHEL K.H., MORETTO H.-H., WODITSCH P., *Industrial Inorganic Chemistry*, Wiley-VCH Verlag GmbH, 2nd ed., 2000.

[BUL 08] BULGAREVICH D.S., HORIKAWA Y., SAKO T., "ATR FT-IR studies of supercritical methanol", *J. Supercrit. Fluids*, no. 46, pp. 206–210, 2008.

[BUM 11] BUMAJDAD A., SAMI ALI S., MATHEW A., "Characterization of iron hydroxide/oxide nanoparticles prepared in microemulsions stabilized with cationic/non-ionic surfactant mixtures", *J. Colloid Interface Sci.*, no. 355, pp. 282–292, 2011.

[BUN 62] BUNDY F.P., "Direct conversion of graphite to diamond in static pressure apparatus", *Science*, vol. 37, no. 3537, pp. 1057–1058, 1962.

[BUR 05] BURDA C., CHEN X., NARAYANAN R., EL-SAYED M.A., "Chemistry and properties of nanocrystals of different shapes", *Chem. Rev.*, no. 105, pp. 1025–1102, 2005.

[BUS 10] BUSKY R.T., BOTCHER T.R., SANDSTROM J. *et al.*, Nontoxic, noncorrosive phosphorus-based primer compositions, US Patent no. 7,857,921, 2010.

[BUSS 97] BUSS R.J., RF-plasma synthesis of nanosize silicon carbide and nitride, Final Report, SAND97-0039, 1997.

[BUZE 07] BUZEA C., PACHECO BLANDINO I.I., ROBBIE K., Nanomaterials and nanoparticles: Sources and toxicity, *Biointerphases*, vol. 2, no. 4, pp. MR17–MR172, 2007.

[CAB 47a] CABRERA N., TERRIEN J., HAMON J., "Sur l'oxydation de l'aluminium en atmosphère sèche", *C. R. Acad. Sci.*, Paris, vol. 224, pp. 1558–1560, 1947.

[CAB 47b] CABRERA N., HAMON J., "Sur l'oxydation de l'aluminium à haute température", *C. R. Acad. Sci.*, Paris, vol. 224, pp. 1713–1715, 1947.

[CAI 12] CAI Y.-G., MA Q.-L., HUANG Y.M., "Effect of sodium chloride on the sol-gel synthesized silica colloidal particles", *Solid State Phenom.*, nos. 181–182, pp. 417–421, 2012.

[CAM 08] CAMPONESCHI E., WALKER J., GARMESTANI H. *et al.*, "Surfactant effects on the particle size of iron (III) oxides formed by sol–gel synthesis", *J. Non-Cryst. Solids*, no. 354, pp. 4063–4069, 2008.

[CAM 99] CAMPBELL T., KALIA R.K., NAKANO A. *et al.*, "Dynamics of oxidation of aluminum nanoclusters using variable charge – molecular-dynamics simulations on parallel computers", *Phys. Rev. Lett.*, vol. 82, no. 24, pp. 4866–4869, 1999.

[CFH 07] CFHA, "Les poussières fines en Suisse", available at: www.bafu.admin.ch/div-5013-f, Berne, 2007.

[CHA 06] CHAKKA V.M., ALTUNCEVAHIR B., JIN Z.Q. *et al.*, "Magnetic nanoparticles produced by surfactant-assisted ball milling", *J. Appl. Phys.*, vol. 99, no. 8, pp. 08E912-1–08E912-3, 2006.

[CHA 96a] CHAMPION Y., BIGOT J, "Characterization of nanocrystalline copper powders prepared by melting in a cryogenic liquid", *Mat. Sci. Eng. A*, nos. 217–218, pp. 58–63, 1996.

[CHA 96b] CHAMPION Y., BIGOT J., "Preparation and characterization of nanocrystalline copper powders", *Scripta Mater.*, vol. 35, No. 4, pp. 517–522, 1996.

[CHA 98] CHAMPION Y., BIGOT J., "Synthesis and structural analysis of aluminum nanocrystalline powders", *Nanostruct. Mater.*, no. 10, pp. 1097–1110, 1998.

[CHA 11a] CHANG X., SUN S., YIN Y., "Green synthesis of tungsten trioxide monohydrate nanosheets as gas sensor", *Mater. Chem. Phys.*, no. 126, pp. 717–721, 2011.

[CHA 11b] CHANG K.-H., HU C.-C., HUANG C.-M. *et al.*, "Microwave-assisted hydrothermal synthesis of crystalline WO_3–WO_3 $0.5H_2O$ mixtures for pseudocapacitors of the asymmetric type", *J. Power Sources*, no. 196, pp. 2387–2392, 2011.

[CHA 96] CHANG J. J., WARNER B. E., "Laser-plasma interaction during visible-laser ablation of methods", *Appl. Phys. Lett.*, no. 69, pp. 473–475, 1996.

[CHA 05] CHAU J.L.H., HSU M.-K., HSIEH C.-C. *et al.*, "Microwave plasma synthesis of silver nanopowders", *Mater. Lett.*, no. 59, pp. 905–908, 2005.

[CHA 06] CHAU J.L.H., HSU M.-K., KAO C.-C., "Microwave plasma synthesis of Co and SiC-coated Co nanopowders", *Mater. Lett.*, no. 60, pp. 947–951, 2006.

[CHA 07] CHAU J. L. H. "Synthesis of Ni and bimetallic FeNi nanopowders by microwave plasma method", *Mater. Lett.*, no. 61, pp. 2753–2756, 2007.

[CHA 14] CHAU J.L.H., YANG C.-C., SHIH H.-H., "Microwave plasma production of metal nanopowders", *Inorganics*, no. 2, pp. 278–290, 2014.

[CHA 14] CHAUDHARY A.-L., SHEPPARD D.A., PASKEVICIUS M. *et al.*, "Mechanochemical synthesis of amorphous silicon nanoparticles", *RSC Adv.*, no. 4, pp. 21979–21983, 2014.

[CHA 06] CHAZELAS C., COUDERT J.F, JARRIGE J. *et al.*, "Synthesis of ultra fine particles by plasma transferred arc: Influence of anode material on particle properties", *J. Eur. Ceram. Soc.*, no. 6, pp. 3499–3507, 2006.

[CHE 04] CHEN L.C., PAI S.-H., "In-situ measurement and control of electric discharge on submerged arc synthesis process for continuous TiO_2 nanoparticle fabrication", *Mater. Trans.*, vol. 45, no. 10, pp. 3071–3078, 2004.

[CHE 05a] CHEN P., JI Y., FENG S., "Preparation of spherical Bi_20_3 powder by plasma and precipitation processes", *Plasma Sci. Technol.*, vo1.7, no. 6, pp. 3139–3142, 2005.

[CHE 05b] CHEN X., WANG X., AN C. *et al.*, "Synthesis of Sb_2O_3 nanorods under hydrothermal conditions", *Mater. Res. Bull.*, no. 40, pp. 469–474, 2005.

[CHE 08] CHEN L., YOKEL R.A., HENNIG B. *et al.*, "Manufactured aluminum oxide nanoparticles decrease expression of tight junction proteins in brain vasculature", *J. Neuroimmune Pharm.*, vol. 3, no. 4, pp. 286–295, 2008.

[CHE 10] CHENG J.L., HNG H.H., LEE Y.W. *et al.*, "Kinetic study of thermal- and impact-initiated reactions in Al-Fe_2O_3 nanothermite", *Combust. Flame*, vol. 157, no. 12, pp. 2241–2249, 2010.

[CHE 11] CHENG X., BAOSHAN WU B., YANG Y. *et al.*, "Synthesis of iron nanoparticles in water-in-oil microemulsions for liquid-phase Fischer–Tropsch synthesis in polyethylene glycol", *Catal. Commun.*, no. 12, pp. 431–435, 2011.

[CHE 12] CHEN Q., KANEKO T., HATAKEYAMA R., "Synthesis of superfine ethanol-soluble CoO nanoparticles via discharge plasma in liquid", *Appl. Phys. Express*, no. 5, pp. 096201-1–096201-3, 2012.

[CHE 14] CHEN L.,MASHIMO T.,OKUDERA H. *et al.*, "Synthesis of $WO_3.H_2O$ nanoparticles by pulsed plasma in liquid", *RSC Adv.*, no. 4, 28673–28677, 2014.

[CHE 92] CHEN Y., Contribution à la physique du procédé de mécanosynthèse, PhD Thesis University of Paris XI, 1992.

[CHE 99] CHENG Z.-H., MACKAY G.R., SMALL D.A. *et al.*, "Phase development in titanium by mechanical alloying under hydrogen atmosphere", *J. Phys. D: Appl. Phys.*, no. 32, pp. 1934–1937, 1999.

[CHO 07] CHO C.H., PARK S.H., CHOI Y.W. *et al.*, "Production of nanopowders by wire explosion in liquid media", *Surf. Coat. Tech.*, no. 201, pp. 4847–4849, 2007.

[CHO 10] CHO C., HA Y.-C., KANG C. *et al.*, "Preparation of a Nickel nanopowder by wire explosion in liquid Media", *J. Korean Phys. Soc.*, vol. 57, no. 6, pp. 1807–1810, 2010.

[CHO 10] CHOI H., VERIANSYAH B., KIM J. *et al.*, "Continuous synthesis of metal nanoparticles in supercritical methanol", *J. Supercrit. Fluids*, no. 52, pp. 285–291, 2010.

[CHO 92] CHOU C.H., PHILLIPS J., "Plasma production of metallic nanoparticles", *J. Mater. Res.*, vol. 7, no. 8, pp. 2107–2113, 1992.

[CHO 97] CHOULIER D., RAHOUADJ R., GAFFET E., "Mécanique de ma mécanosynthèse: Bilan et perspectives", *Ann. Chim. Sci. Mat.*, no. 22, pp. 351–361, 1997.

[CHO 10] CHOWDHURY S., SULLIVAN K., PIEKIEL N. *et al.*, "Diffusive vs. explosive reaction at the nanoscale", *J. Phys. Chem., C*, vol. 114, no. 20, pp. 9191–9195, 2010.

[CIS 12] CISZEK F., COMET M., SOURGEN F. *et al.*, Dispositif de pilotage d'un missile ou d'un projectile, European patent no. 2,226,605, 2012.

[CLA 03] CLAPSADDLE B.J., ZHAO L., GASH A.E. *et al.*, "Synthesis and characterization of mixed metal oxide nanocomposite energetic materials", *Materials Research Society Fall Meeting*, Boston, USA, November 30 – December 4, 2003.

[CLA 10] CLARK B.R., PANTOYA M.L., "The aluminium and iodine pentoxide reaction for the destruction of spore forming bacteria", *Phys. Chem. Chem. Phys.*, vol. 12, no. 39, pp. 12653–12657, 2010.

[COM 02] COMET M., SCHREYECK-REINERT L., LOUIS C. *et al.*, "Synthesis and characterization of high surface area aluminium and alumina microtubes from carbonaceous materials", *J. Mater. Chem.*, vol. 12, no. 3, pp. 754–757, 2002.

[COM 06a] COMET M., SPITZER D., "Des thermites classiques aux composites interstitiels métastables", *Actual. Chimique*, vol. 299, pp. 20–25, 2006.

[COM 06b] COMET M., SPITZER D., "Elaboration and characterization of nano-sized $Al_xMo_yO_z$ thermites", *33rd International Pyrotechnics Seminar*, Fort Collins, USA, July 2006.

[COM 08a] COMET M., SPITZER D., Procédé de fabrication de micro et/ou nanothermites et nanothermites associées, Brevet FR2905882, 2008.

[COM 08b] COMET M., SIEGERT B., PICHOT V. *et al.*, "Preparation of explosive nanoparticles in a porous chromium(III) oxide matrix: a first attempt to control the reactivity of explosives", *Nanotechnology*, vol. 19, no. 28, pp. 285716, 2008.

[COM 08c] COMET M., PICHOT V., SPITZER D. *et al.*, "Elaboration and characterization of manganese oxide (MnO_2) based 'green' nanothermites", *39th International Annual Conference of ICT, Karlsruhe*, Germany, June 2008.

[COM 09] COMET M., PICHOT V., SIEGERT B. *et al.*, "Use of nanodiamond as a reducing agent in a chlorate-based energetic composition", *Propell. Explos. Pyrot.*, vol. 34, no. 2, pp. 166–173, 2009.

[COM 10a] COMET M., PICHOT V., SIEGERT B. *et al.*, "Phosphorus-based nanothermites: a new generation of energetic materials", *J. Phys. Chem. Solids*, vol. 71, no. 2, pp. 64–68, 2010.

[COM 10b] COMET M., SIEGERT B., SCHNELL F. *et al.*, "Phosphorus-based nanothermites: a new generation of pyrotecnics illustrated by the example of n-CuO/Red P mixtures", *Propell. Explos. Pyrot.*, vol. 35, no. 3, pp. 220–225, 2010.

[COM 11a] COMET M., SIEGERT B., SCHNELL F. *et al.*, "Experimental strategies for the desensitization of nanothermites", *International Pyrotechnic Safety Symposium*, Bordeaux, France, 15–16 November 2011.

[COM 11b] COMET M., SIEGERT B., SCHNELL F. *et al.*, "Control of the reactivity of phosphorus based nanothermites", *37th International Pyrotechnics Seminar*, Reims, France, May 2011.

[COM 11c] COMET M., PICHOT V., SIEGERT B. *et al.*, "Preparation of Cr_2O_3 nanoparticles for superthermites by the detonation of an explosive nanocomposite material", *J. Nanopart. Res.*, vol. 13, no. 5, pp. 1961–1969, 2011.

[COM 12] COMET M., SCHNELL F., SIEGERT B. *et al.*, "Nanothermites at the ISL: preparation, desensitization, applications", *38th International Pyrotechnics Seminar*, Denver, Colorado, USA, June 2012.

[COM 13a] COMET M., SIEGERT B., PICHOT V. *et al.*, "Reactive characterization of nanothermites", *J. Therm. Anal. Calorim.*, vol. 111, no. 1, pp. 431–436, 2013.

[COM 13b] COMET M., Les nanomatériaux énergétiques: préparation, caractérisation, propriétés et utilisations, Thesis, University of Strasbourg, 2013.

[COM 13c] COMET M., SCHNELL F., PICHOT V. *et al.*, "La sécurité pyrotechnique: fait scientifique ou réglementaire?", *Actual. Chimique*, vol. 373, pp. 31–36, 2013.

[COM 13d] COMET M., SPITZER D., HASSLER D. *et al.*, Activation of energetic compositions by magnetic mixing, European patent no. 2,615,077, 2013.

[COM 14a] COMET M., PICHOT V., SCHNELL F. *et al.*, "Oxidation of detonation nanodiaonds in a reactive formulation", *Diamond Relat. Mater.*, no. 47, pp. 35–39, 2014.

[COM 14b] COMET M., SCHNELL F., PICHOT V. *et al.*, "Boron as fuel for ceramic thermites", *Energ. Fuel*, vol. 28, no. 6, pp. 4139–4148, 2014.

[COM 15a] COMET M., VIDICK G., SCHNELL F. *et al.*, "Sulfates-based nanothermites: an expanding horizon for metastable interstitial composites", *Angew. Chem. Int. Ed.*, vol. 54, no. 15, pp. 4458–4462, 2015.

[COM 15b] COMET M., MARTIN C., KLAUMÜNZER M. *et al.*, "Energetic nanocomposites for detonation initiation in high explosives without primary explosive", *Appl. Phys. Lett.*, vol. 107, no. 243 p. 108, 2015.

[COO 95] COOK T.M., COURTNEY T.H., "The effect of ball size distribution on attritor efficiency", *Metall. Mater.Trans., A*, no. 26, pp. 2389–2397, 1995.

[COU 96] COURTNEY T.H., MAURICE D., "Process modeling of the mechanics of mechanical alloying", *Scripta Mater.*, vol. 34, no. 1, pp. 5–11, 1996.

[CRA 10] CRANE C. A., PANTOYA M. L., DUNN J., "Infrared measurements of energy transfer from energetic materials to steel substrates", *Int. J. Therm. Sci.*, vol. 49, no. 10, pp. 1877–1885, 2010.

[CRA 11] CRANE C.A., COLLINS E.S., PANTOYA M.L. *ET AL.*, "Nanoscale investigation of surfaces exposed to a thermite spray", *Appl. Therm. Eng.*, vol.31, no. 6–7, pp. 1286–1292, 2011.

[CUK 01] CUKROV L.M., TSUZUKI T., MCCORMICK P.G., "SnO_2 nanoparticles prepared by mechanical processing", *Scripta Mater.*, no. 44, pp. 1787–1790, 2001.

[DAL 96] DALLIMORE M.P., MCCORMICK P.G., "Dynamics of planetary ball milling:a comparison of computer simulated processing parameters with CuO/Ni displacement reaction milling kinetics", *Mater. Trans.*, vol. 37, no. 5, pp. 1091–1098, 1996.

[DAN 04] DANILENKO V.V., "On the history of the discovery of nanodiamond synthesis", *Phys. Solid State*, vol. 46, no. 4, pp. 595–599, 2004.

[DAS12] DAS R., DAS B.K., SHUKLA R. *et al.*, "Analysis of electrical explosion of wire systems for the production of nanopowder", *Sadhana*, vol. 37, no. 5, pp. 629–635, 2012.

[DAV 13] DAVID B., SCHNEEWEISS O., PIZUROVA N. *et al.*, "Atmospheric-pressure microwave torch discharge generated gamma-Fe_2O_3 nanopowder", *Phys. Procedia*, no. 44, pp. 206–212, 2013.

[DAV 14] DAVID B., PIZUROVA N., SYNE P. *et al.*, "ε-Fe_2O_3 nanoparticles synthesized in atmospheric-pressure microwave torch", *Mater. Lett.*, no. 116, pp. 370–373, 2014.

[DAV 99] DAVIES N., Red phosphorus for use in screening smoke compositions, Pentagon Reports, available at: www.dtic.mil/dtic/tr/fulltext/u2/a372367.pdf, 1999.

[DE 05] DE DIOS M., BARROSO F., TOJO C. *et al.*, "Effects of the reaction rate on the size control of nanoparticles synthesized in microemulsions", *Colloids Surf., A*, no. 270–271, pp. 83–87, 2005.

[DEA 10] DEAN S.W., PANTOYA M.L., GASH A.E. *et al.*, "Enhanced convective heat transfer in nongas generating nanoparticle thermites", *J. Heat Transf.*, vol. 132, no. 11, pp. 111201.1–111201.7, 2010.

[DEL 97] DEL BIANCO L., HERNANDO A., BONETTI E. *et al.*, "Grain-boundary structure and magnetic behavior in nanocrystalline ball-milled iron", *Phys. Rev., B*, vol. 56, no. 14, pp. 8894–8901, 1997.

[DEM 10] DEMASKE B.J., ZHAKHOVSKY V.V., INOGAMOV N.A. *et al.*, "Ablation and spallation of gold films irradiated by ultrashort laser pulses", *Phys. Rev., B*, no. 82, pp. 064113(1)–064113(5), 2010.

[DEM 15] DEMIRCI S., OZTUŸRK B., YILDIRIM S. *et al.*, "Synthesis and comparison of the photocatalytic activities of flame spray pyrolysis and sol–gel derived magnesium oxide nano-scale particles", *Mater. Sci. Semicond. Process.*, no. 34, pp. 154–161, 2015.

[DEM 14] DEMOISSON F., PIOLET R., ARIANE M. *et al.*, "Influence of the pH on the ZnO nanoparticle growth in supercritical water:experimental and simulation approaches", *J. Supercrit. Fluids*, no. 95, pp. 75–83, 2014.

[DEN 11] DENSMORE J.M., BISS M.M., MCNESBY K.L. *et al.*, "High-speed digital color imaging pyrometry", *Appl. Optics*, vol. 50, no. 17, pp. 2659–2665, 2011.

[DEN 14] DENSMORE J.M., SULLIVAN K.T., GASH A.E. *et al.*, "Expansion behavior and temperature mapping of thermites in burn tubes as a function of fill length", *Propell. Explos. Pyrot.*, vol. 39, no. 3, pp. 416–422, 2014.

[DES 04] DESHPANDE K., MUKASYAN A., VARMA A., "Direct synthesis of iron oxide nanopowders by the combustion approach: reaction mechanism and properties", *Chem. Mater.*, vol. 16, no. 24, pp. 4896–4904, 2004.

[DES 06] DESTRÉE C., NAGY J.B., "Mechanism of formation of inorganic and organic nanoparticles from microemulsions", *Adv. Colloid Interface Sci.*, no. 123–126, pp. 353–357, 2006.

[DHA 97] DHAS N.A., KOLTYPIN Y., GEDANKEN A., "Sonochemical preparation and characterization of ultrafine chromium oxide and manganese oxide powders", *Chem. Mater.*, no. 9, pp. 3159–3163, 1997.

[DHA 98] DHAS N. A.,RAJ C.P., GEDANKEN A., "Synthesis, characterization, and properties of metallic copper nanoparticles", *Chem. Mater.*, no. 10, pp. 1446–1452, 1998.

[DI 15] DI CARLO G., LUALDI M., VENEZIA A.M. *et al.*, "Design of cobalt nanoparticles with tailored structural and morphological properties via O/W and W/O microemulsions and their deposition onto silica", *Catalysts*, no. 5, pp. 442–459,2015.

[DIK 10] DIKICI B., PANTOYA M.L., LEVITAS V., "The effect of pre-heating on flame propagation in nanocomposite thermites", *Combust.Flame*, vol. 157, no. 8, pp. 1581–1585, 2010.

[DIN 14] DING Y., YAMADA R., GRESBACK R. *et al.*, "A parametric study of non-thermal plasma synthesis of silicon nanoparticles from a chlorinated precursor", *J. Phys. D: Appl. Phys.*, no. 47, pp. 485202, 2014.

[DIN 96a] DING J., TSUZUKI T., MCCORMICK P.G. *et al.*, "Ultrafine Cu particles prepared by mechanochemical process", *J. Alloys Compd.*, no. 234, pp. L1–L3, 1996.

[DIN 96b] DING J., TSUZUKI T., MCCORMICK P.G. *et al.*, "Ultrafine Co and Ni particles prepared by mechanochemical processing", *J. Phys. D: Appl. Phys.*, no. 29, pp. 2365–2369, 1996.

[DIN 96c] DING J., TSUZUKI T., MCCORMICK P.G. *et al.*, "Structure and magnetic properties of ultrafine Fe powders by mechanochemical processing", *J. Magn. Magn. Mater.*, no. 162, pp. 271–276, 1996.

[DIN 96d] DING J., TSUZUKI T., MCCORMICK P.G., "Ultrafine alumina particles prepared by mechanical/thermal proceding", *J. Am. Ceram. Soc.*, vol. 79, no. 11, pp. 2956–58, 1996.

[DIN 97a] DING J., TSUZUKI T., MCCORMICK P.G., "Mechanochemical synthesis ofultrafine ZrO_2 powder", *Nanostruct. Mater.*, vol. 8, no. 1, pp. 75–81, 1997.

[DIN 97b] DING J., TSUZUKI T., MCCORMICK P.G., "Hematite powders synthesized by mechanochemical processing", *Nanostruct. Mater.*, vol. 8, no. 6, pp. 739–747, 1997.

[DIN 99] DING J., TSUZUKI T., MCCORMICK P.G., "Microstructural evolution of Ni-NaCl mixtures during mechanochemical reaction and mechanical milling", *J. Mater. Sci.*, no. 34, pp. 5293–5298, 1999.

[DJU 15] DJURIŠIĆ A.B., LEUNG Y.H., NG A.M.C. *et al.*, "Toxicity of metal oxide nanoparticles: mechanisms, characterization, and avoiding experimental artefacts", *Small*, vol. 11, no. 1, pp. 26–44, 2015.

[DOD 01a] DODD A.C., TSUZUKI T., MCCORMICK P.G., "Nanocrystalline zirconia powders synthesised by mechanochemical processing", *Mater. Sci. Eng., A*, no. 301, pp. 54–58, 2001.

[DOD 01b] DODD A.C. , RAVIPRASAD K., MCCORMICK P.G., "Synthesis of ultrafine zirconia powders by mechanical processing", *Scripta Mater.*, no. 44, pp. 689–694, 2001.

[DOD 01c] DODD A.C., MCCORMICK P.G., "Solid-state chemical synthesis of nanoparticulate zirconia", *Acta Mater.*, no. 49, pp. 4215–4220, 2001.

[DOL 05] DOLGOBORODOV A.Y., MAKHOV M.N., KOLBANEV I.V. *et al.*, "Detonation in an aluminum-teflon mixture", *J. Expp.Theor. Phys. Lett.*, vol. 81, no. 7, pp. 311–314, 2005.

[DON 13] DONG Z., AL-SHARAB J.F., KEAr B.H. *et al.*, "Combined flame and electrodeposition synthesis of energetic coaxial tungsten-oxide/aluminum nanowire arrays", *Nano Lett.*, vol. 13, no. 9, pp. 4346–4350, 2013.

[DOS 14] DOS SANTOS V., DA SILVEIRA N.P., BERGMANN C.P., "In-situ evaluation of particle size distribution of ZrO_2-nanoparticles obtained by sol–gel", *Powder Technol.*, no. 267, pp. 392–397, 2014.

[DRE 09] DREIZIN E.L., "Metal-based reactive nanomaterials", *Prog. Energy Combust. Sci.*, no. 35, pp. 141–167, 2009.

[DUR 06] DURÃES L., CAMPOS J., PORTUGAL A., "Radial combustion propagation in iron (III) oxide/aluminum thermite mixtures", *Propell. Explos. Pyrot.*, vol. 31, no. 1, pp. 42–49, 2006.

[DUR 09] DURÃES L., PLAKSIN I., ANTUNES J. *et al.*, "Radial combustion dynamics in Fe_2O_3/Al thermite: variability of the flame propagation profiles", *AIP Conf. Proc.*, vol. 1195, pp. 428–431, 2009.

[DUT 09] DUTRO G. M., YETTER R. A., RISHA G.A. *et al.*, "The effect of stoichiometry on the combustion behavior of a nanoscale Al/MoO_3 thermite", *P. Combust. Inst.*, vol. 32, no. 2, pp. 1921–1928, 2009.

[EAP 04] EAPEN B.Z., HOFFMANN V.K., SCHOENITZ M. *et al.*, "Combustion of aerosolized spherical aluminum powders and flakes in air", *Combust. Sci. Technol.*, vol. 176, no. 7, pp. 1055–1069, 2004.

[EBE 46] EBELMEN J.-J., "Recherches sur les combinaisons des acides boriques et siliciques avec les ethers", *Ann. Chim. Phys.*, Serie 3, no. 16, pp. 129–166, 1846.

[ECK 92] ECKERT J., HOLZER J.C., KRILL IIIC.E. *et al.*, "Structural and thermodynamic properties of nanocrystalline fee metals prepared by mechanical attrition", *J. Mater. Res.*, vol. 7, no. 7, pp. 1751–1761, 1992.

[EGA 14] EGAN G.C., SULLIVAN K.T., LAGRANGE T., REED B.W. *et al.*, "In situ imaging of ultra-fast loss of nanostructure in nanoparticle aggregates", *J. Appl. Phys.*, vol. 115, no. 084903, 2014.

[EGA 15] EGAN G.C., ZACHARIAH M.R., "Commentary on the heat transfer mechanisms controlling propagation in nanothermites", *Combust. Flame*, vol. 162, no. 7, pp. 2959–2961, 2015.

[EID 04] EIDEN-ASSMANN S, WIDONIAK J., MARET G., "Synthesis and characterization of porous and nonporous monodisperse colloidal TiO_2 particles", *Chem. Mater.*, no. 16, pp. 6–11, 2004.

[EID 93] EIDELMAN S., ALTSHULER A., "Synthesis of nanoscale materials using detonation of solid explosives", *Nanostruct. Mater.*, vol. 3, no. 1–6, pp. 31–41, 1993.

[EL 05] EL-ESKANDARANY M.S., "Fabrication and characterizations of new nanocomposite WC/Al_2O_3 materials by room temperature ball milling and subsequent consolidation", *J. Alloys Compd.*, no. 391, pp. 228–235, 2005.

[EL 15] EL-ESKANDARANY M.S., "Amorphous-i phase-big cube-amorphous cyclic phase transformation of mechanically alloyed Zr 75Ni 20Al 5 system", *Intermetallics*, no. 63, pp. 27–36, 2015.

[ETH 06a] ETHAYARAJA M., BANDYOPADHYAYA R., "Population balance models and Monte Carlo simulation for nanoparticle formation in water-in-oil microemulsions: implications for CdS synthesis", *J. Am. Chem. Soc.*, no. 128, pp. 17102–17113, 2006.

[ETH 06] ETHAYARAJA M., DUTTA K., BANDYOPADHYAYA R., "Mechanism of nanoparticle formation in self-assembled colloidal templates: population balance model and Monte Carlo simulation", *J. Phys. Chem., B*, no. 110, pp. 16471–16481, 2006.

[ETH 07] ETHAYARAJA M., DUTTA K., MUTHUKUMARAN D. *et al.*, "Nanoparticle formation in water-in-oil microemulsions: experiments, mechanism, and Monte Carlo simulation", *Langmuir*, no. 23, pp. 3418–3423, 2007.

[FAB 90] FABBRO R., FOURNIER J., BALLARD P. *et al.*, "Physical study of laser-produced plasma in confined geometry"*J. Appl. Phys.*, vol. 68, no. 2, pp. 775–784, 1990.

[FAN 14] FAN G., REN S., QU S. *et al.*, "Stability and nonlinear optical properties of Cu nanoparticles prepared by femtosecond laser ablation of Cu target in alcohol and water", *Opt. Commun.*, no. 330, pp. 122–130, 2014.

[FAR 10] FARLEY C., PANTOYA M.L., "Reaction kinetics of nanometric aluminum and iodine pentoxide", *J. Therm. Anal. Calorim.*, vol. 102, no. 2, pp. 609–613, 2010.

[FAR 10] FARRUKH M.A. , HENG B.-T., ADNAN R., "Surfactant-controlled aqueous synthesis of SnO_2 nanoparticles via the hydrothermal and conventional heating methods", *Turk J. Chem.*, no. 34, pp. 537–550, 2010.

[FAU 11] FAUCHAIS P., MONTAVON G., LIMA R.S. *et al.*, "Engineering a new class of thermal spray nano-based microstructures from agglomerated nanostructured particles, suspensions and solutions: an invited review", *J. Phys. D: Appl. Phys.*, no. 44 pp. 093001-1–093001-53, 2011.

[FAU 80] FAUCHAIS P., "Utilisation industrielle actuelle et potentielle des plasmas: synthèses, traitement des poudres, traitements métallurgiques, traitements de surface", *Revue Phys. Appl.*, no. 15, pp. 1281–1301, 1980.

[FAU 96a] FAUCHAIS P., COUDERT J.F., PATEYRON B., "La production des plasmas thermiques", *Rev. Gen. Therm.*, no. 35, pp. 543–560, 1996.

[FAU 96b] FAUCHAIS P., COUDERT J.F., PATEYRON B., "Mesure de température dans les plasmas thermiques", *Rev. Gen. Therm.*, no. 35, pp. 324–337–560, 1996.

[FEC 90] FECHT H.J., HELLSTERN E., FU Z. *et al.*, "Nanocrystallin metals prepared by hygh-energy ball milling", *Metall. Trans., A*, no. 21A, pp. 2333–2337, 1990.

[FED 60] FEDOROFF B.T., AARONSON H.A., REESE E.F. *et al.*, Encyclopedia of Explosives and Related Items, New Jersey: US Army Research and Development Command, TACOM, ARDEC, Picatinny Arsenal, 1960.

[FEN 13] FENG J., JIAN G., LIU Q. *et al.*, "Passivated Iodine Pentoxide Oxidizer for Potential Biocidal Nanoenergetic Applications", *ACS Appl. Mater. Interfaces*, vol. 5, no. 18, pp. 8875–8880, 2013.

[FER 05] FERGUSON J.D., BUECHLER K.J., WEIMER A.W. *et al.*, "SnO_2 atomic layer deposition on ZrO_2 and Al nanoparticles: Pathway to enhanced thermite materials", *ACS Appl. Mater. Interfaces*, vol. 5, no. 18, pp. 8875–8880, 2013.

[FER 07] FERNANDEZ C.A., WAI C.M., "Continuous tuning of cadmium sulfide and zinc sulfide nanoparticle size in a water-in-supercritical carbon dioxide microemulsion", *Chem. Eur. J.*, no. 13, pp. 5838–5844, 2007.

[FIE 92] FIELD J.E., *The Properties of Natural and Synthetic Diamond*, Elsevier Academic Press, 1992.

[FIS 96] FISCHER S.H., GRUBELICH M.C., "A survey of combustible metals, thermites, and intermetallics for pyrotechnic applications", *32nd AIAA/ASME/SAE/ASEE Joint Propulsion Conference*, Lake Buena Vista, Florida, USA, July 1996.

[FIS 06] FISHBURN J.M., WITHFORD M.J., COUTTS D.W. *et al.*, "Study of the fluence dependent interplay between laser induced material removal mechanisms in metals: Vaporization, melt displacement and melt ejection", *Appl. Surf. Sci.*, no. 252, pp. 5182–5188, 2006.

[FLE 87] FLETCHER P.D.I., HOWET J.A.M., ROBINSON B.H., "The kinetics of solubilisate exchange between water droplets of a water-in-oil microemulsion", *J. Chem. Soc., Faraday Trans. I*, no. 83, pp. 985–1006, 1987.

[FOL 07] FOLEY T., PACHECO A., MALCHI J. *et al.*, "Development of nanothermite composites with variable electrostatic discharge ignition thresholds", *Propell. Explos. Pyrot.*, vol. 32, no. 6, pp. 431–434, 2007.

[FOR 08] FORKER M., DE LA PRESA P., HOFFBAUER W. *et al.*, "Structure, phase transformations, and defects of HfO2 and ZrO2 nanoparticles studied by 181Ta and 111Cd perturbed angular correlations, 1H magic-angle spinning NMR, XPS, and X-ray and electron diffraction", *Phys. Rev., B*, no. 77, p. 054108, 2008.

[FRA 98] FRANK A.C., STOWASSER F., SUSSEK H. *et al.*, "Detonations of gallium azides: a simple route to hexagonal GaN nanocrystals", *J. Am. Chem. Soc.*, vol. 120, no. 14, pp. 3512–3513, 1998.

[FUJ 14] FUJITA M., KOMATSU N., KIMURA T., "Sonochemical preparation of carbon spheres", *Ultrason. Sonochem.*, no. 21, pp. 943–945, 2014.

[FUK 74] FUKANO Y., "Particles of gamma-iron quenched at room temperature", *Jpn J. Appl. Phys.*, vol. 13, no. 6, pp. 702–713, 1610, 1974.

[FUR 14] FURUKAWA S., REBOUL J., DIRING S. *et al.*, "Structuring of metal-organic frameworks at the mesoscopic/macroscopic scale", *Chem. Soc. Rev.*, vol. 43, pp. 5700–5734, 2014.

[FUZ 00] FUZELLIER H., COMET M., "Étude synoptique des explosifs", *Actual. Chimique*, vol. 233, pp. 4–11, 2000.

[GAF 04] GAFFET E., LE CAËR G., "Mechanical processing for nanomaterials", in Nalwa H.S. (ed.), *Encyclopedia of Nanoscience and Nanotechnology*, American Scientific Publishers, 2004.

[GAF 91] GAFFET E., "Planetary ball-milling: an expérimental parameter phase diagram", *Mater. Sci. Eng., A*, vol. 132, pp. 181–193, 1991.

[GAF 95] GAFFET E., ABDELLAOUI M., MALHOUROUX-GAFFET N., "Formation of nanostructured material induced by mechanical processings", *Mater. Trans. JIM*, vol. 36, no. 2, pp. 198–209, 1995.

[GAG 57] GAGE, R.M., Arc torches and process, U. S. Patent no. 2,806,124, 1957.

[GAN 07] GANGOPADHYAY S., SHENDE R., APPERSON S. *et al.*, "On-chip igniter and method of manufacture", US 2007/0099335, 2007.

[GAS 01] GASH A. E., TILLOTSON T.M., SATCHER J.H. *et al.*, "Use of epoxides in the sol-gel synthesis of porous iron (III) Oxide Monoliths from Fe(III) Salts", *Chem. Mater.*, vol. 13, no. 3, pp. 999–1007, 2001.

[GAS 03] GASH A.E., SATCHER J.H., SIMPSON R.L., "Strong akaganeite aerogel monoliths using epoxides: synthesis and characterization", *Chem. Mater.*, vol. 15, no. 17, pp. 3268–3275, 2003.

[GAS 04] GASH A.E., SATCHER J.H., SIMPSON R.L. *et al.*, "Nanostructured energetic materials with sol-gel methods", *Mater. Res. Soc. Sympp. P.*, vol. 800, pp. 55–66, 2004.

[GAS 11] GASH A.E., SIMPSON R.L., SATCHER J.H., "Aerogels and sol-gel composites as nanostructured energetic materials", in AEGERTER M.A., LEVENTIS N., KOEBEL M.M., *Aerogels Handbook*, Springer, New York, 2011.

[GEF 39] GEFFCKEN W., BERGER E., Verfahren zur Änderung des Reflexionsvermögen optischer Gläser, DE Patent 736411 C, 1939.

[GES 12] GESNEr J., PANTOYA M.L., LEVITAS V.I., "Effect of oxide shell growth on nano-aluminum thermite propagation rates", *Combust. Flame*, vol. 159, no. 11, pp. 3448–3453, 2012.

[GHA 08] GHASEMI S., MOUSAVi M.F., SHAMSIPUR M. *et al.*, "Sonochemical-assisted synthesis of nano-structured lead dioxide", *Ultrason. Sonochem.*, no. 15, pp. 448–455, 2008.

[GIA 10] GIAMMANCO F., GIORGETTi E., MARSILI P. *et al.*, "Experimental and theoretical analysis of photofragmentation of Au nanoparticles by picosecond laser radiation", *J. Phys. Chem., C*, vol. 114, no. 8, pp. 3354–3363, 2010.

[GIB 11a] GIBOT P., COMET M., EICHHORN A. *et al.*, "Highly insensitive/reactive thermite prepared from Cr_2O_3 nanoparticles", *Propell. Explos. Pyrot.*, vol. 36, no. 1, pp. 80–87, 2011.

[GIB 11b] GIBOT P., COMET M., VIDAL L. *et al.*, "Synthesis of WO_3 nanoparticles for superthermites by the template method from silica spheres", *Solid State Sci.*, vol. 13, no. 5, pp. 908–914, 2011.

[GIB 11c] GIBOT P., MORY J., MOITRIER F. *et al.*, "Miniaturization of micrometric SiC from a detonation process of highly energetic material", *Powder Technol.*, vol. 208, no. 2, pp. 324–328, 2011.

[GIE 94] GIESCHE H., "Synthesis of monodispersed silica powders II. Controled growth reaction and continuous production process", *J. Eur. Ceram. Soc.*, no. 14, pp. 205–214, 1994.

[GIL 83] GILMAN P.S., BENJAMIN J.S., "Mechanical alloying", *Ann. Rev. Mater. Sci.*, no. 13, pp. 279–300, 1983.

[GIR 93] GIRSHICK S.L., CHIU C.-P., MUNO R. *et al.*, "Thermal plasma synthesis of ultrafine iron particles", *J. Aerosol Sci.*, vol. 24, no. 3, pp. 367 382, 1993.

[GLA 15] GLAVIER L., TATON G., DUCÉRÉ J.-M. *et al.*, "Nanoenergetics as pressure generator for nontoxic impact primers: Comparison of Al/Bi_2O_3, Al/CuO, Al/MoO_3 nanothermites and Al/PTFE", *Combust. Flame*, vol. 162, pp. 1813–1820, 2015.

[GLE 89] GLEITER H., "Nanocrystalline materials", *Prog. Mater Sci.*, vol. 33, pp. 223–315, 1989.

[GLU 08] GLUSHENKOV A.M., ZHANG H.Z., CHEN Y., "Reactive ball milling to produce nanocrystalline ZnO", *Mater. Lett.*, vol. 62, no. 24, pp. 4047–4049, 2008.

[GOL 97] GOLDSCHMIDT H., Method of producing metals and alloys, US Patent no. 578868 A, 1897.

[GOO 09] GOO W.H., BAC L.H., PARK E.J. *et al.*, "Synthesis and characterization of nano-sized Zn powder by electrical explosion of wire in liquid", *Mod. Phys. Lett., B*, vol. 23, no. 31–32, p 3903–3909, 2009.

[GOO 06] GOORTANI B.M., MENDOZA N., PROULX P., "Synthesis of SiO_2 Nanoparticles in RF Plasma Reactors: Effect of Feed Rate and Quench Gas Injection", *Int. J. Chem. React. Eng.*, vol. 4, Article A33, 2006.

[GOP 00] GOPALAN S., SINGHAL S.C., "Mechanochemical synthesis of nano-sized CeO_2", *Scripta Mater.*, no. 42, pp. 993–996, 2000.

[GOU 10] GOULLÉ J.-P., SAUSSEREAU E., MAHIEU L. *et al.*, "Une nouvelle approche biologique: le profil métallique", *Ann. Biol. Clin.*, vol. 68, no. 4, pp. 429–440, 2010.

[GOU 93] GOURSAT A.G., VERNET G., RIMBERT J-F. *et al.*, Procédé de fabrication de poudres métalliques à partir d'un matériau métallique de fusion, French Patent no. 8307414, 1993.

[GRA 04] GRANIER J.J., PANTOYA M.L., "Laser ignition of nanocomposite thermites", *Combust. Flame*, vol. 138, no. 4, pp. 373–383, 2004.

[GRO 06] GROMOV A.A., FÖRTER-BARTH U., TEIPEL U., "Aluminium nanopowders produced by electrical explosion of wires and passived by non-inert coatings: Characterisation and reactivity with air and water", *Power Technol.*, vol. 164, pp. 111–115, 2006.

[HA 09] HA J.-H., MURALIDHARAN P., KIM D.K., "Hydrothermal synthesis and characterization of self-assembled h-WO_3 nanowires/nanorods using EDTA salts", *J. Alloys Compd.*, no. 475, pp. 446–451, 2009.

[HA 10] HA Y.-C., CHO C., "Production of highly-dispersed nano-sized sn powders in a liquid medium by using a high-energy electrical explosion", *J. Korean Phys. Soc.*, vol. 57, no. 6, pp. 1574–1576, 2010.

[HAF 14] HAFEEZULLAH, YAMANI Z. H., IQBAL J. *et al.*, "Rapid sonochemical synthesis of In_2O_3 nanoparticles their doping optical, electrical and hydrogen gas sensing properties", *J. Alloys Compd.*, no. 616, pp. 76–80, 2014

[HAI 05] HAIBO Z., WEIPING C., YUE L. *et al.*, "Composition/structural evolution and optical properties of ZnO/Zn nanoparticles by laser ablation in liquid media", *J. Phys. Chem., B*, vol. 109, no. 39, pp. 18260–18266, 2005.

[HAL 49] HALE G.C., HART D., Fuse powder composition, US Patent no. 2,468,061, 1949.

[HAM 96] HAMADA K., SENNA M., "Mechanochemical effects on the properties of starting mixtures for $PbTiO_3$ ceramics by using a novel grinding equipment", *J. Mater. Sci.*, no. 31, pp. 1725–1728, 1996.

[HAM 11] HAMMERSTROEM D.W., BURGERS M.A., CHUNG S.W. *et al.*, "Aluminum nanoparticles capped by polymerization of alkyl-substituted epoxides: ratio-dependent stability and particle size", *Inorg. Chem.*, vol. 50, no. 11, pp. 5054–5059, 2011.

[HAN 11] HAN N.S., SHIM H.S., SEO J.H. *et al.*, "Optical properties and lasing of ZnO nanoparticles synthesized continuously in supercritical fluids", *Chem. Phys. Lett.*, no. 505, pp. 51–56, 2011.

[HAN 14] HAN Z.-W., HAN Y.-C., XU S., "Preparation of nano-cerium dioxide and its effect on the thermal decomposition of ammonium perchlorate", *J. Therm. Anal. Calorim.*, vol. 116, pp. 273–278, 2014.

[HAN 11] HANAOr D.A.H., SORRELL C.C., "Review of the anatase to rutile phase transformation", *J. Mater. Sci.*, vol. 46, pp. 855–874, 2011.

[HAR 03] HARILAL S.S., BINDHU C.V., TILLACK M.S. *et al.*, "Internal structure and expansion dynamics of laser ablation plumes into ambient gases", *J. Appl. Phys.*, no. 93, pp. 2380–2388, 2003.

[HAS 11] HASEGAWA G., SHIMONAKA M., ISHIHARA Y., "Differential genotoxicity of chemical properties and particle size of rare metal and metal oxide nanoparticles", *J. Appl. Toxicol.*, vol. 32, no. 1, pp. 72–80, 2012.

[HAS 90] HASHIMOTO H., WATANABE R., "Model simulation of energy consumption during vibratory ball milling of metal powder", *Mater. Trans., JIM*, vol. 31, no. 3, pp. 219–224, 1990.

[HAT 13] HATTORI Y., NOMURA S., MUKASA S. *et al.*, "Synthesis of tungsten trioxide nanoparticles by microwave plasma in liquid and analysis of physical properties", *J. Alloys Compd.*, no. 560, pp. 105–110, 2013.

[HAY 90] HAYAKAWA K., IWAMA S., "Preparation of ultrafine gamma Fe particles by microwave plasma processing" *J. Cryst. Growth*, no. 99, pp. 188–191, 1990.

[HE 07] HE C., SASAKI T., USUI H. *et al.*, "Fabrication of ZnO nanoparticles by pulsed laser ablation in aqueous media and pH-dependent particle size: An approach to study the mechanism of enhanced green photoluminescence", *J. Photochem. Photobiol. A*, no. 191, pp. 66–73, 2007.

[HE 08] HE C., SASAKI T., SHIMIZU Y. *et al.*, "Synthesis of ZnO nanoparticles using nanosecond pulsed laser ablation in aqueous media and their self-assembly towards spindle-like ZnO aggregates", *Appl. Surf. Sci.*, no. 254, pp. 2196–2202, 2008.

[HEN 93] HENGLEIN A., "Physicochemical properties of small metal particles in solution: 'microelectrode' reactions, chemisorption, composite metal particles, and the atom-to-metal transition", *J. Phys. Chem.*, vol. 97, no. 21, pp. 5457–5471, 1993.

[HEN 10] HENZ B.J., HAWA T., ZACHARIAH M.R., "On the role of built-in electric fields on the ignition of oxide coated nanoaluminum: Ion mobility versus Fickian diffusion", *J. Appl. Phys.*, vol. 107, no. 024901, 2010.

[HEO 15] HEON L., KIM H.G., KIM B.H. *et al.*, "Investigation on sized-regulated iron nanoparticles prepared by liquid Phase plasma reduction process", *J. Nanosci. Nanotechnol.*, no. 15, pp. 518–521, 2015.

[HER 14] HERNANDEZ-URESTI D.B., SANCHEZ-MARTINEZ D., MARTINEZ-DE LA CRUZ A. *et al.*, "Characterization and photocatalytic properties of hexagonal and monoclinic WO_3 prepared via microwave-assisted hydrothermal synthesis", *Ceram. Int.*, no. 40, pp. 4767–4775, 2014.

[HON 06] HONG Y.C., UHM H.S., "Synthesis of MgO nanopowder in atmospheric microwave plasma torch", *Chem. Phys. Lett.*, no. 422, pp. 174–178, 2006.

[HON 11] HONG Y.C., LHO T., LEE B.J. *et al.*, "Synthesis of titanium dioxide in $O_2/Ar/SO_2/TiCl_4$ microwave torch plasma and its band gap narrowing", *Curr. Appl. Phys.*, no. 11, pp. 517–520, 2011.

[HON 94] HONG L.B., BANSAL C., FULTZ B., "Steady state grain size and thermal stability of nanophase Ni_3Fe and Fe_3X (X = Si, Zn, Sn) synthesized by ball milling at elevated temperatures", *Nanostruct. Mater.*, vol. 4, no. 8, pp. 949–956, 1994.

[HOR 10] HORCAJADA P., CHALATI T., SERRE C. *et al.*, "Porous metal-organic-framework nanoscale carriers as a potential platform for drug delivery and imaging", *Nat. Mater.*, vol. 9, pp. 172–178, 2010.

[HU 07] HU X., YU J.C., GONG J. *et al.*, "α-Fe_2O_3 nanorings prepared by a microwave-assisted hydrothermal process and their sensing properties", *Adv. Mater.*, no. 19, pp. 2324–2329, 2007.

[HU 14a] HU X., ZHANG X., SHEN X. *et al.*, "Plasma-induced synthesis of CuO nanofibers and ZnO nanoflowers in water", *Plasma Chem. Plasma Process.*, no. 34, pp. 1129–1139, 2014.

[HU 14b] HU S., MELTON C., MUKHERJEE D., "A facile route for the synthesis of nanostructured oxides and hydroxides of cobalt using laser ablation synthesis in solution (LASIS)", *Phys. Chem. Chem. Phys.*, no. 16, pp. 24034–24044, 2014.

[HU 15] HU X., LIAO X., XIAO L. *et al.*, "High-energy pollen-like porous Fe2O3/Al thermite: synthesis and properties", *Propell. Explos. Pyrot.*, vol. 40, no. 6, p. 867–872, 2015.

[HUA 10a] HUANG Z., WU Q., LI X. *et al.* "Synthesis and characterization of nano-sized boron powder prepared by plasma torch", *Plasma Sci. Technol.*, vol. 12, no. 5, pp. 577–580, 2010.

[HUA 10b] HUANG Y.-W., WU C.-H., ARONSTAM R.S., "Toxicity of transition metal oxide nanoparticles: recent insights from in vitro studies", *Materials*, vol. 3, no. 10, pp. 4842–4859, 2010.

[HUA 97a] HUANG H., PAN J., MCCORMICK P.G., "On the dynamics of mechanical milling in a vibratory mill", *Mater. Sci. Eng. A*, vol. 232, pp. 55–62, 1997.

[HUA 97b] HUANG H., PAN J. MCCORMICK P.G., "Prediction of impact forces in a vibratory ball mill using an inverse technique", *Int. J. Impact Eng.*, vol. 19, no. 2, pp. 117–126, 1997.

[HUA 97c] HUANG B., ISHIHARA K.N., SHINGU P.H. "Metastable phases of Al–Fe system by mechanical alloying", *Mater. Sci. and Eng., A*, no. 231, pp. 72–79, 1997.

[HUS 05] HUSSAIN S.M., HESS K.L., GEARHART J.M. *et al.*, "In vitro toxicity of nanoparticles in BRL 3A rat liver cells", *Toxicol. In Vitro*, vol. 19, no. 7, pp. 975–983, 2005.

[ILE 52] ILER R. K., Aqueous dispersion of water-insoluble organic polymers containing colloïdal silica, US Patent no. 2,597,872, 1952.

[IM 08] IM J.-H., LEE J.-H., PARK D.-W., "Synthesis of nano-sized tin oxide powder by argon plasma jet at atmospheric pressure", *Surf. Coat. Technol.*, no. 202, pp. 5471–5475, 2008.

[IMA 96] IMAMURA H., SAKASAI N., KAJII Y., "Hydrogen absorption of Mg-based composites prepared by mechanical milling: Factors affecting its characteristics", *J. Alloys Compd.*, no. 232, pp. 218–223, 1996.

[INA 05] INASAWA S., SUGIYAMA M., YAMAGUCHI Y., "Bimodal size distribution of gold nanoparticles under picosecond laser pulses", *J. Phys. Chem., B*, vol. 109, no. 19, pp. 9404–9410, 2005.

[INO 10] INOGAMOV N.A., ZHAKHOVSKY V.V., FAENOV A.Y. *et al.*, "Spallative ablation of dielectrics by X-ray laser", *Appl. Phys., A*, vol. 101, no. 1, p 87–96, 2010.

[INT 14] INTARTAGLIA R., KOMAL BAGGA K., BRANDI F., "Study on the productivity of silicon nanoparticles by picosecond laser ablation in water: towards gram per hour yield", *Opt. Express*, vol. 22, no. 3, pp. 3117–3127, 2014.

[ION 11] IONIN A.A., KUDRYASHOV S.I., SELEZNEV L.V. *et al.*, "Dynamics of the spallative ablation of a GaAs surface irradiated by femtosecond laser pulses", *J. Expp. Theor. Phys. lett.*, vol. 94, no. 10, pp. 753–758, 2011.

[ION 13] IONIN A.A., KUDRYASHOV S.I., SELEZNEV L.V. *et al.*, "Thermal melting and ablation of silicon by femtosecond laser radiation", *J. Expp. Theor. Phys.*, vol. 116, no. 3, pp. 347–362, 2013.

[IPE 14] IPEKSAC T., KAYA F., KAYA C., "Template-free hydrothermal method for the synthesis of multi-walled CuO nanotubes", *Mater. Lett.*, no. 130, pp. 68–70, 2014.

[ISA 33] ISABEY J., "Cuivre", in PASCAL P., BAUD P. (ed), *Traité de chimie minérale*, vol. 8, Masson et Cie, Paris, 1933.

[ISH 06] ISHIKAWA Y., SHIMIZU Y., SASAKI T. *et al.*, "Preparation of zinc oxide nanorods using pulsed laser ablation in water media at high temperature", *J. Colloid Interface Sci.*, no. 300, pp. 612–615, 2006.

[ISM 14] ISMAIL R.A., FADHIL F.A., "Effect of electric field on the properties of bismuth oxide nanoparticles prepared by laser ablation in water", *J. Mater. Sci.: Mater. Electron.*, no. 25, pp. 1435–1440, 2014.

[IVA 03] IVANOV Y.F.,. OSMONOLIEV M.N, SEDOI V.S. *et al.*, "Productions of ultra-fine powders and their use in high energetic compositions", *Prop., Explosives, Pyrot.*, vol. 28, no. 6, pp. 319–333, 2003.

[IWA 92] IWAMA S., HAYAKAWA K., "Vaporization and condensation of metals in a flowing gas with high velocity", *Nanostruct. Mater.*, no. 1, pp. 113–118, 1992.

[IWA 95] IWAMA S., MIHAMA K., "Nanometer-sized beta-Mn and amorphous Sb particles formed by the flowing gas evaporation technique", *NanoStruct. Mater.*, vol. 6, pp. 305–308, 1995.

[JAI 06] JAIN K., SRIVASTAVA A., RASHMI, "Synthesis and controlling the morphology of SnO2 nanocrystals via hydrothermal treatment", *ECS Trans.*, vol. 1, no. 21, pp. 1–7, 2006.

[JAN 11] JANKAUSKAS V., PADGURSKAS J., ZUNDA A. *et al.*, "Research into nanoparticles obtained by electric explosion of conductive materials", *Surf. Eng. Appl. Electrochem.*, vol. 47, no. 2, pp. 170–175, 2011.

[JAN 02] JANZEN C., KLEINWECHTER H., KNIPPING J. *et al.*, "Size analysis in low-pressure nanoparticle reactors: comparison of particle mass spectrometry with in situ probing transmission electron microscopy", *J. Aerosol Sci.*, no. 33, pp. 833–841, 2002.

[JIA 13a] JIAN G., FENG J., JACOB R.J. *et al.*, "Super-reactive Nanoenergetic Gas Generators Based on Periodate Salts", *Angew. Chem. Int. Ed.*, vol. 52, no. 37, pp. 9743–9746, 2013.

[JIA 13b] JIAN G., LIU L., ZACHARIAH M. R., "Facile aerosol route to hollow CuO spheres and its superior performance as an oxidizer in nanoenergetic gas generators", *Adv. Funct. Mater.*, vol. 23, pp. 1341–1346, 2013.

[JIA 13c] JIAN G., CHOWDHURY S., SULLIVAN K. *et al.*, "Nanothermite reactions: Is gas phase oxygen generation from the oxygen carrier an essential prerequisite to ignition?", *Combust. Flame*, vol. 160, no. 2, pp. 432–437, 2013.

[JIA 98] JIANG W.H., YATSUI K., "Pulsed wire discharge for nanosize powder synthesis", *IEEE T. Plasma Sci.*, vol. 26, no. 5, pp. 1498–1501, 1998.

[JOL 04] JOLIVET J.-P., FROIDEFOND C., POTTIER A. *et al.*, "Size tailoring of oxide nanoparticles by precipitation in aqueous medium. A semi-quantitative modelling", *J. Mater. Chem.*, no. 14, pp. 877–882, 2004.

[JON 04] JONES J.W., Energy dense explosives, US Patent no. 6,679,960, 2004.

[JOU 05] JOUET R.J., WARREN A.D., ROSENBERG D.M. *et al.*, "Surface passivation of bare aluminum nanoparticles using perfluoroalkyl carboxylic acids", *Chem. Mater.*, vol. 17, no. 11, pp. 2987–2996, 2005.

[JUN 01] JUNG J., PERRUT M., "Particle design using supercritical fluids: Literature and patent survey", *J. Supercrit. Fluids*, no. 20, pp. 179–219, 2001.

[KAB 03] KABASHIN A.V,. MEUNIER M., "Synthesis of colloidal nanoparticles during femtosecond laser ablation of gold in water", *J. Appl. Phys.*, vol. 94, no. 12, pp. 7941–7943, 2003.

[KAL 98] KALYANARAMAN R., SANG YOO, KRUPASHANKARA M.S. *et al.*, "Synthesis and consolidation of iron nanopowders" *Nanostruct. Mater.*, vol. 10, no. 8, pp. 1379–1392, 1998.

[KAM 98] KAMAT P.V. FLUMIANI M., HARTLAND G.V., "Picosecond dynamics of silver nanoclusters. photoejection of electrons and fragmentation", *J. Phys. Chem. B*, vol. 102, no. 17, pp. 3123–3128, 1998.

[KAP 13] KAPLOWITZ D.A., JIAN G., GASKELL K. *et al.*, "Aerosol synthesis and reactivity of thin oxide shell aluminum nanoparticles via fluorocarboxylic acid functional coating", *Part. Part. Syst. Charact.*, vol. 30, no. 10, pp. 881–887, 2013.

[KAP 14] KAPLOWITZ D.A., JIAN G., GASKELL K. *et al.*, "Synthesis and reactive properties of Iron Oxide-Coated Nanoaluminum", *J. Energ. Mater.*, vol. 32, no. 2, pp. 95–105, 2014.

[KAR 62] KARIORIS F.G., FISH B.R., "An exploding wire aerosol generator", *J. Col. Sci.*, no. 17, pp. 155–161, 1962.

[KAW 06] KAWASAKI M., NISHIMURA N., "1064-nm laser fragmentation of thin Au and Ag flakes in acetone for highly productive pathway to stable metal nanoparticles", *Appl. Surf. Sci.*, vol. 253, no. 4, pp. 2208–2216, 2006.

[KAW 09] KAWASAKI S.-I., YAN XIUYI Y., SUE K. *et al.*, "Continuous supercritical hydrothermal synthesis of controlled size and highly crystalline anatase TiO_2 nanoparticles", *J. Supercrit. Fluids*, no. 50, pp. 276–282, 2009.

[KAW 10] KAWASAKI S.-I., SUE K., OOKAWARA R. *et al.*, "Engineering study of continuous supercritical hydrothermal method using a T-shaped mixer: Experimental synthesis of NiO nanoparticles and CFD simulation", *J. Supercrit. Fluids*, no. 54, pp. 96–102, 2010.

[KEL 98] KELLY R., DREYFUS R.W., "On the effect of Knudsen-layer formation on studies of vaporization, sputtering and desorption", *Surf. Sci.*, no. 198, pp. 263–276, 1988.

[KEN 13] KENNEDY A.J., MELBY N.L., MOSER R.D. *et al.*, "Fate and Toxicity of CuO Nanospheres and Nanorods used in Al/CuO Nanothermites Before and After Combustion", *Environ. Sci. Technol.*, vol. 47, no. 19, pp. 11258–11267, 2013.

[KHA 06] KHAN A.S., SUH Y. S., CHEN X. *et al.*, "Nanocrystalline aluminum and iron: Mechanical behavior at quasi-static and high strain rates, and constitutive modeling", *Int. J. Plast.*, no. 22, pp. 195–209, 2006.

[KHA 08] KHAN A.S., FARROKH B., TAKACS L., "Effect of grain refinement on mechanical properties of ball-milled bulk aluminum", *Mater. Sci. Eng., A*, no. 489, pp. 77–84, 2008.

[KHA 10] KHATIRKAR R.K., MURTY B.S., "Structural changes in iron powder during ball milling", *Mater. Chem. Phys.*, no. 123, pp. 247–253, 2010.

[KIM 01] KIM E.J., HAHN S.H., "Microstructure and photoactivity of titania nanoparticles prepared in nonionic W/O microemulsions", *Mater. Sci. Eng., A*, no. 303, pp. 24–29, 2001.

[KIM 04] KIM S.H., ZACHARIAH M.R., "Enhancing the rate of energy release from nanoenergetic materials by electrostatically enhanced assembly", *Adv. Mater.*, vol. 16, no. 20, pp. 1821–1825, 2004.

[KIM 07] KIM J.H., HONG Y.C., UHM H.S., "Synthesis of oxide nanoparticles via microwave plasma decomposition of initial materials", *Surf. Coat. Technol.*, no. 201, pp. 5114–5120, 2007.

[KIM 09] KIM J., KIM D., VERIANSYAH B. *et al.*, "Metal nanoparticle synthesis using supercritical alcohol", *Mater. Lett.*, no. 63, pp. 1880–1882, 2009.

[KIM 11] KIM D.K., BAE J.H., KANG M.K. *et al.*, "Analysis on thermite reactions of CuO nanowires and nanopowders coated with Al", *Curr. Appl. Phys.*, vol. 11, no. 4, pp. 1067–1070, 2011.

[KIM 13a] KIM H.G., LEE H., KIM B.H. *et al.*, "Synthesis process of cobalt nanoparticles in liquid-phase plasma", *Jpn. J. Appl. Phys.*, no. 52, pp. 01AN03-1 – 01AN03-4, 2013.

[KIM 13b] KIM H.G., LEE H., KIM S.J. *et al.*, "Synthesis of manganese nanoparticles in the liquid phase plasma", *J. Nanosci. Nanotechnol.*, no. 13, pp. 6103–6108, 2013.

[KIM 14] KIM M., SON W.S., AHN K.H. *et al.*, "Hydrothermal synthesis of metal nanoparticles using glycerol as a reducing agent", *J. Supercrit. Fluids*, no. 90, pp. 53–59, 2014.

[KIM 15a] KIM S.C., KIM B.H., CHUNG M. *et al.*, "Preparation of aluminum nanoparticles using bipolar pulsed electrical discharge in water", *J. Nanosci. Nanotechnol.*, no. 15, pp. 5350–5353, 2015.

[KIM 15b] KIM S.J., KIM B.H., CHUNG M. *et al.*, "The synthesis of nickel nanoparticles by liquid phase plasma processing", *J. Nanosci. Nanotechnol.*, no. 13, pp. 1997–2000, 2015.

[KIM 15c] KIM J., HONG S.A., YOO J., "Continuous synthesis of hierarchical porous ZnO microspheres in supercritical methanol and their enhanced electrochemical performance in lithium ion batteries", *Chem. Eng. J.*, no. 266, pp. 179–188, 2015.

[KIM 15d] KIM J.H., KIM S.B., CHOI M.G. *et al.*, "Flash-ignitable nanoenergetic materials with tunable underwater explosion reactivity: The role of sea urchin-like carbon nanotubes", *Combust. Flame*, vol. 162, no. 4, pp. 1448–1454, 2015.

[KIM 63] KIMOTO K., KAMIYA Y., NONOYAMA M. *et al.*, "An electron microscope study on fine metal particles prepared by evaporation in argon gas at low pressure", Jpn *J. Appl. Phys.*, vol. 2, no. 11, pp. 702–713, 1963.

[KIM 87] KIMURA K., "Metal colloïds produced by means of gas evaporation technique. IV Size distribution of small Mg and In particles", *Bull. Chem. Soc. Jpn.*, no. 60, pp. 3093–3097, 1987.

[KIM 95] KIMURA Y., TAKAKI S., "Microstructural changes during annealing of work-hardened mecanically milled metallic powders (Overview)", *Mater. Trans., JIM*, vol. 36, no. 2 pp. 289–296, 1995.

[KIN 03] KINEMUCHI Y., MURAI K., SANGURAI C. *et al.*, "Nanosize powders vof aluminium nitride synthesized by pulsed wire discharge", *J. Am. Ceram. Soc.*, no. 86, pp. 420–424, 2003.

[KIR 63] KIRSHENBAUM A.D., GROSSE A.V., Basic research in inorganic and physical chemistry at high temperatures, 5th Annual Report, United States, OSTI identifier: 4686208, no. TID–18951, 1963.

[KIR 08] KIRILLOV P.L., BOBKOV V.P., FOKIN L.R. *et al.*, "Structural materials", in KIRILLOV P.L., STANCULESCU A. (ed.), *Thermophysical Properties of Materials for Nuclear Engineering: A Tutorial and Collection of Data*, IAEA, Vienna, 2008.

[KLE 02] KLEINWECHTER H., JANZEN C., KNIPPINg J. *et al.*, "Formation and properties of ZnO nano-particles from gas phase synthesis processes", *J. Mater. Sci.*, no. 37, pp. 4349–4360, 2002.

[KLE 05] KLEINER S., BERTOCCO F., KHALID F.A. *et al.*, "Decomposition of process control agent during mechanical milling and its influence on displacement reactions in the Al–TiO_2 system", *Mater. Chem. Phys.*, no. 89, pp. 362–366, 2005.

[KLIN 12] KLINBUMRUNG A., THONGTEM T., THONGTEM S., "Characterization of orthorhombic α-MoO_3 microplates produced by a microwave plasma process", *J. Nano Mat.*, vol. 2012.

[KO 06] KO T.S., YANG S., HSU H.C. *et al.*, "ZnO nanopowders fabricated by DC thermal plasma synthesis", *Mater. Sci. and Eng. B*, no. 134, pp. 54–58, 2006.

[KOB 08] KOBAYASHI N., KAWAKAMI Y., KAMADA K. *et al.*, "Spherical submicron-size copper powders coagulated from a vapor phase in RF induction thermal plasma", *Thin Solid Films*, no. 516, pp. 4402–4406, 2008.

[KOC 08] KOCH E.C., "Special materials in pyrotechnics: V. military applications of phosphorus and its compounds", *Propell. Explos. Pyrot.*, vol. 33, no. 3, pp. 165–176, 2008.

[KOC 11] KOCH E.C., WEISER V., ROTH E. *et al.*, "Consideration of some 4f-metals as new flare fuels-Europium, samarium, thulium and ytterbium", *42nd International Annual Conference of the Fraunhofer ICT*, Karlsruhe, Germany, 2011.

[KOC 12] KOCH E.C., WEISER V., ROTH E. *et al.*, "Combustion of ytterbium metal", *Propell. Explos. Pyrot.*, vol. 37, no. 1, pp. 9–11, 2012.

[KOC 93] KOCH C.C., "The synthesis and structure of nanocrystalline materials produced by mechanical attrition: a review", *Nanostruct. Mater.*, no. 2, pp. 109–129, 1993.

[KOC 96] KOCH C.C., WHITTENBERGER J.D., "Review: mechanical milling/alloying of intermetallics", *Intermetallics*, no. 4, pp. 339–355, 1996.

[KOJ 08] KOJIMA T., SUGIMOTO T., "Formation mechanism of amorphous TiO_2 spheres in organic solvents 3. Effects of water, temperature, and solvent composition", *J. Phys. Chem. C*, no. 112, pp. 18445–18454, 2008.

[KOL 03] KOLEN'KO Y.V., BURUKHIN A.A., CHURAGULOV B.R. *et al.*, "Synthesis of nanocrystalline TiO_2 powders from aqueous $TiOSO_4$ solutions under hydrothermal conditions", *Mater. Lett.*, no. 57, pp. 1124–1129, 2003.

[KOL 12] KOLTYPIN Y., NIKITENKO S.I., GEDANKEN A., "The sonochemical preparation of tungsten oxide nanoparticles", *J. Mater. Chem.*, no. 12, pp. 1107–1110, 2012.

[KOL 96] KOLTYPIN Y., KATABI G., CAO X. *et al.*, "Sonochemical preparation of amorphous nickel", *J. Non-Cryst. Solids*, no. 201, pp. 159–162, 1996.

[KOR 05] KOROGODOV V.S., KIRDYASHKIN A.I., MAKSIMOV Y.M. *et al.*, "Microwave radiation from combustion of an Iron-aluminum thermite mixture", *Combust. Explos. Shock Waves*, vol. 41, no. 4, pp. 481–483, 2005.

[KOT 03] KOTOV Y.A., "Electric explosion of wires as a method for preparation of nanopowders", *J. Nanopart. Res.*, 5, pp. 539–550, 2003.

[KOT 99] KOTOV Y. A., SAMATOV O.M., "Production of nanometer sized AlN powders by the exploding wire method", *NanoStruct. Mater.*, no. 12, pp. 119–122, 1999.

[KRE 13] KREYLING W.G., "Toxicokinetics of inhaled nanoparticles", in *Nanomaterials*, Wiley-VCH, Weinheim, 2013.

[KRI 15] KRIEDEMANN B., FESTER V., "Critical process parameters and their interactions on the continuous hydrothermal synthesis of iron oxide nanoparticles", *Chem. Eng. J.*, no. 281, pp. 312–321, 2015.

[KUB 14] KUBOTA S., MORIOKA T., TAKESUE M. *et al.*, "Continuous supercritical hydrothermal synthesis of dispersiblezero-valent copper nanoparticles for ink applications in printedelectronics", *J. Supercrit. Fluids*, no. 86, pp. 33–40, 2014.

[KUM 03] KUMAR V.G., AURBUCH D., GEDANKEN A., "A comparison between hot-hydrolysis and sonolysis of various Mn (II) salts", *Ultrason. Sonochem.*, no. 10, pp. 17–23, 2003.

[KUM 12] KUMAR S.M., MURUGAN K., CHANDRASEKHAR S.B. *et al.*, "Synthesis and characterization of nano silicon and titanium nitride powders using atmospheric microwave plasma technique", *J. Chem. Sci.*, vol. 124, no. 3, pp. 557–563, 2012.

[KUN 10] KUNTZ J.D., CERVANTES O.G., GASH A.E. *et al.*, "Tantalum-tungsten oxide thermite composites prepared by sol-gel synthesis and spark plasma sintering", *Combust. Flame*, vol. 157, no. 8, pp. 1566–1571, 2010.

[KUR 82] KURASAWA T., TAKAHASHI T., NODA K. *et al.*, "Thermal expansion of lithium oxide", *J. Nucl. Mater.*, vol. 107, pp. 334–336, 1982.

[KWO 01] KWON Y.S., JUNG Y.H., YAVOROVSKY N.A. *et al.*, "Ultra-fine powder by wire explosion method", *Scripta Mater.*, no. 44, pp. 2247–2251, 2001.

[KWO 02] KWON Y.S., GROMOV A.A., ILYIN A.P., "Reactivity of superfine aluminium poders stabilized by aluminium diboride", *Combust. Flame*, no. 131, pp. 349–352, 2002.

[KWO 03] KWON Y.S., GROMOV A.A., ILYIN A.P. *et al.*, "Passivation process for superfine aluminum powders obtained by electrical explosion of wires", *Appl. Surf. Sci.*, vol. 211, no. 1–4, pp. 57–67, 2003.

[KWO 07] KWON Y.S., GROMOV A.A., STROKOVA J.I., "Passivation of the surface of aluminum nanopowders by protective coatings of the different chemical origin", *Appl. Surf. Sci.*, vol. 253, no. 12, pp. 5558–5564, 2007.

[LAG 08] LAGRANGE T., CAMPBELL G.H., REED B.W. *et al.*, "Nanosecond time-resolved investigations using the in situ of dynamic transmission electron microscope (DTEM)", *Ultramicroscopy*, vol. 108, no. 11, pp. 1441–1449, 2008.

[LAM 00] LAM C., ZHANG Y.F., TANG Y.H. *et al.*, "Large-scale synthesis of ultrafine Si nanoparticles by ball milling", *J. Cryst. Growth*, no. 220, pp. 466–479, 2000.

[LAM 50] LAMER V.K., DINEGAR R., "Theory, production and mechanism of formation of monodispersed hydrosols", *J. Am. Chem. Soc.*, vol. 72, no. 11, pp. 4847–4854, 1950.

[LAN 03] LANDSTRÖM L., MARTON ZS., ARNOLD N. *et al.*, "In situ monitoring of size distributions and characterization of nanoparticles during W ablation in N_2 atmosphere", *J. Appl. Phys.*, vol. 94, no. 3, pp. 2011–2017, 2003.

[LAU 04] LAUCHT H., BARTUCH H., KOVALEV D., "Silicon initiator, from the idea to functional tests", *7th International Symposium & Exhibition on Sophisticated Car Occupant Safety Systems*, Karlsruhe, Germany, 2004.

[LEE 07] LEE S.H., OH S.M., PARK D.W., "Preparation of silver nanopowder by thermal plasma", *Mater. Sci. Eng., C*, no. 27, pp. 1286–290, 2007.

[LEE 10] LEE S.B., JUNG J.H., BAE G.N. *et al.*, "in-situ characterization of metal nanopowders manufactured by the wire electrical explosion process", *Aerosol Sci. Technol.*, no. 44, pp. 1131–1139, 2010.

[LEE 14a] LEE D.J., KIM S.J., LEE J. *et al.*, "Bipolar pulsed electrical discharge for synthesis of tungsten nanoparticles in the aqueous solutions", *Sci. Adv. Mater.*, no. 6, pp. 1599–1604, 2014.

[LEE 14b] LEE H., PARK S.H., SEO S.G. *et al.*, "Preparation and characterization of copper nanoparticles via the liquid phase plasma method", *Curr. Nanosci.*, no. 10, pp. 7–10, 2014.

[LEE 96] LEE W., KWUN S.I., "The effects of process control agents on mechanical alloying mechanisms in the Ti-Al system", *J. Alloys Compd.*, no. 240, pp. 193–199, 1996.

[LEE 98] LEE P.Y., YANG J.L., LIN H.M., "Amorphization behaviour in mechanically alloyed Ni–Ta powders", *J. Mater. Sci.*, no. 33, pp. 235–239, 1998.

[LEI 12] LEI P., BOIES A.M., CALDER S. *et al.*, "Thermal plasma synthesis of superparamagnetic iron oxide nanoparticles" *Plasma Chem. Plasma Process.*, no. 32, pp. 519–531, 2012.

[LEP 08] LEPAROUX M., LOHER M., SCHREUDERS C. *et al.*, "Neural network modelling of the inductively coupled RF plasma synthesis of silicon nanoparticles", *PowderTechnol.*, no. 185, pp. 109–115, 2008.

[LES 12] LESTER E., AKSOMAITYTE G., LI J. *et al.*, "Controlled continuous hydrothermal synthesis of cobalt oxide (Co_3O_4) nanoparticles", *Prog. Cryst. Growth Charact. Mater.*, no. 58, pp. 3–13, 2012.

[LEU 10] LEUNG A.M., PEARCE E.N., BRAVERMAN L.E., "Perchlorate, iodine and the thyroid", *Best. Pract. Res. Cl. En.*, vol. 24, no. 1, pp. 133–141, 2010.

[LEV 04] LEVEUGLE E., IVANOV D.S., ZHIGILEI L.V., "Photomechanical spallation of molecular and metal targets: molecular dynamics study", *Appl. Phys. A*, no. 79, pp. 1643–1655, 2004.

[LEV 06] LEVITAS V.I., ASAY B.W., SON S.F. *et al.*, "Melt dispersion mechanism for fast reaction of nanothermites", *Appl. Phys. Lett.*, vol. 89, no. 071909, 2006.

[LEV 07] LEVITAS V.I., ASAY B.W., SON S.F. *et al.*, "Mechanochemical mechanism for fast reaction of metastable intermolecular composites based on dispersion of liquid metal", *J. Appl. Phys.*, vol. 101, no. 083524, 2007.

[LEV 08a] LEVITAS V.I., PANTOYA M.L., DIKICI B., "Melt dispersion versus diffusive oxidation mechanism for aluminum nanoparticles: Critical experiments and controlling parameters", *Appl. Phys. Lett.*, vol. 92, no. 011921, 2008.

[LEV 08b] LEVITAS V.I., PANTOYA M.L., WATSON K.W., "Melt-dispersion mechanism for fast reaction of aluminum particles: Extension for micron scale particles and fluorination", *Appl. Phys. Lett.*, vol. 92, no. 201917, 2008.

[LEV 09] LEVITAS V.I., "Burn time of aluminum nanoparticles: strong effect of the heating rate and melt-dispersion mechanism", *Combust. Flame*, vol. 156, no. 2, pp. 543–546, 2009.

[LEV 11] LEVITAS V.I., DIKICI B., PANTOYA M.L., "Toward design of the pre-stressed nano- and microscale aluminum particles covered by oxide shell", *Combust. Flame*, vol. 158, no. 7, pp. 1413–1417, 2011.

[LI 03] LI Y.X., ZHOU X.Z., WANG Y. *et al.*, "Preparation of nano-sized CeO_2 by mechanochemical reaction of cerium carbonate with sodium hydroxide", *Mater. Lett.*, no. 58, pp. 245–249, 2003.

[LI 05] LI Y.X., CHEN W.F., ZHOU X.Z. *et al.*, "Synthesis of CeO_2 nanoparticles by mechanochemical processing and the inhibiting action of NaCl on particle agglomeration", *Mater. Lett.*, no. 59, pp. 48–52, 2005.

[LI 07] LI Y.Y., YANG C., CHEN W.P. *et al.*, "Oxygen-induced amorphization of titanium by ball milling", *J. Mater.Res.*, vol. 22, no. 7, pp. 1927–1932, 2007.

[LI 13] LI R., XU H., HU H. *et al.*, "Microstructured Al/Fe_2O_3/nitrocellulose energetic fibers realized by electrospinning", *J. Energ. Mater.*, vol. 32, no. 1, pp. 50–59, 2013.

[LIA 03] LIANG C., SHIMIZU Y., SASAKI T. *et al.*, "Synthesis of ultrafine SnO_{2-x}nanocrystals by pulsed laser-induced reactive quenching in liquid medium", *J. Phys. Chem., B*, vol. 107, no. 35, pp. 9220–9225, 2003.

[LIA 04] LIANG C., SASAKI T., SHIMIZU Y. *et al.*, "Pulsed-laser ablation of Mg in liquids: surfactant-directing nanoparticle assembly for magnesium hydroxide nanostructures", *Chem. Phys. Lett.*, no. 389, p . 58–63, 2004

[LID 05] LIDE D.R., "Properties of the elements and inorganic compounds", *CRC Handbook of Chemistry and Physics*, Internet Version 2005, CRC Press, Boca Raton, FL, 2005, available at http://www.hbcpnetbase.com.

[LIM 06] LIMA M.D., BONADIMANN R., DE ANDRADE M.J. *et al.*, "Nanocrystalline $Cr_2O_3$and amorphous CrO_3 produced by solution combustion synthesis", *J. Eur. Ceram. Soc.*, vol. 26, no. 7, pp. 1213–1220, 2006.

[LIN 09a] LIN C.H., CHEN S.Y., SHEN P., "Defects, lattice correspondence, and optical properties of spinel-like Cr_3O_4 condensates by pulsed laser qblation in water", *J. Phys. Chem., C*, vol. 113, no. 37, pp. 16356–16363, 2009.

[LIN 09b] LIN C.H., CHEN S.Y., HO N.J. *et al.*, "Shape-dependent local internal stress of α-Cr_2O_3 nanocrystal fabricated by pulsed laser ablation", *J. Phys. Chem. Solids*, no. 70, pp. 1505–1510, 2009.

[LIN 10] LIN G., TAN D., LUO F. *et al.*, "Fabrication and photocatalytic property of α-Bi_2O_3 nanoparticles by femtosecond laser ablation in liquid", *J. Alloys Compd.*, no. 507, pp. L43–L46, 2010.

[LIN 15] LIN S., GUO Y., LI X. *et al.*, "Glycine acid-assisted green hydrothermal synthesis and controlled growth of WO_3 nanowires", *Mater. Lett.*, no. 152, pp. 102–104, 2015.

[LIU 09] LIU Z., YUAN Y., KHAN S. *et al.*, "Generation of metal-oxide nanoparticles using continuous-wave fibre laser ablation in liquid", *J. Micromech. Microeng.*, no. 19, pp. 054008 (1)–054008(7), 2009

[LIU 10] LIU L., HU Z., CUI Y. *et al.*, "A facile route to the fabrication of morphology-controlled Sb_2O_3 nanostructures", *Solid State Sci.*, no. 12, pp. 882–886, 2010.

[LIU 11] LIU P., LI Z., CAI W. *et al.*, "Fabrication of cuprous oxide nanoparticles by laser ablation in PVP aqueous solution", *RSC Advances*, no. 1, pp. 847–851, 2011.

[LIU 13] LIU L.C., ZHANG Q.G., ZHAO J.P. *et al.*, "Study on characteristics of nanopowders synthesized by nanosecond electrical explosion of thin aluminum wire in the argon gas", *IEEE T. Plasma Sci.*, vol. 41, no. 8, Part 2 SI, pp. 2221–2226, 2013.

[LIU 15] LIU J., KHAN A.S., TAKACS L. *et al.*, "Mechanical behavior of ultrafine-grained/nanocrystalline titanium synthesized by mechanical milling plus consolidation: Experiments, modeling and simulation", *Int. J. Plast.*, no. 64, pp. 151–163, 2015.

[LIU 99] LIU L., LU L., CHEN L. *et al.*, "Solid-gas reactions driven by mechanical alloying of niobium and tantalum in nitrogen", *Metall. Mater. Trans. A*, no. 30, pp. 1097–1100, 1999.

[LOM 09] LOMBARD A., "Toxicologie industrielle", *Techniques de l'Ingénieur*, vol. TIB568DUO, no. SE 1605, pp. 1–21, 2009.

[LOP 03] LOPEZ-QUINTELA M.A., "Synthesis of nanomaterials in microemulsions: formation mechanisms and growth control", *Curr. Opin. Colloid Interface Sci.*, no. 8, pp. 137–144, 2003.

[LOP 14] LOPES P.A.L., SANTOS M.B., MASCARENHAS A.J.S. *et al.*, "Synthesis of CdS nano-sphere by a simple and fast sonochemical method at room temperature", *Mater. Lett.*, no. 136, pp. 111–113, 2014.

[LOP 93] LOPEZ QUINTELA M.A., RIVAS J., "Chemical reactions in microemulsions: a powerful method to obtain ultrafine particles", *J. Colloid Interface Sci.*, no. 158, pp. 446–451, 1993.

[LOR 03] LORAZO P., LEWIS L., MEUNIER M., "Short-pulse laser ablation of solids: from phase explosion to fragmentation", *Phys. Rev. Lett.*, vol. 91, no. 22, pp. 1–4, 2003.

[LOR 06] LORAZO P., LEWIS L., MEUNIER M., "Thermodynamic pathways to melting, ablation, and solidification in absorbing solids under pulsed laser irradiation", *Phys. Rev., B*, vol. 73, no. 13, pp. 1–22,. 2006.

[LU 02] LU Y., ZHU Z., WU W. *et al.*, "Detonation chemistry of a CHNO explosive: catalytic assembling of carbon nanotubes at low pressure and temperature state", *Chem. Commun.*, vol. 22, pp. 2740–2741, 2002.

[LU 03] LU Y., ZHU Z., WU W. *et al.*, "Catalytic formation of carbon nanotubes during detonation of m-dinitrobenzene", *Carbon*, vol. 41, no. , pp. 179–198, 2003.

[LU 04] LU Y., ZHU Z., LIU Z., "Catalytic growth of carbon nanotubes through CHNO explosive detonation", *Carbon*, vol. 42, no. 2, pp. 361–370, 2004.

[LU 05] LU Y., ZHU Z., LIU Z., "Carbon-encapsulated Fe nanoparticles from detonation-induced pyrolysis of ferrocène", *Carbon*, vol. 43, no. 2, pp. 369–374, 2005.

[LU 15] LU L., AN X., "Silver nanoparticles synthesis using H_2 as reducing agent intoluene–supercritical CO_2 microemulsion", *J. Supercrit. Fluids*, no. 99, pp. 29–37, 2015.

[LUN 05] LUND G.K., HANKS J.M., JOHNSTON H.E., Method for the production of α-alane, Patent WO 2005/102919, 2005.

[LUN 07] LUNG J.K., HUANG J.C., TIEN D.C. *et al.*, "Preparation of gold nanoparticles by arc discharge in water", *J. Alloys Compd.*, no. 434–435, pp. 655–658, 2007.

[LUO 10a] LUO N., LI X.J., WANG X.H. *et al.*, "Synthesis of carbon-encapsulated metal nanoparticles by a detonation method", *Combust. Explos. Shock Waves*, vol. 46, no. 5, pp. 609–615, 2010.

[LUO 10b] LUO N., LI X., WANG X. *et al.*, "Synthesis and characterization of carbon-encapsulated iron/iron carbide nanoparticles by a detonation method", *Carbon*, vol. 48, no. 13, pp. 3858–3863, 2010.

[LUO 12a] LUO N., LIU K.X., LI X. *et al.*, "Systematic study of detonation synthesis of Ni-based nanoparticles", *Chem. Eng. J.*, vol. 210, pp. 114–119, 2012.

[LUO 12b] LUO N., LIU K.X., LIU Z.Y. *et al.*, "Controllable synthesis of carbon coated iron-based composite nanoparticles", *Nanotechnology*, vol.23, no. 475603, 2012.

[MA 10] MA J., LIAN J., DUAN X. *et al.*, "α-Fe2O3: hydrothermal synthesis, magnetic and electrochemical properties", *J. Phys. Chem., C*, no. 114, pp. 10671–10676, 2010.

[MAF 00a] MAFUNE F., KOHNO J., TAKEDA Y. *et al.*, "Formation and size control of sliver nanoparticles by laser ablation in aqueous solution", *J. Phys. Chem., B*, vol. 104, no. 39, pp. 9111–9117, 2000.

[MAF 00b] MAFUNE F., KOHNO J.Y., TAKEDA Y. *et al.*, "Structure and stability of silver nanoparticles in aqueous solution produced by laser ablation", *J. Phys. Chem., B*, vol. 104, no. 35, pp. 8333–8337, 2000.

[MAF 01] MAFUNE F., KOHNO J., TAKEDA Y. *et al.*, "Formation of gold nanoparticles by laser ablation in aqueous solution of surfactant", *J. Phys. Chem.*, B, no. 105, pp. 5114–5120, 2001.

[MAF 03] MAFUNE F., KOHNO J.Y., TAKEDA Y. *et al.*, "Formation of stable platinum nanoparticles by laser ablation in water", *J. Phys. Chem.*, B, no. 107, pp. 4218–4223, 2003.

[MAH 02] MAHADEVAN R., LEE D., SAKURAI H. *et al.*, "Measurement of condensed-phase reaction kinetics in the aerosol phase using single particle mass spectrometry", *J. Phys. Chem., A*, vol. 106, no. 46, pp. 11083–11092, 2002.

[MAH 08] MAHFOUZ R., CADETE SANTOS AIRES F.J., BRENIER A. *et al.*, "Synthesis and physico-chemical characteristics of nanosized particles produced by laser ablation of a nickel target in water", *Appl. Surf. Sci.*, no. 254, pp. 5181–5190, 2008.

[MAK 04] MAKIMURA T., MIZUTA T., TAKAHASHI T. *et al.*, "In situ size measurement of Si nanoparticles and formation dynamics after laser ablation", *Appl. Phys., A*, no. 79, pp. 819–821, 2004.

[MAL 08] MALCHI J.Y., YETTER R.A., FOLEY T.J. *et al.*, "The effect of added Al_2O_3 on the propagation behavior of an Al/CuO nanoscale thermite", *Combust. Sci. Technol.*, vol. 180, no. 7, pp. 1278–1294, 2008.

[MAL 09] MALCHI J.Y., FOLEY T.J., YETTER R.A., "Electrostatically self-assembled nanocomposite reactive microspheres", *ACS Appl. Mater. Inter.*, vol. 1, no. 11, pp. 2420–2423, 2009.

[MAN 13] MANEERATANASARN P., KHAI T.V., KIM S.Y. *et al.*, "Synthesis of phase-controlled iron oxide nanoparticles by pulsed laser ablation in different liquid media", *Phys. Status Solidi A*, vol. 210, no. 3, pp. 563–569, 2013.

[MANG 05] MANGOLINI L., THIMSEN E., KORTSHAGEN U., "High-yield plasma synthesis of luminescent silicon nanocrystals", *Nano Lett.*, vol. 5, no. 4, pp. 655–659, 2005.

[MAO 07] MAO X., WEN S.B., RUSSO R.E., "Time resolved laser-induced plasma dynamics", *Appl. Surf. Sci.*, no. 253, pp. 6316–6321, 2007.

[MAO 97] MAO X., RUSSO R.E., "Observation of plasma shielding by measuring transmitted and reflected laser pulse temporal profiles", *Appl. Phys. A Mater. Sci. Process.*, no. 64, pp. 1–6, 1997.

[MAR 09a] MARTIROSYAN K.S., WANG L., VICENT A. *et al.*, "Synthesis and performances of bismuth trioxide nanoparticles for high energy gas generator use", *Nanotechnology*, vol. 20, pp. 405609 (8), 2009.

[MAR 09b] MARTIROSYAN K.S., WANG L., LUSS D., "Novel nanoenergetic system based on iodine pentoxide", *Chem. Phys. Lett.*, vol. 483, no. 1–3, pp. 107-110, 2009.

[MAR 09c] MARTIROSYAN K.S., WANG L., VICENT A. *et al.*, "Nanoenergetic Gas-Generators: Design and Performance", *Propell. Explos. Pyrot.*, vol. 34, no. 6, pp. 532–538, 2009.

[MAR 11] MARTIROSYAN K.S., "Nanoenergetic gas-generators: principles and applications", *J. Mater. Chem.*, vol. 21, pp. 9400–9405, 2011.

[MAR 12] MARTIROSYAN K.S., HOBOSYAN M., LYSHEVSKI S.E., "Enabling nanoenergetic materials with integrated microelectronics and MEMS platforms", *12th IEEE International Conference on Nanotechnology (IEEE-NANO)*, Birmingham, United Kingdom, August 2012.

[MAR 05] MARZIK J.V., SUPLINSKAS R.J., WILKE R.H.T. *et al.*, " Plasma synthesized doped B powders for MgB_2 superconductors", *Physica C*, no. 423, pp. 83–88, 2005.

[MAS 14] MASJEDI-ARANI M., SALAVATI-NIASARI M., GHANBARI D. *et al.*, "A sonochemical-assisted synthesis of spherical silica nanostructures by using a new capping agent", *Ceram. int.*, no. 40, pp. 495–499, 2014.

[MAS 16] MASJEDI-ARANI M., GHANBARI D., SALAVATI-NIASARI M. *et al.*, "Sonochemical synthesis of spherical silica nanoparticles and polymeric nanocomposites", *J. Clust. Sci.*, vol. 27, no. 1, pp. 39–53, 2016.

[MAT 06] MATSUI I., "Preparation of magnetic nanoparticles by pulsed plasma chemical vapor synthesis", *J. Nanopart. Res.*, no. 8, pp. 429–443, 2006.

[MAU 92] MAURICE D., COURTNEY T.H., "Modeling of the mechanical alloying process", *JOM*, vol. 44, no. 8, pp. 10–14, 1992.

[MAU 94] MAURICE D., COURTNEY T.H., "Modeling of mechanical alloying: part I,deformation, coalescence, and fragmentation mechanisms", *Metall. Mater. Trans., A*, no. 25, pp. 147–158, 1994.

[MAU 95a] MAURICE D., COURTNEY T.H., "Modeling of mechanical alloying: part II,development of computational modeling programs", *Metall. Mater. Trans.*, A, no. 26, pp. 2431–2435, 1995.

[MAU 95b] MAURICe D., COURTNEY T.H., "Modeling of mechanical alloying: part III,applications of computational programs", *Metall. Mater. Trans., A*, no. 26, pp. 2437–2444, 1995.

[MAU 96a] MAURICE D., COURTNEY T.H., "Milling dynamics: part II,dynamics of a SPEX mill on a one-dimentional mill", *Metall. Mater. Trans., A*, no. 27, pp. 1973–1979, 1996.

[MAU 96b] MAURICE D., COURTNEY T.H., "Milling dynamics: part III,integration of local and global modeling of mechanical alloying devices", *Metall. Mater. Trans., A*, no. 27, pp. 1981–1986, 1996.

[MBA 15] MBARKI R., MNIF A., HAMZAOUI A.H., "Structural dielectric relaxation and electrical conductivity behavior in MgO powders synthesized by sol-gel", *Mater. Sci. Semicond. Process.*, no. 29, pp. 300–306, 2015.

[MCC 01] MCCORMICK P.G., TSUZUKI T., ROBINSON J.S. *et al.*, "Nanopowders synthesized by mechanochemical processing", *Adv. Mater.*, vol. 13, no. 12–13, pp. 1008–1010, 2001.

[MCM 14] MCMAHON B.W., PEREZ J.P.L., YU J. *et al.*, "Synthesis of nanoparticles from malleable and ductile metals Using powder-free, reactant-assisted mechanical attrition", *ACS Appl. Mater. Interfaces*, no. 6, pp. 19579–19591, 2014.

[MEH 91] MEHTA P., SINGH A.K., KINGON A.I., "Nonthermal microwave plasma synthesis of crystalline titanium oxide and titanium nitride nanoparticles", *Mat. Res. Soc. Sympp. Proc.*, no. 249, pp. 153–159. 1991.

[MEI 05] MEI Q.S., WANG S.C., CONG H.T. *et al.*, "Pressure-induced superheating of Al nanoparticles encapsulated in Al_2O_3 shells without epitaxial interface", *Acta Mater.*, vol. 53, no. 4, pp. 1059–1066, 2005.

[MEI 03] MEILINGER M., "Promoted ignition-combustion tests with structured aluminum packings in gaseous oxygen with argon or nitrogen dilution at 0.1 and 0.6 MPa", *Flammability and sensitivity of materials in oxygen-enriched atmospheres: tenth volume*, vol. 1454, pp. 137–150, 2003.

[MEI 12] MEIR Y., JERBY E., "Thermite powder ignition by localized microwaves", *Combust. Flame*, vol. 159, no. 7, pp. 2474–2479, 2012.

[MEN 98] MENCH M.M., KUO K.K., YEH C.L. *et al.*, "Comparison of thermal behavior of regular and ultra-fine aluminum powders (Alex) made from plasma explosion process", *Combust. Sci. Technol.*, vol. 135, no. 1–6, pp. 269–292, 1998.

[MEN 13] MENDIVIL M.I., KRISHNAN B., SANCHEZ F.A. *et al.*, "Synthesis of silver nanoparticles and antimony oxide nanocrystals by pulsed laser ablation in liquid media", *Appl. Phys., A*, no. 110, pp. 809–816, 2013.

[MEN 04] MENON L., PATIBANDLA S., BHARGAVA. *et al.*, "Ignition studies of Al/Fe_2O_3 energetic nanocomposites", *Appl. Phys. Lett.*, vol. 84, no. 23, pp. 4735–4737, 2004.

[MEZ 05] MEZIANI M.J., PATHAK P., BEACHAM F. *et al.*, "Nanoparticle formation in rapid expansion of water-in-supercritical carbon dioxide microemulsion into liquid solution", *J. Supercrit. Fluids*, vol. 34, no. 1, pp. 91–97, 2005.

[MIA 15] MIAO B., ZENG W., HUSSAIN S. *et al.*, "Large scale hydrothermal synthesis of monodisperse hexagonal WO_3 nanowire and the growth mechanism", *Mater. Lett.*, no. 147, pp. 12–15, 2015.

[MIK 92] MIKI M., YAMASAKI T., OGINO Y., "Preparation of nanocrystalline NbN and (Nb, Al)N powders by mechanical alloying under nitrogen atmosphere", *Mater. Trans., JIM*, vol. 33, no. 9, pp. 839–844, 1992.

[MIL 12] MILLET Y., "Corrosion du titane et de ses alliages", *Techniques de l'Ingénieur*, vol. TIB373DUO, no. COR 320, pp. 1–15, 2012.

[MIO 02] MIO H., KANO J., SAITO F. *et al.*, "Effects of rotational direction and rotation-to-revolution speed ratio in planetary ball milling", *Mater. Sci. Eng., A*, no. 332, pp. 75–80, 2002.

[MIO 95] MIOTELLO A., PERLONGO A., KELLY R., "Laser-Pulse sputtering of aluminium: gas-dynamic effects with recondensation and reflection conditions at the Knudsen layer", *Nucl. Instr. Meth. Phys. Res. Sect. B*, no. 101, pp. 148–155, 1995

[MIR 10] MIRJALILI F., HASMALIZA M., CHUAH ABDULLAH L., "Size-controlled synthesis of nano a-alumina particles through the sol–gel method" *Ceram. Int.*, no. 36, pp. 1253–1257, 2010.

[MOO 06] MOORE K., PANTOYA M.L., "Combustion of environmentally altered molybdenum trioxide nanocomposites", *Propell. Explos. Pyrot.*, vol. 31, no. 3, pp. 182–187, 2006.

[MOO 07] MOORE K., PANTOYA M.L., SON S.F., "Combustion behaviors resulting from bimodal aluminum size distributions in thermites", *J. Propul. Power*, vol. 23, no. 1, pp. 181–185, 2007.

[MOO 95] MOORE J.J., FENG H.J., "Combustion synthesis of advanced materials: part I, reaction parameters", *Prog. Mater. Sci.*, vol. 39, no. 4–5, pp. 243–273, 1995.

[MOR 90] MORRIS D.G., MORRIS M.A., "Mechanical alloying of Fe-Al with oxide and carbide dispersoids", *Mater. Sci. Eng.*, A, no. 125, pp. 97–106, 1990.

[MOT 39] MOTT N.F., "A theory of the formation of protective oxide films on metals", *Trans. Faraday Soc.*, vol. 35, pp. 1175–1177, 1939.

[MOT 40] MOTT N.F., "Oxidation of metals and the formation of protective films", *Nature*, vol. 145, pp. 996–1000, 1940.

[MOT 47] MOTT N.F., "The theory of the formation of protective oxide films on metals.-III", *Trans. Faraday Soc.*, vol. 43, pp. 429–434, 1947.

[MUK 07] MUKASYAN A.S., EPSTEIN P., DINKA P., "Solution combustion synthesis of nanomaterials", *Proc. Combust. Inst.*, vol. 31, no. 2, pp. 1789–1795, 2007.

[MUT 08] MUTO H., MIYAJIMA K., MAFUNE F., "Mechanism of laser-induced size reduction of gold nanoparticles as studied by single and double laser pulse excitation", *J. Phys. Chem., C*, vol. 112, no. 15, pp. 5810–5815, 2008.

[NAG 14] NAGABHUSHANA G.P., SAMRAT D., CHANDRAPPA G.T., "α-MoO_3 nanoparticles: solution combustion synthesis, photocatalytic and electrochemical properties", *RSC Adv.*, vol. 4, pp. 56784–56790, 2014.

[NAI 74] NAIRNE E., "Electrical Experiments by Mr. EDWARD NAIRNE, of London, Mathematical Instrument-Maker, Made with a Machine of His Own Workmanship, A description of Which is Prefixed", *Phil. Trans.*, no. 64, pp. 79–89, 1774.

[NAK 07] NAKAMURA R., TOKOZAKURA D., NAKAJIMA H. *et al.*, "Hollow oxide formation by oxidation of Al and Cu nanoparticles", *J. Appl. Phys.*, vol. 101, no. 074303, 2007.

[NAT 11] NATH A., KHARE A., "Size induced structural modifications in copper oxide nanoparticles synthesized via laser ablation in liquids", *J. Appl. Phys.*, no. 110, pp. 043111-1–043111-6, 2011.

[NAZ 06] NAZARENKO O., "Nanopowders produced by electric explosion of wires", *Proceeding of European Congress of Chemical Engeneering (ECC-6)*, Copenhagen, Denmark, September 2006.

[NED 93] NEDDERSEN J., CHUMANOV G., COTTON T.M., "Laser-ablation of metals – a new method for preparing SERS active colloids", *Appl. Spectrosc.*, vol. 47, no. 12, pp. 1959–1964, 1993.

[NEV 14] NEVES N., LAGOA A., CALADO J. *et al.*, "Al-doped ZnO nanostructured powders by emulsion detonation synthesis – improving materials for high quality sputtering targets manufacturing", *J. Eur. Ceram. Soc.*, vol. 34, no. 10, pp. 2325–2338, 2014.

[NIA 09] NIAZI J.H., GU M.B., "Toxicity of Metallic Nanoparticles in Microorganisms- a Review", DANS Y.J. KIM, U. PLATT, M.B. GU, IWAHASHI H. (ed.), *Atmospheric and Biological Environmental Monitoring*, Springer, Netherlands, 2009.

[NIC 06a] NICHOLS W.T., SASAKI T., KOSHIZAKI N., "Laser ablation of a platinum target in water, II,ablation rate and nanoparticle size distributions", *J. Appl. Phys.*, vol. 100, no. 11, pp. 114912(1)–114912(6), 2006.

[NIC 06b] NICHOLS W.T., SASAKI T., KOSHIZAKI N., "Laser ablation of a platinum target in water. III. Laser-induced reactions", *J. Appl. Phys.*, vol. 100, no. 11, pp. 114913(1)–114913(7), 2006.

[NIE 91] NIEH T.G., WADSWORTH J., "Hall-Petch relation in nanocrystalline solids", *Scripta Mater.*, no. 25, pp. 955–958, 1991.

[NIS 74] NISHIDA I., KIMOTO K., "Crystal habit and crystal structure of fine chromium particles: an electron microscope and electron diffraction study of fine metallic particles prepared by evaporation in argon at low pressure (III)", *Thin Solid Films*, no. 23, pp. 179–189, 1974.

[NOE 07] NOËL S., HERMANN J., ITINA T., "Investigation of nanoparticle generation during femtosecond laser ablation of metals", *Appl. Surf. Sci.*, no. 253, pp. 6310–6315, 2007.

[NOG 08] NOGUCHI T., MATSUI K., ISLAM N.M. *et al.*, "Rapid synthesis of -Al_2O_3 nanoparticles in supercritical water by continuous hydrothermal flow reaction system", *J. Supercrit. Fluids*, no. 46, pp. 129–136, 2008.

[NOO 15] NOOR F., VOROZHTSOV A., LERNER M. *et al.*, "Exothermic characteristics of aluminum based nanomaterials", *Powder Technol.*, vol. 282, pp. 19–24, 2015.

[NOZ 07] NOZAKI T., SASAKI K., OGINO T. *et al.*, "Silicon nanocrystal synthesis in microplasma reactor", *J. Therm. Sci. Tech. Jpn*, vol. 2, no. 2, pp. 192–199, 2007.

[OGA 88] OGALE S.B., "Pulsed-laser induces and ion-beam-induced surface synthesis and modification of oxides, nitrides and carbides", *Thin Solid Films*, no. 163, pp. 215–227, 1988.

[OGI 90] OGINO Y., YAMASAKI T., MURAYAMA S. *et al.*, "Non-equilibrium phases formed by mechanical alloying of CrCu alloys", *J. Non-Cryst. Solids*, nos. 117–118, pp. 737–740, 1990.

[OH 05] OH S.-M., LI J.-G., ISHIGAKI T., "Nanocrystalline TiO_2 powders synthesized by in-flight oxidation of TiN in thermal plasma: mechanisms of phase selection and particle morphology evolution", *J. Mater. Res.*, vol. 20, no. 2, pp. 529–537, 2005.

[OHD 01] OHDE H., HUNT F., WAI C.M., "Synthesis of silver and copper nanoparticles in a water-in-supercritical-carbon dioxide microemulsion", *Chem. Mater.*, no. 13, pp. 4130–4135, 2001.

[OHK 11a] OHKURA Y., RAO P.M., ZHENG X., "Flash ignition of Al nanoparticles: mechanism and applications", *Combust. Flame*, vol. 158, no. 12, pp. 2544–2548, 2011.

[OHK 11b] OHKURA Y., LIU S.-Y., RAO P.M., ZHENG X., "Synthesis and ignition of energetic CuO/Al core/shell nanowires", *Proc. Combust. Inst.*, vol. 33, no. 2, pp. 1909–1915, 2011.

[OLE 96] OLESZAK D., SHINGU P.H., "Nanocrystalline metals prepared by low energy ball milling", *J. Appl. Phys.*, vol. 79, no. 6, pp. 2975–2980, 1996.

[OLM 33] OLMER L.-J., "Argent", DANS P. PASCAL, P. BAUD (ed.), *Traité de chimie minérale* – Tome VIII, Masson et Cie, Paris, 1933.

[OMU 07] OMURZAK E., JASNAKUNOV J., MAIRYKOVA N. *et al.*, "Synthesis method of nanomaterials by pulsed plasma in liquid", *J. Nanosci. Nanotechnol.*, no. 7, pp. 3157–3159, 2007.

[OUY 08] OUYANG X., LI X., YAN H. *et al.*, "Synthesis of TiO_2 nanoparticles from sprayed droplets of titanium tetrachloride by the gas-phase detonation method", *Combust. Explos. Shock Waves*, vol. 44, no. 5, pp. 597–600, 2008.

[OZA 01] OZAWA E., KAWAKAMI Y., SETO T., "Formation and size control of tungsten nano particles produced by Nd:YAG laser irradiation", *Scripta mater.*, no. 44, pp. 2279–2283, 2001.

[PAB 96] PABI S.K., MURTY B.S., "Mechanism of mechanical alloying in Ni-A1 and Cu-Zn systems", *Mater. Sci. Eng.*, A, no. 214, pp. 146–52, 1996.

[PAB 99] PABI S.K. , MANNA I., MURTY B.S., "Alloying behaviour in nanocrystalline materials during mechanical alloying", *Bull. Mater. Sci.*, vol. 22, no. 3, pp. 321–327, 1999.

[PAN 95] PANDA S., PRATSINIS S.E., "Modeling the synthesis of aluminum particles by evaporation -condensation in an aerosol flow reactor", *Nanostruct. Mater.*, no. 5, pp. 755–767, 1995.

[PAN 14] PANDEY B.K., SHAHI A.K., SHAH J. *et al.*, "Optical and magnetic properties of Fe_2O_3 nanoparticles synthesized by laser ablation/fragmentation technique in different liquid media", *Appl. Surf. Sci.*, no. 289, pp. 462–471, 2014.

[PAN 02] PANG Y.-X, BAO X., "Aluminium oxide nanoparticles prepared by water-in-oil microemulsions", *J. Mater. Chem.*, no. 12, pp. 3699–3704, 2002.

[PAN 07] PANG S., WANG L.G., ZHANG Z., " Preparation of manganese and their derivate compounds by arc plasma method", *Surf. Coat. Technol.*, no. 201, pp. 5451–5453, 2007.

[PAN 05] PANTOYA M.L., SON S.F., DANEN W.C. *et al.*, "Characterization of Metastable Intermolecular Composites", *ACS Sym. Ser.*, vol. 891, no. 16, pp. 227–240, 2005.

[PAN 06] PANTOYA M.L., GRANIER J.J., "The effect of slow heating rates on the reaction mechanisms of nano and micron composite thermite reactions", *J. Therm. Anal. Calorim.*, vol. 85, no. 1, pp. 37–43, 2006.

[PAR 05a] PARK K., LEE D., RAI A. *et al.*, "Size-resolved kinetic measurements of aluminum nanoparticle oxidation with single particle mass spectrometry", *J. Phys. Chem., B*, vol. 109, no. 15, pp. 7290–7299, 2005.

[PAR 05b] PARK Y.K., TADD E.H., ZUBRIS M. *et al.*, "Size-controlled synthesis of alumina nanoparticles from aluminum alkoxides", *Mater. Res. Bull.*, no. 40, pp. 1506–1512, 2005.

[PAR 07] PARK S., LEE Y.K., JUNG M. *et al.*, "Cellular Toxicity of Various Inhalable Metal Nanoparticles on Human Alveolar Epithelial Cell", *Inhal. Toxicol.*, vol. 19, pp. 59–65, 2007.

[PAR 11] PARK E.-J., KIM H., KIM Y. *et al.*, "Repeated-dose toxicity attributed to aluminum nanoparticles following 28-day oral administration, particularly on gene expression in mouse brain", *Toxicol. Environ. Chem.*, vol. 93, no. 1, pp. 120–133, 2011.

[PAR 96] PARK H.K., GRIGOROPOULOS C.P., POON C.C. *et al.*, "Optical probing of the temperature transients during pulsed-laser induced boiling of liquids", *Appl. Phys. Lett.*, vol. 68, no. 5, pp. 596–598, 1996.

[PAS 31] PASCAL P., "Iode", in PASCAL P., BAUD P. (eds.), *Traité de chimie minérale - Tome I*, Masson et Cie, Paris, 1931.

[PAS 09] PASKEVICIUS M., WEBB J., PITT M.P. *et al*, "Mechanochemical synthesis of aluminium nanoparticles and their deuterium sorption properties to 2 kbar", *J. Alloys Compd.*, no. 481, pp. 595–599, 2009.

[PAT 13] PATEL V.K., BHATTACHARYA S., "High-Performance Nanothermite Composites Based on Aloe-Vera-Directed CuO Nanorods", *ACS Appl. Mater. Interfaces*, vol. 5, no. 24, pp. 13364–13374, 2013.

[PAT 02] PATIL K.C., ARUNA S.T., MIMANI T., "Combustion Synthesis: an update", *Curr. Opin. Solid Sate Mater. Sci.*, vol. 6, no. 6, pp. 507–512, 2002.

[PAT 87] PATIL P.P., PHASE D.M., KULKARNI S.A. *et al.*, "Pulsed-laser-induced reactive quenching at liquid-solid interface: aqueous oxidation of iron", *Phys. Rev. Lett.*, no. 58, pp. 238–241, 1987.

[PAT 97] PATIL K.C., ARUNA S.T., EKAMBARAM S., "Combustion synthesis", *Curr. Opin. Solid State Mater. Sci.*, vol. 2, no. 2, pp. 158–165, 1997.

[PAT 02] PATZKE G.R., KRUMEICH F., NESPER R., "Oxidic nanotubes and nanorods-anisotropic modules for a future nanotechnology", *Angew. Chem. Int. Ed.*, vol. 41, no. 14, pp. 2446–2461, 2002.

[PAT 04] PATZKE G.R., MICHAILOVSKI A., KRUMEICH F. *et al.*, "One-step synthesis of submicrometer fibers of MoO_3", *Chem. Mater.*, no. 16, pp. 1126–1134, 2004.

[PEI 09] PEI Z., XU H., ZHANG Y., "Preparation of Cr_2O_3 nanoparticles via C_2H_5OH hydrothermal reduction", *J. Alloys Compd.*, no. 468, pp. L5–L8, 2009.

[PEN 14] PENDASHTEH A., RAHMANIFAR M.S., MOUSAVI M.F. "Morphologically controlled preparation of CuO nanostructures under ultrasound irradiation and their evaluation as pseudocapacitor materials", *Ultrason. Sonochem.*, no. 21, pp. 643–652, 2014.

[PER 04] PERRY W.L., SMITH B.L., BULIAN C.J. *et al.*, "Nano-scale tungsten oxides for metastable intermolecular composites", *Propell. Explos. Pyrot.*, vol. 29, no. 2, pp. 99–105, 2004.

[PER 07] PERRY W.L., TAPPAN B.C., REARDON B.L. *et al.*, "Energy release characteristics of the nanoscale aluminum-tungsten oxide hydrate metastable intermolecular composite", *J. Appl. Phys.*, vol. 101, no. 064313, 2007.

[PES 04] PESIRI D., AUMANN C.E., BILGER L. *et al.*, "Industrial Scale Nano-Aluminum Powder Manufacturing", *J. Pyro.*, vol. 19, pp. 19–31, 2004.

[PET 09] PETERSEN S., JAKOBI J., BARCIKOWSKI S., "In situ bioconjugation – Novel laser based approach to pure nanoparticle-conjugates", *Appl. Surf. Sci.*, vol. 255, no. 10, pp. 5435–5438, 2009.

[PET 11] PETERMANN N., STEIN N., SCHIERNING G. *et al.*, "Plasma synthesis of nanostructures for improved thermoelectric properties", *J. Phys., D*, vol. 44, pp. 174034, 2011.

[PET 10] PETRANTONI M., ROSSI C., CONÉDÉRA V. *et al.*, "Synthesis process of nanowired Al/CuO thermite", *J. Phys. Chem. Solids*, vol. 71, no. 2, pp. 80–83, 2010.

[PHI 04] PHILLIPS J., PERRY W.L., KROENKE W.J., Method for producing metallic nanoparticles, US Patent 6,689,192 B1, 2004.

[PHU 08] PHUOC T.X., HOWARD B.H., MARTELLO D.V. *et al.*, "Synthesis of $Mg(OH)_2$, MgO, and Mg nanoparticles using laser ablation of magnesium in water and solvents", *Opt. Laser Eng.*, no. 46, pp. 829–834, 2008.

[PIC 13] PICHOT V., RISSE B., SCHNELL F. *et al.*, "Understanding ultrafine nanodiamond formation using nanostructured explosives", *Sci. Rep.*, vol. 3, no. 2159, 2013.

[PIC 15] PICHOT V., COMET M., MIESCH J. *et al.*, "Nanodiamond for tuning the properties of energetic composites", *Journal of Hazardous Materials*, vol. 300, pp. 194–201, 2015.

[PIE 10] PIERCEY D.G., KLAPÖTKE T.M., "Nanoscale Aluminum – Metal oxide (thermite) reactions for application in energetic materials", *Cent. Eur. J. Energ. Mater.*, vol. 7, no. 2, pp. 115–129, 2010.

[PIL 03] PILENI M.-P. "The role of soft colloidal templates in controlling the size and shape of inorganic nanocrystals", *Nat. Mater.*, no. 2, pp. 145–150, 2003.

[PIN 08] PINKAS J., VENDULA R.V., ZBORIL R. *et al.*, "Sonochemical synthesis of amorphous nanoscopic iron (III) oxide from $Fe(acac)_3$", *Ultrason. Sonochem.*, no. 15, pp. 257–264, 2008.

[SAG 03] SAGMEISTER M., POSTL M., BROSSMAN U. *et al.*, "Structure and electrical properties of nanoparticulate tungsten oxide prepared by microwave plasma synthesis", *J. Phys.: Condens. Matter.*, no. 23, pp. 334206 (7pp), 2003.

[SAI 02] SAITO K., TAKATANI K., SAKKA T. *et al.*, "Observation of the light emitting region produced by pulsed laser irradiation to a solid-liquid interface", *Appl. Surf. Sci.*, vol. 197, pp. 56–60, 2002.

[SAI 08] SAITO M., YASUKAWA K., UMEDA T. *et al.*, "Copper nanoparticles fabricated by laser ablation in polysiloxane", *Opt. Mater.*, no. 30, pp. 1201–1204, 2008.

[SAI 11] SAITO G., HOSOKAI S., TSUBOTA M. *et al.*, "Synthesis of copper/copper oxide nanoparticles by solution plasma", *J. Appl. Phys.*, no. 110, pp. 023302-1– 23302-6, 2011.

[SAI 80] SAITO Y., MUHAMA K., UYEDA R., "Formation of ultrafine metal particles by gas-evaporation VI. Bcc metals, Fe, V, Nb, Ta, Cr, Mo and W", *Jpn J. Appl. Phys.*, vol. 19, no. 9, pp. 1603–1610, 1980.

[SAJ 10a] SAJTI C.L., SATTARI R., CHICHKOV B. *et al.*, "Ablation efficiency of α-Al_2O_3 in liquid phase and ambient air by nanosecond laser irradiation", *Appl. Phys., A*, no. 100, pp. 203–206, 2010.

[SAJ 10b] SAJTI C.L., SATTARI R., CHICHKOV B.N. *et al.*, "Gram scale synthesis of pure ceramic nanoparticles by laser ablation in liquid", *J. Phys. Chem., C*, vol. 114, no. 6, pp. 2421–2427, 2010.

[SAN 98] SANCHEZ-LOPEZ J.C., CABALLERO A., FERNANDEZ A., "Characterisation of passivated aluminium nanopowders: an XPS and TEM / EELS study", *J. Eur Ceram. Soc.*, no. 18, pp. 1195–1200, 1998.

[SAN 07] SANDERS V.E., ASAY B.W., FOLEY T.J. *et al.*, "Reaction Propagation of Four Nanoscale Energetic Composites (Al/MoO_3, Al/WO_3, Al/CuO, and Bi_2O_3)", *J. Propul. Power*, vol. 23, no. 4, pp. 707–714, 2007.

[SAN 07] SANDU I., MOREAU P., GUYOMARD D. *et al.*, "Synthesis of nanosized Si particles via a mechanochemical solid–liquid reaction and application in Li-ion batteries", *Solid State Ionics*, no. 178, pp. 1297–1303, 2007.

[SAR 07a] SARATHI R., SINDHU T.K., CHAKRAVARTHY S.R., "Generation of nano aluminium powder through wire explosion process and its characterization", *Mater. Charact.*, no. 58, pp. 148–155, 2007.

[SAR 07b] SARATHI R., SINDHU T.K., CHAKRAVARTHY S.R., "Generation and characterization of nano aluminium powder obtained through wire explosion process", *Bull. Mater. Sci.*, vol. 30, no. 2, pp. 187–195, 2007.

[SAR 10] SARATHI R., SANKAR B., CHAKRAVARTHY S.R., "Influence of nano aluminium powder produced by wire explosion process at different ambience on hydrogen generation", *J. Electr. Eng. Slovak*, vol. 61, no. 4, pp. 215–221, 2010.

[SAS 04] SASAKI T., LIANG C., NICHOLS W.T. *et al.*, "Fabrication of oxide base nanostructures using pulsed laser ablation in aqueous solutions", *Appl. Phys., A*, no. 79, pp. 1489–1492, 2004.

[SAS 05] SASOH A., WATANABE K., SANO Y. *et al.*, "Behavior of bubbles induced by the interaction of a laser pulse with a metal plate in water", *Appl. Phys., A*, vol. 80, no. 7, pp. 1497–1500, 2005.

[SAT 03] SATO T., TANIGAKI T., SUZUKIA H. *et al.*, "Structure and optical spectrum of ZnO nanoparticles produced in RF plasma", *J. Cryst.Growth*, no. 255, pp. 313–316, 2003.

[SCH 45] SCHAFTHAUL K.F.E., "Die neuesten geologischen hypothesen und ihr verhältniβ zur naturwissentschaft über haupt", *Gelehrte Anzeigen Bayer Akad.*, Vol 20, pp. 557–594, 1845.

[SCH 98] SCHITTENHEIM H., CALLIES G., STRAUB A. *et al.*, "Measurements of wavelength-dependent transmission in excimer laser-induced plasma plumes and their interpretation", *J. Phys. D: Appl. Phys.*, no. 31, pp. 418–427, 1998.

[SCH 07] SCHLABACH S., SZABO D.V., VOLLATH D. P. *et al.*, "Zirconia and titania nanoparticles studied by electric hyperfine interactions, XRD and TEM", *J. Alloys Compd.*, no. 434–435, pp. 590–593, 2007.

[SCH 12] SCHMIDLIN L., PICHOT V., COMET M. *et al.*, "Identification, quantification and modification of detonation nanodiamond functional groups", *Diam. Relat. Mater.*, vol. 22, pp. 113–117, 2012.

[SCH 07] SCHUMACHER B., OCHS R., TROβE H. *et al.*, "Nanogranular SnO_2 layers for gas sensing applications by in situ deposition of nanoparticles produced by the Karlsruhe microwave plasma process", *Plasma Process. Polym.*, no. 4, pp. S865–S870, 2007.

[SCH 59] SCHULMAN J. H., STOECKENIUS W., PRINCE L.M., "Mechanism of formation and structure of micro emulsions by electron microscopy", *J. Phys. Chem.*, vol. 63, no. 10, pp. 1677–1680, 1959.

[SCH 92] SCHULZ O., HAUSNER H., "Plasma synthesis of silicon nitride powders. I. RF-plasrna system for the synthesis of ceramic powders", *Ceram. Int.*, no. 18, pp. 177–183, 1992.

[SEA 15] SEABRA A.B., DURÁN N., "Nanotoxicology of Metal Oxide Nanoparticles", *Metals*, vol. 5, no. 2, pp. 934–975, 2015.

[SEL 14] SELVAKUMAR D., DHARMARAJ N., KADIRVELU K. *et al.*, "Effect of sintering temperature on structural and optical properties of indium(III) oxide nanoparticles prepared with Triton X-100 by hydrothermal method", *Spectrochim. Acta, Part A*, no. 133, pp. 335–339, 2014.

[SEM 10] SEMALTIANOS N.G.,LOGOTHETIDIS S., PERRIE W. *et al.*, "Silicon nanoparticles generated by femtosecond laser ablation in a liquid environment", *J. Nanopart. Res.*, no. 12, pp. 573–580, 2010.

[SEM 14] SEMALTIANOS N.G., HENDRY E., CHANG H. *et al.*, "Laser ablation of a bulk Cr target in liquids for nanoparticle synthesis", *RSC Adv.*, no. 4, pp. 50406–50411, 2014.

[SEO 11] SEONG G., TAKAMI S., ARITA T. *et al.*, "Supercritical hydrothermal synthesis of metallic cobalt nanoparticles and its thermodynamic analysis", *J. Supercrit. Fluids*, no. 60, pp. 113–120, 2011.

[SÉV 12] SÉVERAC F., ALPHONSE P., ESTÈVE A. *et al.*, "High-energy Al/CuO nanocomposites obtained by DNA-Directed assembly", *Adv. Funct. Mater.*, vol. 22, no. 2, pp. 323–329, 2012.

[SHE 08] SHENDE R., SUBRAMANIAN S., HASAN S. *et al.*, "Nanoenergetic composites of CuO nanorods, nanowires, and al-nanoparticles", *Propell. Explos. Pyrot.*, vol. 33, no. 2, pp. 122–130, 2008.

[SHI 99] SHI Y., DING J., LIU X. *et al.*, "$NiFe_2O_4$ ultrafine particles prepared by co-precipitation/mechanical alloying", *J. Magn. Magn. Mater.*, no. 205, pp. 249–254, 1999.

[SHI 86] SHIMIZU T., "A Concept and the Use of Negative Explosives", *11th International Pyrotechnics Seminar*, Vail, Colorado, USA, July 1986.

[SHI 10] SHIN N.C., LEE Y.-H., SHIN Y.H. *et al.*, "Synthesis of cobalt nanoparticles in supercritical methanol", *Mater. Chem. Phys.*, no. 124, pp. 140–144, 2010.

[SHI 12] SHIN M.-S., KIM J.-W., MENDES MORAES C.A. *et al.*, "Reaction characteristics of Al/Fe2O3 nanocomposites", *J. Ind. Eng. Chem.*, vol. 18, no. 5, pp. 1768–1773, 2012.

[SHI 13] SHIN H., LEE S., JUNG H.S. *et al.*, "Effect of ball size and powder loading on the milling efficiency of a laboratory-scale wet ball mill", *Ceram. Int.*, no. 39, pp. 8963–8968, 2013.

[SHU 13] SHUI A., ZHU W., XU L. *et al.*, "Green sonochemical synthesis of cupric and cuprous oxides nanoparticles and their optical properties", *Ceram. Int.*, no. 39, pp. 8715–8722, 2013.

[SIE 10] SIEGERT B., COMET M., MULLER O. *et al.*, "Reduced-sensitivity nanothermites containing manganese oxide filled carbon nanofibers", *J. Phys. Chem., C*, vol. 114, no. 46, pp. 19562–19568, 2010.

[SIE 11] SIEGERT B., COMET M., SPITZER D., "Safer energetic materials by a nanotechnological approach", *Nanoscale*, vol. 3, pp. 3534–3544, 2011.

[SIM 04] SIMAKIN A.V., VORONOV V.V., KIRICHENKO N.A. *et al.*, "Nanoparticles produced by laser ablation of solids in liquid environment", *Appl. Phys., A*, no. 79, pp. 1127–1132, 2004.

[SIN 07] SINGH S.C., GOPAL R., "Zinc nanoparticles in solution by laser ablation technique", *Bull. Mater. Sci.*, vol. 30, no. 3, pp. 291–293, 2007.

[SIN 08] SINGH S.C., GOPAL R., "Synthesis of colloidal zinc oxide nanoparticles by pulsed laser ablation in aqueous media", *Physica E*, no. 40, pp. 724–730, 2008.

[SIN 90] SINGH R.K., NARAYAN J., "Pulsed-laser evaporation technique for deposition of thin films : Physics and theoretical model", *Phys. Rev. B*, vol. 41, no. 13, pp. 8843–8859, 1990.

[SIR 11] SIRAJ K., SOHAIL Y., TABASSUM A., "Metals and metal oxides particles produced by pulsed laser ablation under high vacuum", *Turk. J. Phys.*, no. 35, pp. 179–183, 2011.

[SIV 13] SIVAKUMAR G., DUSANE R.O. , JOSHI S.V., "A novel approach to process phase pure -Al_2O_3 coatings by solution precursor plasma spraying", *J. Eur. Ceram. Soc.*, no. 33, pp. 2823– 2829, 2013.

[SLO 13] SLOCIK J.M., CROUSE C.A., SPOWART J.E. *et al.*, "Biologically tunable reactivity of energetic nanomaterials using protein cages", *Nano Lett.*, vol. 13, no. 6, pp. 2535–2540, 2013.

[SMI 60] SMITH W.E., ANDERSON A.R., Processes for preparing metal alkyls and alkoxides, US Patent no. 2,965,663, 1960.

[SMI 65] SMITH H.M., TURNER F., "Vacuum deposited thin film using a rubis laser", *Appl. Opt.*, no. 4, pp. 147–148, 1965.

[SON 07] SON S.F., ASAY B.W., FOLEY T.J. *et al.*, "Combustion of nanoscale Al/MoO_3 thermite in microchannels", *J. Propul.Power*, vol. 23, no. 4, pp. 715–721, 2007.

[SON 09] SON S.F. DUTRO G.M., ZASECK K.M. *et al.*, "Combustion modes of nanoscale energetic composites", *Int. J. Energ. Mater. Chem. Propul.*, vol. 8, no. 4, pp. 309–319, 2009.

[SON 11] SØNDERGAARD M., BØJESEN E.D., CHRISTENSEN M. *et al.*, "Size and morphology dependence of ZnO nanoparticles synthesized by a fast continuous flow hydrothermal method", *Cryst. Growth Des.*, no. 11, pp. 4027–4033, 2011.

[SON 15] SONG J.-C., XUE F.-F., LU Z.-Y. *et al.*, "Controllable synthesis of hollow mesoporous silica particles by a facile one-pot sol–gel method", *Chem. Commun.*, no. 51, pp. 10517–10520, 2015.

[SON 15] SONIA S., POONGODI S., KUMAR P.S. *et al.*, "Hydrothermal synthesis of highly stable CuO nanostructures for efficient photocatalytic degradation of organic dyes", *Mater. Sci. Semicond. Process.*, no. 30, pp. 585–591, 2015.

[SPI 07] SPITZER D., COMET M., "Synthesis, structural and reactive characterization of miscellaneous nanothermites", *J. Pyrot.*, vol. 26, pp. 60–64, 2007.

[SPI 10] SPITZER D., COMET M., BARAS C. *et al.*, "Energetic nano-materials: Opportunities for enhanced performances", *J. Phys. Chem. Solids*, vol. 71, no. 2, pp. 100–108, 2010.

[SPI 13] SPITZER D., RISSE B., HASSLER D., Preparation of nanoparticles by flash evaporation, Patent WO 2013/117671, 2013.

[SPI 14] SPITZER D., RISSE B., SCHNELL F. *et al.*, "Continuous engineering of nano-cocrystals for medical and energetic applications", *Sci. Rep.*, vol.4, no. 6575, 2014.

[STA 13] STACY S.C., PANTOYA M.L., "Laser ignition of nano-composite energetic loose powders", *Propell. Explos. Pyrot.*, vol. 38, no. 3, pp. 441–447, 2013.

[STA 03] STALLING W.E., LAMB H.H., "Synthesis of nanostructured titania powders via hydrolysis of titanium isopropoxide in supercritical carbon dioxide", *Langmuir*, no. 19, pp. 2989–2994, 2003.

[STA 15] STARIK A.M., SAVEL'EV A.M., TITOVA N.S., "Specific features of ignition and combustion of composite fuels containing aluminum nanoparticles (review)", *Combust. Explos. Shock Waves*, vol. 51, no. 2, pp. 197–222, 2015.

[STO 02] STOIAN R., ROSENFELD A., ASHKENASI D. *et al.*, "Surface charging and impulsive ion ejection during ultrashort pulsed laser ablation", *Phys. Rev. Lett.*, vol. 88, no. 9, pp. 097603(1)–097603(4), 2002.

[STR 09] STRATAKIS E., BARBEROGLOU M., FOTAKIS C. *et al.*, "Generation of Al nanoparticles via ablation of bulk Al in liquids with short laser pulses", *Opt. Express*, vol. 17, no. 15 , pp. 12650–12659, 2009.

[STR 09] STRIGUL N., VACCARI L., GALDUN C. *et al.*, "Acute toxicity of boron, titanium dioxide, and aluminum nanoparticles to Daphnia magna and Vibrio fischeri", *Desalination*, vol. 248, no. 1–3, pp. 771–782, 2009.

[STU 01] STUBICAR M., BERMANEC V., STUBICAR N. *et al.*, "Microstructure evolution of an equimolar powder mixture of ZrO_2–TiO_2 during high-energy ball-milling and post-annealing" , *J. Alloys Compd*, no. 316, pp. 316–320, 2001.

[SU 05] SU C.-Y., LIN C.-K., CHENG C.-W., "A modified plasma arc gas condensation technique to synthesize nanocrystalline tungsten oxide powders", *Mater. Trans.*, vol. 46, no. 5 pp. 1016–1020, 2005.

[SU 08] SU C.-Y., LIN C.-K., YANG T.-K. *et al.*, "Oxygen partial pressure effect on the preparation of nanocrystalline tungsten oxide powders by a plasma arc gas condensation technique", *Int. J. Refract. Met. Hard Mater.*, no. 26, pp. 423–428, 2008.

[SUE 11] SUE K., KAWASAKI S.-I, SUZUKI M. *et al.*, "Continuous hydrothermal synthesis of Fe_2O_3, NiO, and CuO nanoparticles by super rapid heating using a T-type micro mixer at 673K and 30MPa", *Chem. Eng. J.*, no. 166, pp. 947–953, 2011.

[SUE 07] SUESMATSU H., NISHIMURA S., MURAI K. *et al.*, "Pulsed wire discharge apparatus for mass production of copper nanopowders", *Rev. Scientific Instrum.*, vol. 78, no. 5, pp. 056105–056105-3, 2007.

[SUL 09] SULAIMANKULOVA S., OMURZAK E., JASNAKUNOV J. *et al.*, New Preparation Method of Nanocrystalline Materials by Impulse Plasma in Liquid", *J. Clust. Sci.*, no. 20, pp. 37–49, 2009.

[SUL 09] SULLIVAN K., YOUNG G., ZACHARIAH M.R., "Enhanced reactivity of nano-B/Al/CuO MIC's", *Combust. Flame*, vol. 156, no. 2, pp. 302–309, 2009.

[SUL 10a] SULLIVAN K.T., PIEKIEL N.W., CHOWDHURY S. *et al.*, "Ignition and Combustion Characteristics of Nanoscale Al/$AgIO_3$: A Potential Energetic Biocidal System", *Combust. Sci. Technol.*, vol. 183, no. 3, pp. 285–302, 2010.

[SUL 10b] SULLIVAN K.T., CHIOU W.-A., FIORE R. *et al.*, "In situ microscopy of rapidly heated nano-Al and nano-Al/WO_3 thermites", *Appl. Phys. Lett.*, vol. 97, no. 133104, 2010.

[SUL 12a] SULLIVAN K.T., WORSLEY M.A., KUNTZ J.D. *et al.*, "Electrophoretic deposition of binary energetic composites", *Combust. Flame*, vol. 159, no. 6, pp. 2210–2218, 2012.

[SUL 12b] SULLIVAN K.T., KUNTZ J.D., GASH A.E., "Electrophoretic deposition and mechanistic studies of nano-Al/CuO thermites", *J. Appl. Phys.*, vol. 112, no. 024316, 2012.

[SUL 12c] SULLIVAN K.T., PIEKIEL N.W., WU C. *et al.*, "Reactive sintering: An important component in the combustion of nanocomposite thermites", *Combust. Flame*, vol. 159, no. 1, pp. 2–15, 2012.

[SUL 13] SULLIVAN K.T., WU C., PIEKIEL N.W. *et al.*, "Synthesis and reactivity of nano-Ag_2O as an oxidizer for energetic systems yielding antimicrobial products", *Combust.Flame*, vol. 160, no. 2, pp. 438–446, 2013.

[SUL 14] SULLIVAN K.T., KUNTZ J.D., GASH A.E., "The Role of Fuel Particle Size on Flame Propagation Velocity in Thermites with a Nanoscale Oxidizer", *Propell. Explos. Pyrot.*, vol. 39, no. 3, pp. 407–415, 2014.

[SUN 01] SUNDERMAN F.W., "Nasal Toxicity, Carcinogenicity, and Olfactory Uptake of Metals", *Ann. Clin. Lab. Sci.*, vol. 31, no. 1, pp. 3–24, 2001.

[SUN 06] SUN J., PANTOYA M.L., SIMON S.L., "Dependence of size and size distribution on reactivity of aluminum nanoparticles in reactions with oxygen and MoO_3", *Thermochim. Acta.*, vol. 444, no. 2, pp. 117–127, 2006.

[SUN 07] SUN J., SIMON S.L., "The melting behavior of aluminum nanoparticles", *Thermochim.Acta*, vol. 463, no. 1–2, pp. 32–40, 2007.

[SUN 09] SUN G., LI X., ZHANG Y. *et al.*, "A simple detonation technique to synthesize carbon-coated cobalt", *J. Alloy Compd.*, vol. 473, no. 1–2, pp. 212–214, 2009.

[SUN 15] SUNDARAM D.S., YANG V., ZARKO V.E., "Combustion of Nano Aluminum Particles (Review)", *Combust. Explos. Shock Waves*, vol. 51, no. 2, pp. 173–196, 2015.

[SUR 01] SURYANARAYANA C., "Mechanical alloying and milling,"*Prog. Mater. Sci.*, no. 46, pp. 1–184, 2001.

[SUR 95] SURYANARAYANA C., "Does a disordered y- TiAl phase exist in mechanically alloyed Ti-Al powders?", *Intermetallics*, no. 3, pp. 153–160, 1995.

[SUS 89] SUSLICK K.S., "The chemical effects of ultrasound", *Sci. Am.*, pp. 80–86, February 1989.

[SUS 91] SUSLICK K.S., CHOE S.-B., CICHOWLAS A.A. *et al.*, "Sonochemical synthesis of amorphous iron", *Nature*, no. 353, pp. 414–416, 1991.

[SUS 96] SUSLICK K.S., HYEON T., FANG M., "Nanostructured materials generated by high-intensity ultrasound: Sonochemical synthesis and catalytic studies", *Chem. Mater.*, no. 8, pp. 2172–2179, 1996.

[SUS 96] SUSLICK K.S., FANG M., HYEON T., "Sonochemical synthesis of iron colloids", *J. Am. Chem. Soc.*, no. 118, pp. 11960–11961, 1996.

[SUZ 12] SUZUKI K., TANAKA N., ANDO A. *et al.*, "Size-selected copper oxide nanoparticles synthesized by laser ablation", *J. Nanopart. Res.*, no. 14, pp. 863(1)–863(11), 2012.

[SUZ 95] SUZUKI T.S., NAGUMO M., "Metastable intermediate phase formation at reaction milling of titanium and n-heptane", *Scr. Metall. Mater.*, vol. 32, no. 8, p 1215–1225, 1995.

[SVE 00] SVERGUN D.I. , KONAREV P.V., VOLKOV V.V. *et al.*, "A small angle x-ray scattering study of the droplet–cylinder transition in oil-rich sodium bis(2-ethylhexyl) sulfosuccinate microemulsions", *J. Chem. Phys.*, vol. 113, no. 4, pp. 1651–1664, 2000.

[SVR 05] SVRCEK V., REHSPRINGER J.-L., GAFFET E. *et al.*, "Unaggregated silicon nanocrystals obtained by ball milling", *J. Cryst. Growth*, no. 275, pp. 589–597, 2005.

[SVR 11] SVRCEK V., MARIOTTI D., NAGAI T. *et al.*, "Photovoltaic applications of silicon nanocrystal based nanostructures induced by nanosecond laser fragmentation in liquid media", *J. Phys. Chem., C*, vol. 115, no. 12, pp. 5084–5093, 2011.

[SYL 04a] SYLVESTRE J.-P., POULIN S., KABASHIN A.V. *et al.*, "Surface chemistry of gold nanoparticles produced by laser ablation in aqueous media", *J. Phys. Chem. B*, vol. 108, no. 43, pp. 16864–16869, 2004.

[SYL 04b] SYLVESTRE J.-P., KABASHIN A.V., SACHER E. *et al.*, "Stabilization and size control of gold nanoparticles during laser ablation in aqueous cyclodextrins", *J. Am. Chem. Soc.*, no. , 126, pp. 7176–7177, 2004.

[SYL 05] SYLVESTRE J.-P., KABASHIN A.V., SACHER E. *et al.*, "Femtosecond laser ablation of gold in water: influence of the laser-produced plasma on the nanoparticle size distribution", *Appl. Phys. A*, no. 80, pp. 753–758, 2005.

[SYN 11] SYNEK P., JASEK O., ZAJICKOVA L. *et al.*, "Plasmachemical synthesis of maghemite nanoparticles in atmospheric pressure microwave torch", *Mater. Lett.*, no. 65, pp. 982–984, 2011.

[SZA 07] SZABO D.V., SCHLABACH S., OCHS R., "Analytical TEM investigation of size effects in SnO_2 nanoparticles produced by microwave plasma synthesis", *Microsc. Microanal*, no. 13, pp. 430–431, 2007.

[SZA 12] SZABO D.V., KILIBARDA G., SCHLABACH S. *et al.*, "Structural and chemical characterization of SnO_2-based nanoparticles as electrode material in Li-ion batteries", *J. Mater.Sci.*, no. 47, pp. 4383–4391, 2012.

[TAK 94] TAKACS L., PARDAVI-HORVATH M., "Nanocomposite formation in the Fe_3O_4-Zn system by reaction milling", *J. Appl. Phys.*, Vol 75, no. 10, pp. 5864–5866, 1994.

[TAK 99] TAKAMI A., KURITA H., KODA S., "Laser-induced size reduction of noble metal particles", *J. Phys. Chem., B*, vol. 103, no. 8, pp. 1226–1232, 1999.

[TAP 14] TAPPAN B.C., DIRMYER M.R., RISHA G.A., "Evidence of a kinetic isotope effect in nanoaluminum and water combustion", *Angew. Chem. Int. Ed.*, vol. 53, no. 35, pp. 9218–9221, 2014.

[TAR 10] TARASENKO N., NEVAR A., NEDELKO M., "Properties of zinc-oxide nanoparticles synthesized by electrical-discharge technique in liquids", *Phys. Status Solidi A*, vol. 207, no. 10, pp. 2319–2322, 2010.

[TAT 13] TATON G., LAGRANGE D., CONEDERA V. *et al.*, "Micro-chip initiator realized by integrating Al/CuO multilayer nanothermite on polymeric membrane", *J. Micromech. Microeng.*, vol. 23, no. 105009, 2013.

[TEN 06] TENDERO C., TIXIER C., TRISTANT P. *et al.*, "Atmospheric pressure plasmas: A review", *Spectrochimica Acta Part B*, no. 61, pp. 2–30, 2006.

[THI 11] THIRUVENGADATHAN R., BEZMELNITSYN A., APPERSON S. *et al.*, "Combustion characteristics of novel hybrid nanoenergetic formulations", *Combust.Flame*, vol. 158, no. 5, pp. 964–978, 2011.

[THI 14] THIRUVENGADATHAN R., CHUNG S.W., BASURAY S. *et al.*, "A versatile self-assembly approach toward high performance nanoenergetic composite using functionalized graphene", *Langmuir*, vol. 30, no. 22, pp. 6556–6564, 2014.

[THI 15] THIRUVENGADATHAN R., STALEY C., GEESON J.M. *et al.*, "Enhanced combustion characteristics of bismuth trioxide-aluminum nanocomposites prepared through graphene oxide directed self-assembly", *Propell. Explos. Pyrot.*, vol. 40, no. 5, pp. 729–734, 2015.

[TIA 14] TIAN J.X., ZHANG Z.Y, YAN J.F. *et al.*, "Hydrothermal synthesis and infrared emissivity property of flower-like SnO_2 particles", *AIP Adv.*, no. 4, pp. 047131-1–047131-8, 2014.

[TIE 09] TIEN D.-C., TSENG K.-H., LIAO C.-H. *et al.*, "Identification and quantification of ionic silver from colloidal silver prepared by electric spark discharge system and its antimicrobial potency study", *J. Alloys Compd.*, no. 473, pp. 298–302, 2009.

[TIL 01] TILLOTSON T.M., GASH A.E., SIMPSON R.L. *et al.*, "Nanostructured energetic materials using sol-gel methodologies", *J. Non-Cryst.Solids*, vol. 285, no. 1–3, pp. 338–345, 2001.

[TOB 94] TOBA M., MIZUKAMI F., NIWA S. *et al.*, "Effect of preparation methods on properties of amorphous alumina / silicas", *J. Mater.Chem.*, vol. 4, no. 4, pp. 1131–1135, 1994.

[TOJ 97] TOJO C., BLANCO M.C., RIVADULLA F. *et al.*, "Kinetics of the formation of particles in microemulsions", *Langmuir*, no. 13, pp. 1970–1977, 1997.

[TON 91] TONEJC A., DUZEVIK D., TONEJC A.M. "Effects of ball milling on pure antimony, on Ga-Sb alloy and on Ga + Sb powder mixture; oxidation, glass formation and crystallization", *Mater. Sci. Eng., A*, no. 134, pp. 1372–1375, 1991.

[TOU 14] TOUFIQ A.M., WANG F., JAVED Q. *et al.*, "Hydrothermal synthesis of MnO_2 nanowires: structural characterizations, optical and magnetic properties", *Appl. Phys., A*, no. 116, pp. 1127–1132, 2014.

[TOU 75] TOULOUKIAN Y.S., KIRBY R.K., TAYLOR R. E. *et al.*, "Thermal expansion – Metallic Elements and Alloys", in TOULOUKIAN Y.S. (ed.), *Thermophysical properties of matter* (Volume 12), Plenum Publishing Corporation, New York, 1975.

[TOU 77] TOULOUKIAN Y.S., KIRBY R.K., TAYLOR R.E. *et al.*, "Thermal expansion – Nonmetallic Solids", in TOULOUKIAN Y.S. (ed.), *Thermophysical properties of matter* (Volume 13), Plenum Publishing Corporation, New York, 1977.

[TRE 10] TREBS A., FOLEY T.J., "Semi-empirical model for reaction progress in nanothermite", *J. Propul.Power*, vol. 26, no. 4, pp. 772–775, 2010.

[TSE 09] TSENG K.-H., HUANG J.-C., LIAO C.-Y. *et al.*, "Preparation of gold ethanol colloid by the arc discharge method", *J. Alloys Compd.*, no. 472, pp. 446–450, 2009.

[TSU 01] TSUJI T., IRYO K., NISHIMURA Y. *et al.*, "Preparation of metal colloids by a laser ablation technique in solution: influence of laser wavelength on the ablation efficiency (II)", *J. Photochem. Photobiol. A*, no. 145, pp. 201–207, 2001.

[TSU 02] TSUJI T., IRYO K., WATANABE N. *et al.*, "Preparation of silver nanoparticles by laser ablation in solution: influence of laser wavelength on particle size", *Appl. Surf. Sci.*, no. 202, pp. 80–85, 2002.

[TSU 03] TSUJI T., IRYO K., WATANABE N. *et al.*, "Preparation of silver nanoparticles by laser ablation in solution: influence of laser wavelength on particle size", *Appl. Surf. Sci.*, no. 202, pp. 80–85, 2003.

[TSU 07] TSUJI T., OKAZAKI Y., TSUBOI Y. *et al.*, "Nanosecond time-resolved observations of laser ablation of silver in water", *Jpn. J. Appl. Phys.*, vol. 46, no. 4A, pp. 1533–1535, 2007.

[TSU 00] TSUZUKI T., MCCORMICK P.G., "Synthesis of Cr_2O_3 nanoparticles by mechanical processing", *Acta Mater.*, no. 48, pp. 2795–2801, 2000.

[TSU 01a] TSUZUKI T., MCCORMICK P.G., "Mechanochemical synthesis of niobium pentoxide nanoparticles", *Mater. Trans, JIM*, vol. 42, no. 8, pp. 1623–1628, 2001.

[TSU 01b] TSUZUKI T., MCCORMICK P.G., "ZnO nanoparticles synthesis by mechanical processing", *Scripta Mater.*, no. 44, pp. 1731–1734, 2001.

[TSU 01c] TSUZUKI T., MCCORMICK P.G., "Synthesis of ultrafine ceria powders by mechanochemical processing", *J. Am. Ceram.Soc .*, vol. 84, no. 7, pp. 1453–1458, 2001.

[TSU 11] TSUZUKI T, SCHÄFFEL F., MUROI M. *et al.*, "α-Fe_2O_3 nano-platelets prepared by mechanochemical/thermal processing", *Power Technol.*, no. 210, pp. 198–202, 2011.

[TSU 98] TSUZUKI T., HARRISON W.T.A., MCCORMICK P.G., "Synthesis of ultrafine gadolinium oxide powder by mechanochemical processing"*J. Alloys Compd.*, no. 281, pp. 146–151, 1998.

[TSU 99] TSUZUKI T., PIRAULT E., MCCORMICK P.G., "Mechanochemical synthesis of gadolinium oxide nanoparticles", *Nanostruct. Mater.*, vol. 11 no. 1, pp. 125–131, 1999.

[TUR 14] TURKEZ H., SONMEZ E., TURKEZ O. *et al.*, "The risk evaluation of tungsten oxide nanoparticles in cultured rat liver cells for its safe applications in nanotechnology", *Braz. Arch. Biol. Technol.*, vol. 57, no. 4, pp. 532–541, 2014.

[VAE 12] VAEZI M.R., GHASSEMI S. H.M.S., SHOKUHFAR A., "Effect of different sizes of balls on crystalline size, strain, and atomic diffusion on Cu-Fe nanocrystals produced by mechanical alloying", *J. Theor. Appl. Phys.*, vol. 6, no. 29, p 1–7, 2012.

[VAL 03] VALLIAPAN S., PUSZYNSKI J. A., "Combustion characteristics of metal-based nanoenergetic systems", *Proc. SD Acad. Science*, vol. 82, pp. 97–101, 2003.

[VAL 05] VALLIAPPAN S., SWIATKIEWICZ J.J. *et al.*, "Reactivity of aluminum nanopowders with metal oxides", *Powder Technol.*, vol. 156, no. 2–3, pp. 164–169, 2005.

[VAN 13] VAN OVERSCHELDE O., DERVAUX J., YONGE L. *et al.*, "Screening effect in gold nanoparticles generated in liquid by KrF ablation", *Laser Phys.*, vol. 23, no. 5, pp. 055901(1)– 055901(7), (2013).

[VAY 98] VAYSSIERES L., CHANEAC C., TRONC E. *et al.*, "Size tailoring of magnetite particles formed by aqueous precipitation: An example of thermodynamic stability of nanometric oxide particles", *J. Colloid Interface Sci.*, no. 205, pp. 205–212, 1998.

[VEI 32] VEIL S., "Aluminium", in PASCAL P., BAUD P. (eds.), *Traité de chimie minérale – Tome VII*, Masson et Cie, Paris, 1932.

[VEN 11] VENNEKAMP M., BAUER I., GROH M. *et al.*, "Formation of SiC nanoparticles in an atmospheric microwave plasma", *Beilstein J. Nanotechnol.*, no. 2, 665–673, 2011.

[VER 09] VERIANSYAH B., PARK H., KIM J.-D. *et al.*, "Characterization of surface-modified ceria oxide nanoparticles synthesized continuously in supercritical methanol", *J. Supercrit. Fluids*, no. 50, pp. 283–291, 2009.

[VER 10a] VERIANSYAH B., KIM J.-D., MIN B.K. *et al.*, "Continuous synthesis of magnetite nanoparticles in supercritical methanol", *Matter. Lett.*, no. 64, p . 2197–2200, 2010.

[VER 10b] VERIANSYAH B., KIM J.-D, MIN B.K. *et al.*, "Continuous synthesis of surface-modified zinc oxide nanoparticles in supercritical methanol", *J. Supercrit. Fluids*, no. 52, pp. 76–83, 2010.

[VIA 10] VIAU G.,COLLIERE V., LACROIX L.M. *et al.*, "Internal structure of nanoparticles of Al hollow nanoparticles generated by laser ablation in liquid ethanol", *Chem. Phys. Lett.*, vol. 501, no. 4–6, pp. 419–422, 2011.

[VLA 08] VLADOIU I., STAFE M., NEGUTU C., POPESCU I. M., "Nanopulsed ablation rate of metals dependence on the laser fluence and wavelengh in atmospheric air", *U. P. B., Sci. Bull.A*, vol. 70, no. 4, pp. 119–126, 2008.

[VOL 07] VOLLATH D. "Plasma synthesis of nanoparticles", *KONA, Powders and particules*, no. 25, pp. 39–55, 2007.

[VOL 08] VOLLATH D., "Plasma synthesis of nanopowders", *J Nanopart Res.*, no. 10, pp. 39–57, 2008.

[VOL 93a] VOLLATH D., SICKAFUS K.E., "Synthesis of ceramic oxide powders by microwave plasma pyrolysis", *J. Mater.Sci.*, no. 28, pp. 5943–5948, 1993.

[VOL 93b] VOLLATH D., SICKAFUS K.E., Synthesis of ceramic oxide powders in a microwave plasma device", *J. Mater. Res.*, vol. 8, no. 11, pp. 2978–2984, 1993.

[VOL 95] VOLLATH D., SZABO D.V., TAYLOR R.D. *et al.*, "Synthesis and properties of nanocrystalline superparamagnetic gamma-Fe_2O_3". *Nanostruct. Mater.*, no. 6, pp. 941–944, 1995.

[VOL 96] VOLLATH D., SZABO D.V., WILLIS J.O., "Magnetic properties of nanocrystalline Cr_2O_3 synthesized in a microwave plasma", *Mater. Lett.*, no. 29, pp. 271–279, 1996.

[VON 96] VON DER LINDE D., SCHYLER H., "Breakdown threshold and plasma formation in femtosecond laser–solid interaction", *J. Opt.Soc. Am. B*, vol. 13, no. 1, pp. 216–222, 1996.

[VOS 10] VOSTRIKOV A.A., FEDYAEVA O.N., "Mechanism and kinetics of Al_2O_3 nanoparticles formation by reaction of bulk Al with H_2O and CO_2 at sub- and supercritical conditions"*J. Supercrit.Fluids*, no. 55, pp. 307–315, 2010.

[WAD 67] WADA N., "Preparation of fine metal particles by means of evaporation in helium gas,", *Jpn J. Appl. Phys.*, vol. 6, no. 5, pp. 553–556, 1967.

[WAD 68] WADA N., "Preparation of fine metal particles by means of evaporation in xenon gas"*Jpn J. Appl. Phys.*, vol. 7, no. 10, pp. 1287–1293, 1968.

[WAG 06] WAGNER A., BLECKMANN C., ENGLAND E. *et al.*, "In vitro toxicity of aluminum nanoparticles in rat alveolar macrophages", Air Force Research Laboratory, no. AFRL-HE-WP-TP-2006-0022, 2001.

[WAG 33] WAGNER C., "Beitrag zur Theorie des Anlaufvorgangs", *Z. Phys. Chem. B-Chem. E.*, vol. B21, pp. 25–41, 1933.

[WAH 07] WAHAB R., ANSARI S.G., DAR M.A. *et al.*, "Synthesis of magnesium oxide nanoparticles by sol-gel process", *Mater. Sci. Forum*, no. 558–559, pp. 983–986, 2007.

[WAL 04] WALTER K.C., AUMANN C.E., CARPENTER R.D. *et al.*, "Energetic materials development at technanogy materials development", *Mater. Res. Soc. Sympp. Proc.*, vol. 800, pp. 27–37, 2004.

[WAN 01] WANG C.-C., CHEN D.-H., HUANG T.-C., "Synthesis of palladium nanoparticles in water-in-oil microemulsions", *Colloids Surf.*, A, no. 189, pp. 145–154, 2001.

[WAN 05] WANG S., YANG Y., YU H. *et al.*, "Dynamical effects of the oxide layer in aluminum nanoenergetic materials", *Propell. Explos. Pyrot.*, vol. 30, no. 2, pp. 148–155, 2005.

[WAN 06a] WANG Z., LIU Y., ZENG X., "One-step synthesis of γ-Fe_2O_3 nanoparticles by laser ablation", *Powder Technol.*, no. 161, pp. 65–68, 2006.

[WAN 06b] WANG Z.L., SONG J., "Piezoelectric nanogenerators based on zinc oxide nanowire arrays", *Science*, vol. 312, no. 5771, pp. 242–246, 2006.

[WAN 07] WANG Y., JIANG W., CHENG Z. *et al.*, "Thermite reactions of Al/CuO core-shell nanocomposites with WO_3", *Thermochim.Acta*, vol. 463, no. 1–2, pp. 69–76, 2007.

[WAN 08] WANG X.H., LI X.J., YAN H.H. *et al.*, "Nano-$MnFe_2O_4$ powder synthesis by detonation of emulsion explosive", *Appl. Phys. A*, vol. 90, no. 3, pp. 417–422, 2008.

[WAN 09] WANG W., CLARK R., NAKANO A. *et al.*, "Fast reaction mechanism of a core(Al)-shell (Al_2O_3) nanoparticles in oxygen", *Appl. Phys. Lett.*, vol. 95, no. 261901, 2009.

[WAN 10] WANG H., PYATENKO A., KAWAGUCHI K. *et al.*, "Selective pulsed heating for the synthesis of semiconductor and metal submicrometer spheres", *Angew. Chem. Int. Ed.*, no. 49, pp. 6361 6364, 2010.

[WAN 11] WANG L., LUSS D., MARTIROSYAN K.S., "The behavior of nanothermite reaction based on Bi_2O_3/Al", *J. Appl. Phys.*, vol. 110, no. 074311, 2011.

[WAN 13a] WANG H., JIAN G., YAN S. *et al.*, "Electrospray formation of gelled nano-aluminum microspheres with superior reactivity", *ACS Appl. Mater.Interfaces*, vol. 5, no. 15, pp. 6797–6801, 2013.

[WAN 13b] WANG Y., JIANG W., SONG X.L. *et al.*, "Depositing of amorphous ferri-oxide nanoparticles on surfaces of submicron aluminum powder", *Appl. Mech. Mater.*, vol. 320, pp. 451–455, 2013.

[WAN 13c] WANG Y., ZHU Y., XING Z. *et al.*, "Hydrothermal synthesis of α-MoO_3 and the influence of later heat treatment on its electrochemical properties", *Int. J. Electrochem. Sci.*, no. 8, pp. 9851–9857, 2013.

[WAN 14a] WANG A., CHEN L., XU F. *et al.*, "In situ synthesis of copper nanoparticles within ionic liquid-in-vegetable oil microemulsions and their direct use as high efficient nanolubricants", *RSC Adv.*, no. 4, pp. 45251–45257, 2014.

[WAN 14b] WANG X., ZHANG H., LIU L. *et al.*, "Controlled morphologies and growth direction of WO_3 nanostructures hydrothermally synthesized with citric acid", *Mater. Lett.*, no. 130, pp. 248–251, 2014.

[WAN 14c] WANG L., CAI Y., SONG L.-Y. *et al.*, "High efficient photocatalyst of spherical TiO_2 particles synthesized by a sol–gel method modified with glycol", *Colloids Surf.*, A, no. 461, pp. 195–201, 2014.

[WAN 14d] WANG H., JIAN G., EGAN G.C. *et al.*, "Assembly and reactive properties of Al/CuO based nanothermite microparticles", *Combust. Flame*, vol. 161, pp. 2203–2208, 2014.

[WAN 15a] WANG Y.-F., LI X.-F., LI D.-J. *et al.*, "Controllable synthesis of hierarchical SnO_2 microspheres for dye sensitized solar cells", *J. Power Sources*, no. 280, pp. 476–482, 2015.

[WAN 15b] WANG H., JIAN G., ZHOU W. *et al.*, "Metal iodate-based energetic composites and their combustion and biocidal performances", *ACS Appl. Mater. Interfaces*, vol. 7, no. 31, pp. 17363–17370, 2015.

[WAN 16] WANG J., QIAO Z., YANG Y. *et al.*, "Core-shell Al-Polytetrafluoroethylene (PTFE) configurations to enhance reaction kinetics and energy performance for nanoenergetic materials", *Chem. Eur. J.*, vol. 22, no. 1, pp. 279–284, 2016.

[WAN 92] WANG X.Y., DOWNER M.C., "Femtosecond time-resolved reflectivity of hydrodynamically expanding metal surfaces", *Opt. Lett.*, vol. 17, no. 20, pp. 1450–1452, 1992.

[WAN 93] WANG L.L., MUNIR Z.A., MAXIMOV Y.M., "Thermite reactions: their utilization in the synthesis and processing of materials", *J. Mater. Sci.*, vol. 28, no. 14, pp. 3693–3708, 1993.

[WAT 04] WATANABE H., YAMADA N., OKAJI M., "Linear thermal expansion coefficient of silicon from 293 to 1000 K", *Int. J. Thermophys.*, vol. 25, no. 1, pp. 221–236, 2004.

[WAT 95] WATANABE R., HASHIMOTO H., LEE G.G., "Computer simulation of milling ball motion in mechanical alloying (Overview)", *Mater. Trans., JIM*, vol. 36, no. 2, pp. 102–109, 1995.

[WEE 88] WEEBER A.W., BAKKER H., "Amorphization by ball milling. A review", *Physica B*, vol. 153, pp. 93–135, 1988.

[WEI 05] WEI Z.Q., QIAO H.X., DAI J.F. *et al.*, "Preparation and characterization of Ni nanopowders prepared by anodic arc plasma", *Trans.Nonferrous Met. Soc. China*, vol. 15, no. 1, pp. 51–56, 2005.

[WEI 06] WEI Z.Q., XIA T.-D., MA J. *et al.*, "Growth mechanism of Cu nanopowders prepared by anodic arc plasma", *Trans.Nonferrous Met. Soc. China*, no. 16, pp. 168–172, 2006.

[WEI 09] WEI Z.Q., QIAO H.X., YANG H. *et al.*, "Characterization of NiO nanoparticles by anodic arc plasma method", *J. Alloys Compd.*, no. 479, pp. 855–858, 2009.

[WEI 04] WEIGLE J.C., LUHRS C.C., CHEN C.K. *et al.*, "Generation of aluminum nanoparticles using an atmospheric pressure plasma torch", *J. Phys. Chem., B*, no. 108, pp. 18601–18607, 2004.

[WEI 13] WEIR C., PANTOYA M.L., DANIELS M.A., "The role of aluminum particle size in electrostatic ignition sensitivity of composite energetic materials", *Combust.Flame*, vol. 160, no. 10, pp. 2279–2281, 2013.

[WEI 09] WEISMILLER M.R., MALCHI J.Y., YETTER R.A. *et al.*, "Dependence of flame propagation on pressure and pressurizing gas for an Al/CuO nanoscale thermite", *Proc. Combust. Inst.*, vol. 32, no. 2, pp. 1895–1903, 2009.

[WEI 11a] WEISMILLER M.R., MALCHI J.Y., LEE J.G. *et al.*, "Effects of fuel and oxidizer particle dimensions on the propagation of aluminum containing thermites", *Proc. Combust. Inst.*, vol. 33, no. 2, pp. 1989–1996, 2011.

[WEI 11b] WEISMILLER M.R., LEE J.G., YETTER R.A., "Temperature measurements of Al containing nano-thermite reactions using multi-wavelength pyrometry", *Proc. Combust. Inst.*, vol. 33, no. 2, pp. 1933–1940, 2011.

[WEN 13] WEN J.Z., RINGUETTE S., BOHLOULI-ZANJANI G. *et al.*, "Characterization of thermochemical properties of Al nanoparticle and NiO nanowire composites", *Nanoscale Res. Lett.*, vol. 8, no. 184, 2013.

[WER 11] WERNER D., HASHIMOTO S., "Improved working model for interpreting the excitation wavelength and fluencedependent response in pulsed laser-induced size reduction of aqueous gold nanoparticles", *J. Phys. Chem., C*, vol. 115, no. 12, pp. 5063–5072, 2011.

[WIL 00] WILLMOTT P.R., HUBER J.R., "Pulsed laser vaporization and deposition", *Rev. Mod. Phys.*, vol. 72, no. 1, pp. 315–328, 2000.

[WIT 12] WITSCHGER O., LE BIHAN O., DURAND C. *et al.*, "Préconisations en matière de caractérisation des potentiels d'émission et d'exposition professionnelle aux aérosols lors d'opérations mettant en œuvre des nanomatériaux", *Hygiène et sécurité du travail*, Institut National de Recherche et Sécurité (INRS), vol. 226, pp. 41–55, 2012.

[WOL 96] WOLSKI K., LE CAËR G., DELCROIX P. *et al.*, "Influence of milling conditions on the FeAl intermetallic formation by mechanical alloying", *Mater. Sci. Eng., A*, no. 207, pp. 97–104, 1996.

[WON 11] WONGPISUTPAISAN N., CHAROONSUK P., NARATIP VITTAYAKORN N. *et al.*, "Sonochemical synthesis and characterization of copper oxide nanoparticles", *Energy Procedia*, no. 9, pp. 404–409, 2011.

[WU 03] WU W., ZHU Z., LIU Z. *et al.*, "Preparation of carbon-encapsulated iron carbide nanoparticles by an explosion method", *Carbon*, vol. 41, no. 2, pp. 317–321, 2003.

[WU 08] WU B.X., "High-intensity nanosecond-pulsed laser-induced plasma in air, water, and vacuum: A comparative study of the early-stage evolution using a physics-based predictive model", *Appl. Phys. Lett.*, vol. 93, no. 10, pp. 101104(1)–101104(3), 2008.

[WU 10] WU L., MA Z., HE A. *et al.*, "Studies on destruction of silicon tetrachloride using microwave plasma jet", *J. Hazard. Mater.*, no. 173, pp. 305–309, 2010.

[WU 12] WU C., SULLIVAN K., CHOWDHURY S. *et al.*, "Encapsulation of perchlorate salts within metal oxides for application as nanoenergetic oxidizers", *Adv. Funct. Mater.*, vol. 22, no. 1, pp. 78–85, 2012.

[WU 14a] WU C., ZHIGILEI L.V., "Microscopic mechanisms of laser spallation and ablation of metal targets from large-scale molecular dynamics simulations", *Appl. Phys., A*, no. 114, pp. 11–32, 2014.

[WU 14b] WU C., SHEN L., HUANG Q. *et al.*, "Hydrothermal synthesis and characterization of Bi_2O_3 nanowires", *Mater. Lett.*, no. 65, pp. 1134–1136, 2014.

[XIA 12] XIAO J., LIU P., LIANG Y., LI H. B., YANG G. W, "Porous tungsten oxide nanoflakes for highly alcohol sensitive performance", *Nanoscale*, no. 4, pp. 7078–7083, 2012.

[XIE 06a] XIE X., LI X., YAN H., "Detonation synthesis of zinc oxide nanometer powders", *Mater. Lett.*, vol. 60, no. 25–26, pp. 3149–3152, 2006.

[XIE 06b] XIE X., LI X., ZHAO Z. *et al.*, "Growth and morphology of nanometer $LiMn_2O_4$ powder", *Powder Technol.*, vol. 169, pp. 143–146, 2006.

[XIE 15] XIE A., WANG S., LIU W. *et al.*, "Rapid hydrothermal synthesis of CeO_2 nanoparticles with (2 2 0)-dominated surface and its CO catalytic performance", *Mater. Res. Bull.*, no. 62, pp. 148–152, 2015.

[XU 12] XU D., YANG Y., CHENG H. *et al.*, "Integration of nano-Al with Co_3O_4 nanorods to realize high-exothermic core-shell nanoenergetic materials on a silicon substrate", *Combust.Flame*, vol. 159, no. 6, pp. 2202–2209, 2012.

[YAH 10] YAHIAOUI K., KERDJA T., MALEK S., "Phase explosion in tungsten target under interaction with Nd : YAG laser tripled in frequency", *Surf. Interface Anal.*, no. 42, pp. 1299–1302, 2010.

[YAM 06] YAMADA K., TOKUMOTO Y., NAGATA T. *et al.*, "Mechanism of laser-induced size-reduction of gold nanoparticles as studied by nanosecond transient absorption spectroscopy", *J. Phys. Chem., B*, vol. 110, no. 24, pp. 11751–11756, 2006.

[YAM 96] YAMADA Y., ORII T., UMEZU I. *et al.*, "Optical properties of silicon nanocrystallites prepared by excimer laser ablation in inert gas", *Jpn. J., Appl. Phys.*, vol. 35, pp. 1361–1365, 1996.

[YAN 12] YAN S., JIAN G., ZACHARIAH M.R., "Electrospun nanofiber-based thermite textiles and their reactive properties", *ACS Appl. Mater. Interfaces*, vol. 4, pp. 6432–6435, 2012.

[YAN 13] YAN H., WU L., LI X., WANG X., "Detonation synthesis of SnO_2 nanoparticles in gas phase", *Rare Metal Mat. Eng.*, vol. 42, no. 7, pp. 1325–1327, 2013.

[YAN 02] YANG X.C., RIEHEMANN W., DUBIEL M. *et al.*, "Nanoscaled ceramic powders produced by laser ablation", *Mater. Sci. Eng., B*, no. 95, pp. 299–307, 2002.

[YAN 07a] YANG G.W., "Laser ablation in liquids: applications in the synthesis of nanocrystals", *Prog. Mater. Sci.*, no. 52, pp. 648–698, 2007.

[YAN 07b] YANG L., MAY P.W., YIN L. *et al.*, "Growth of self-assembled ZnO nanoleaf from aqueous solution by pulse laser ablation", *Nanotechnology*, no. 18, pp. 215602(1)–215602(5), 2007.

[YAN 07c] YANG Y.-P., HUANG K.-L., LIU R.-S. *et al.*, "Shape-controlled synthesis of nanocubic Co_3O_4 by hydrothermal oxidation method", *Trans. Nonferrous Met. Soc. China*, no. 17, pp. 1082–1086, 2007.

[YAN 08] YANG S., CAI W., ZENG H. *et al.*, "Polycrystalline Si nanoparticles and their strong aging enhancement of blue photoluminescence", *J. Appl. Phys.*, no. 104, pp. 023516(1)-023516(5), 2008.

[YAN 12] YANG S.-T., WANG T., DONG E. *et al.*, "Bioavailability and preliminary toxicity evaluations of alumina nanoparticules in vivo after oral exposure", *Toxicol. Res.*, vol. 1, pp. 69–74, 2012.

[YAN 13a] YANG Y., WANG P.-P., ZHANG Z.-C. *et al.*, "Nanowire membrane-based nanothermite: towards processable and tunable interfacial diffusion for solid state reaction", *Sci. Rep.*, vol. 3, no. 1694, 2013.

[YAN 13b] YANG Y., ZHANG Z.-C., WANG P.-P. *et al.*, "Hierarchical MnO_2/SnO_2 heterostructures for a novel free-standing ternary thermite membrane", *Inorg. Chem.*, vol. 52, no. 16, pp. 9449–9455, 2013.

[YAO 05] YAO W.-T., YU S.-H., ZHOU Y. *et al.*, "Formation of Uniform CuO nanorods by spontaneous aggregation: selective synthesis of CuO, Cu_2O, and Cu nanoparticles by a solid-liquid phase Arc discharge process", *J. Phys. Chem., B*, no. 109, pp. 14011–14016, 2005.

[YAT 73] YATSUYA S., KASUKABE S., UYEDA R., "Formation of ultrafine metal particles by gas evaporation technique. I. aluminum in helium", *Jpn J. Appl. Phys.*, vol. 12, no. 11, pp. 1675–1684, 1973.

[YAV 92] YAVARI A.R., DESRÉ P.J., BENAMEUR T., "Mechanically driven alloying of immiscibles elements", *Phys. Rev. Lett.*, vol. 68, no. 14, pp. 2235–2238, 1992.

[YE 07] YE R., LI J.-G., ISHIGAKI T., "Controlled synthesis of alumina nanoparticles using inductively coupled thermal plasma with enhanced quenching", *Thin Solid Films*, no. 515, 4251–4257, 2007.

[YEH 99] YEH M.-S., YANG Y.-S., LEE Y.-P. *et al.*, "Formation and characteristics of Cu colloids from CuO powder by laser irradiation in 2-propanol", *J. Phys. Chem., B*, no. 103, pp. 6851–6857, 1999.

[YEN 96] YEN B.K., AIZAWA T., KIHARA J., "Synthesis and formation mechanisms of molybdenum silicides by mechanical alloying", *Mater. Sci. Eng., A*, no. 220, pp. 8–14, 1996.

[YET 09] YETTER R.A., RISHA G.A., SON S.F., "Metal particle combustion and nanotechnology", *Proc. Combust.Inst.*, vol. 32, no. 2, pp. 1819–1838, 2009.

[YOS 04] YOSHIDA K., WAKAI C., MATUBAYASI N. *et al.*, "NMR spectroscopic evidence for an intermediate of formic acid in the water-gas-shift reaction", *J. Phys. Chem., A*, vol. 108, no. 37, pp. 7479–7482,2004.

[YOU 10] YOUNG G., PIEKIEL N., CHOWDHURY S. *et al.*, "Ignition behavior of α-AlH_3", *Combust. Sci. Technol.*, vol. 182, no. 9, pp. 1341–1359, 2010.

[YOU 14] YOUNG G., SULLIVAN K., PIEKIEL N. *et al.*, "Aluminum hydride as a fuel supplement to nanothermites", *J. Propul.Power*, vol. 30, no. 1, pp. 70–77, 2014.

[YU 10] YU P., ZHANG X., CHEN Y. *et al.*, "Solution-combustion synthesis of ε-MnO_2 for supercapacitors", *Mater. Lett.*, vol. 64, no. 1, pp. 61–64, 2010.

[YU 98] YU J., SAVAGE P. E., "Decomposition of formic acid under hydrothermal conditions", *Ind. Eng. Chem. Res.*, no. 37, pp. 2–10, 1998.

[YUN 11] YUN J.-Y., LEE H.-M., CHOI S.-Y. *et al.*, "Characteristics of Fe-Cr-Al alloy nanopowders prepared by electric wire explosion proces under liquid media", *Mater. Trans.*, vol. 52, no. 2, pp. 250-253, 2011.

[ZEN 04] ZENG D.W., XIE C.S., ZHU B.L. *et al.*, "Characteristics of Sb_2O_3 nanoparticles synthesized from antimony by vapor condensation method", *Mater. Lett.*, no. 58, pp. 312–315, 2004.

[ZEN 13] ZENG X., ZHANG X., YANG M. *et al.*, "A facile hydrothermal method for the fabrication of one-dimensional MoO_3 nanobelts", *Mater. Lett.*, no. 112, pp. 87–89, 2013.

[ZHA 03] ZHANG J., WORLEY J., DENOMMEE S. *et al.*, "Synthesis of metal alloy nanoparticles in solutionby laser irradiation of a metal powder suspension", *J. Phys. Chem., B*, vol. 107, no. 29, pp. 6920–6923, 2003.

[ZHA 06a] ZHANG L., WANG W., YANG J. *et al.*, "Sonochemical synthesis of nanocrystallite Bi_2O_3 as a visible-light-driven photocatalyst", *Appl. Catal., A*, no. 308, pp. 105–110, 2006.

[ZHA 06b] ZHANG J., LIU Z., HAN B. *et al.*, "Preparation of silica and TiO_2–SiO_2 core–shell nanoparticles in water-in-oil microemulsion using compressed CO_2 as reactant and antisolvent", *J. Supercrit. Fluids*, no. 36, pp. 194–201, 2006.

[ZHA 08] ZHANG J., LAN C.Q., "Nickel and cobalt nanoparticles produced by laser ablation of solids in organic solution", *Mater. Lett.*, no. 62, pp. 1521–1524, 2008.

[ZHA 09] ZHANG K., ROSSI C., ALPHONSE P. *et al.*, "Integrating Al with NiO nano honeycomb to realize an energetic material on silicon substrate", *Appl. Phys., A*, vol. 94, no. 4, pp. 957–962, 2009.

[ZHA 12a] ZHANG H., BAI L., HU P. *et al.*, "Single-step pathway for the synthesis of tungsten nanosized powders by RF induction thermal plasma", *Int. J. Refract. Met. Hard Mater.*, no. 31, pp. 33–38, 2012.

[ZHA 12b] ZHANG H., DUAN G., LI Y. *et al.*, "Leaf-like tungsten oxide nanoplatelets induced by laser ablation in liquid and subsequent aging", *Cryst. Growth Des.*, no. 12, pp. 2646–2652, 2012.

[ZHA 14] ZHANG H., LI Y., DUAN G. *et al.*, "Tungsten oxide nanostructures based on laser ablation in water and a hydrothermal route", *CrystEngComm*, no. 16, pp. 2491–2498, 2014.

[ZHA 01] ZHAO Y.H., SHENG H.W., LU K., "Microstructure evolution and thermal properties in nanocrystalline Fe during mechanical attrition", *Acta Mater.*, no. 49, pp. 365–375, 2001.

[ZHE 07] ZHENG X. L., XU W.Q., CORREDOR C. *et al.*, "Laser-induced growth of monodisperse silver nanoparticles with tunable surface plasmon resonance properties and a wavelength self-limiting effect", *J. Phys. Chem., C*, vol. 111, no. 41, ppp. 14962–14967, 2007.

[ZHO 04] ZHOU F., WITKIN D., NUTT S.R. *et al.*, "Formation of nanostructure in Al produced by a low-energy ball milling at cryogenic temperature", *Mater. Sci. Eng., A*, no. 375–377, pp. 917–921, 2004.

[ZHO 09a] ZHOU M., WEI Z.Q., QIAO H. *et al.*, "Particle size and pore structure characterization of silver nanoparticles prepared by confined arc plasma", *J. Nano Mat.*, vol. 2009, Article ID 968058, 2009.

[ZHO 09b] ZHOU L., PIEKIEL N., CHOWDHURY S. *et al.*, "T-jump/time-of-flight mass spectrometry for time-resolved analysis of energetic materials", *Rapid Commun. Mass Spectrom.*, vol. 23, no. 1, pp. 194–202, 2009.

[ZHO 10] ZHOU L., PIEKIEL N., CHOWDHURY S. *et al.*, "Time-resolved mass spectrometry of the exothermic reaction between nanoaluminum and metal oxides: the role of oxygen release", *J. Phys. Chem., C*, vol. 114, no. 33, pp. 14269–14275, 2010.

[ZHO 11] ZHOU X., SHEN R., YE Y. *et al.*, "Influence of Al/CuO reactive multilayer films additives on exploding foil initiator", *J. Appl. Phys.*, vol. 110, no. 094505, 2003.

[ZHO 13] ZHOU X., XU D., ZHANG Q. *et al.*, "Facile green in situ synthesis of Mg/CuO core/shell nanoenergetic arrays with a superior heat-release property and long-term storage stability", *ACS Appl. Mater. Interfaces*, vol. 5, no. 15, pp. 7641–7646, 2013.

[ZHO 14a] ZHOU X., XU D., YANG G. *et al.*, "Highly exothermic and superhydrophobic mg/fluorocarbon core/shell nanoenergetic arrays", *ACS Appl. Mater.Interfaces*, vol. 6, no. 13, pp. 10497–10505, 2014.

[ZHO 14b] ZHOU X., TORABI M., LU J. *et al.*, "Nanostructured energetic composites: synthesis, ignition/combustion modeling, and applications", *ACS Appl. Mater. Interfaces*, vol. 6, no. 5, pp. 3058–3074, 2014.

[ZHO 15a] ZHOU L., WANG S., MA H. *et al.*, "Size-controlled synthesis of copper nanoparticles in supercritical water", *Chem. Eng. Res. Des.*, no. 98, pp. 36–43, 2015.

[ZHO 15b] ZHOU W., DELISIO J.B., LI X. *et al.*, "Persulfate salt as an oxidizer for biocidal energetic nano-thermites", *J. Mater.Chem., A*, vol. 3, no. 22, pp. 11838–11846, 2015.

[ZHU 04] ZHU Z., SU D., LU Y. *et al.*, "Molecular 'glass' blowing: from carbon nanotubes to carbon nanobulbs", *Adv. Mater.*, vol. 16, no. 5, pp. 443–447, 2004.

[ZHU 12] ZHU M., WANG Y. *et al.*, "Hydrothermal synthesis of hematite nanoparticles and their electrochemical properties", *J. Phys. Chem., C*, no. 116, pp. 16276–16285, 2012.

[ZOZ 97] ZOZ H., ERNST D., WEISS H. *et al.*, "Mechanical alloying of Ti-24Al-11Nb (AT%) using the simoloyer (ZOZ- horizontal rotary ball mill)", *Mater. Sci. Forum*, vol. 235–238, pp. 59–66, 1997.

Index

N, O

P, R

S, T

V, W, X

Other titles from

iSTE

in

Nanoscience and Nanotechnology

2016

BASSET Philippe, BLOKHINA Elena, GALAYKO Dimitri
Electrostatic Kinetic Energy Harvesting
(Nanotechnologies for Energy Recovery Set – Volume 3)

2015

LEPRINCE-WANG Yamin
Piezoelectric ZnO Nanostructure for Energy Harvesting
(Nanotechnologies for Energy Recovery Set – Volume 1)

ROSSI Carole
Al-based Energetic Nanomaterials
(Nanotechnologies for Energy Recovery Set – Volume 2)

2014

BALESTRA Francis
Beyond CMOS Nanodevices 1
Beyond CMOS Nanodevices 2

LACAZE Pierre Camille, LACROIX Jean-Christophe
Non-volatile Memories

LATU-ROMAIN Laurence, OLLIVIER Maelig
Silicon Carbide One-dimensional Nanostructures

NICU Liviu, LEÏCHLE Thierry
Micro- and Nanoelectromechanical Biosensors

SENEZ Vincent, THOMY Vincent, DUFOUR Renaud
Nanotechnologies for Synthetic Super Non-wetting Surfaces

TRIPON-CANSELIET Charlotte, CHAZELAS Jean
Nanoscale Microwave Engineering: Optical Control of Nanodevices

2012

BELLEVILLE Marc, CONDEMINE Cyril
Energy Autonomous Micro and Nano Systems

CLAVERIE Alain
Transmission Electron Microscopy in Micro-nanoelectronics

LACAZE PIERRE Camille
Nanotechnologies: Concepts, Processing and Applications

2010

BALESTRA Francis
Nanoscale CMOS / Innovative Materials, Modeling and Characterization

QUERLIOZ Damien, DOLLFUS Philippe
The Wigner Monte Carlo Method for Nanoelectronic Devices: Particle Description of Quantum Transport and Decoherence

TAKADOUM Jamal
Nanomaterials and Surface Engineering

VODA Alina
Micro, Nanosystems and Systems on Chips: Modeling, Control and Estimation

2008

DAVIM J. Paulo, JACKSON J.
Nano and Micromachining

NOUAILLHAT Alain
An Introduction to Nanoscience and Nanotechnology

OUISSE Thierry
Electron Transport in Nanostructures and Mesoscopic Devices

2006

RIGNEAULT Hervé *et al.*
Nanophotonics

Printed and bound by CPI Group (UK) Ltd, Croydon, CR0 4YY

13/09/2026

14961017-0004